ALS
Advances in Life Sciences

Transgenic Organisms: Biological and Social Implications

Edited by
J. Tomiuk
K. Wöhrmann
A. Sentker

Birkhäuser Verlag
Basel · Boston · Berlin

Editors

Agr.
QH
442
.6
.T728
1996

PD Dr. J. Tomiuk
Prof. Dr. K. Wöhrmann
Biologisches Institut
Lehrstuhl für Populationsgenetik
Auf der Morgenstelle 28
D-72076 Tübingen
Germany

Dipl.-Biol. A. Sentker
Eduard-Spranger-Strasse 41
D-72076 Tübingen
Germany

Library of Congress Cataloging-in-Publication Data
Transgenic organisms: biological and social implications / edited by
J. Tomiuk, K. Wöhrmann, A. Sentker.
    p.   cm. – – (Advances in life sciences)
Includes bibliographical references and index.
ISBN 3-7643-5262-0 (hardcover : alk. paper). – – ISBN 0-8176-5262-0
(hardcover : alk. paper)
1. Transgenic organisms– –Risk assessment.   I. Tomiuk, J. (Jürgen)
II. Wöhrmann, K. (Klaus), 1928- .   III. Sentker, A. (Andreas),   IV. Series.
QH442.6.T728   1996
575.1– –dc20

Deutsche Bibliothek Cataloging-in-Publication Data
Transgenic organisms: biological and social implications / ed.
by J. Tomiuk ... - Basel ; Boston ; Berlin : Birkhäuser, 1996
    (Advances in life sciences)
    ISBN 3-7643-5262-0 (Basel ...)
    ISBN 0-8176-5262-0 (Boston)
NE: Tomiuk, Jürgen [Hrsg.]

Contents

VI

Preface

In 1992, a group of scientists including molecular biologists, microbiologists, population biologists, ecologists, human geneticists, moral philosophers and others met discussing the state of affairs regarding the deliberate or unintentional release of genetically modified organisms. The proceedings of this meeting were subsequently published by Birkhäuser Verlag as *Transgenic Organisms: Risk Assessment of Deliberate Release* (K. Wöhrmann and J. Tomiuk).

Since then we have gained many new insights that are also worthy of discussion. And although other equally important scientific views on the release of genetically modified organisms exist, we have mainly concentrated on aspects of population biology and evolution. The results of a second meeting in 1995 are summarized here.

We are grateful to colleagues and friends for their help in the translation, correction and review of the authors' contributions. We especially want to thank Jutta Bachmann, Donna Devine, Diana von Finck, Friedrich Laplace, Volker Loeschcke, Rolf Lorenz, Dave Parker and Trevor Petney.

A grant (BMFT N° 0311035) from the Ministerium für Forschung und Technologie der Bundesrepublik Deutschland again made possible the continuation of this cooperative endeavour.

Jürgen Tomiuk
Klaus Wöhrmann
Andreas Sentker

Tübingen, September 1995

Transgenic Organisms – Biological and Social Implications
J. Tomiuk, K. Wöhrmann & A. Sentker (eds)
© 1996 Birkhäuser Verlag Basel/Switzerland

Prologue

K. Wöhrmann, J. Tomiuk and A. Sentker[1]

Department of Population Genetics, University of Tübingen, Auf der Morgenstelle 28, D-72076 Tübingen, Germany
[1]*Eduard-Spranger-Straße 41, D-72076 Tübingen, Germany*

Controversy about genetically modified organisms (GMOs) concentrates mainly on two problems that have arisen in their application:

(i) When GMOs are produced, new genetic information is intentionally transferred from one organism to another, target organism. Depending on the method applied, the DNA in the transgenic organisms used as marker or vehicle must also be considered. This raises questions concerning the stability of the DNA that codes for the new trait and concerning the possible interactions (pleiotropy, position effects) of the transferred genetic information with the host genome.

It is feared that unstable informative traits and unpredictable pleiotropic and position effects may result in properties leading to a selective advantage of deliberately or unintentionally released organisms in our ecosystem and thus to unpredictable changes. And before their consumption cultured plants might produce substances causing health problems in humans.

(ii) Transgenic organisms might escape human control. This could occur during their handling in laboratories and factories but also after their deliberate release into the environment. The escaped organisms could reproduce as such or transfer their genetic information horizontally to the same or to related species. Not every transgene or transgenic organism need have an effect on the environment, but certain traits, on the other hand, do seem to carry potential dangers. This is basically true for all groups: Microorganisms, plants as well as animals.

In order to clarify and assess these problems, several questions have to be resolved.

(i) What kind of phenomena can occur? Are the genes stable? Are there any pleiotropic effects? Can we expect horizontal gene transfer? Are the transgenic organisms able to exist outside their range of use?

(ii) What is the probability of such phenomena occurring?

(iii) What are the consequences for the organisms, for our environment and for humans?

• The final decision will always be a political and an ethical one. One has to weigh "risk" and "usefulness", and decide whether the so-called progress is really desirable and essential for human existence or whether it simply serves economic interests.

The information gained in the sciences involved (e.g., molecular and classical genetics, population genetics and evolutionary genetics) may provide some answers. It was the aim of the two meetings and the respective publications (Wöhrmann and Tomiuk, 1993 and the present volume) to give a representative summary of the extent of scientific knowledge relating to the release of genetically modified organisms.

The first contributions of this volume deal with molecular aspects. P. Meyer discusses the inactivation of gene expression in transgenic plants and summarizes recent data on homology-dependent gene silencing, considering transcriptional and posttranscriptional silencing. W. J. Miller, L. Kruckenhauser and W. Pinsker, and separately, L. Bachmann analyse the impact of transposable elements on genome evolution and deduce parallels for transgenic organisms. Transposons were discovered by B. McClintock and since then have been described in plants, animals and microorganisms. Transposons behave as "foreign" DNA in host genomes and their effects range from point mutations to chromosomal rearrangements. They have a large impact on the creation of genetic variability and therefore on the evolution of their hosts. Their behaviour is essentially the same as that of foreign DNA transferred to target organisms in gene technology.

Genetic information can spread within a single species group or horizontally through parasexual and sexual systems. Recent intensive research has revealed the extent of an unexpectedly high degree of horizontal gene transfer between microorganisms: M.G. Lorenz and W. Wackernagel deal with the mechanisms and consequences of horizontal gene transfer in natural bacterial populations. Another topic still hotly debated in the context of the release of modified genes concerns the fate of organisms that accidentally escape into the environment. By the same token, numerous examples from biological pest control schemes show that the behaviour of deliberately released organisms cannot always be predicted since our knowledge about ecological relationships and on evolutionary progresses is still limited. K.D. Adam and W.H. Köhler report the possible consequences of the release of transgenic organisms from the point of view of the population geneticist. The interactions between organisms of an ecosystem is complex, and changes in one component may lead to severe consequences for the entire community. Baculoviruses, seen as a benign biological system in biological pest control, are highly specific with respect to their host range and can receive transposable elements from their hosts. Questions and speculations about the biological and evolutionary significance of these observations in connection with transgenes are summarized by J.A. Jehle. In particular, interventions and changes in host-parasite interactions have led to scepticism among ecologists. Apart from baculoviruses, this also concerns *Bacillus thuringiensis* whose toxin should protect the transformed plant against insects. This involves an intervention in host–insect relations and may influence coevolutionary processes of this system (P.W. Braun).

A need for more security against an undesired and uncontrolled spread of genetically modified organisms has led to the development of biological containment systems which include the inability to survive outside the laboratory or the construction of suicide systems to kill the organism after serving the goal of release. T. Schweder summarizes and discusses biological containment systems for *E. coli* and for other bacteria. Despite such systems, an uncontrollable minimal risk still remains. The efforts made to judge this risk are addressed as "risk assessment". "Biosafety research" or "monitoring" fields have been set up in order to develop methods allowing a control of released organisms and their fate. K. Smalla and J.D. van Elsas discuss the monitoring of genetically modified microorganisms and I. Parker and D. Bartsch the prospect for transgenic plants.

Gene technologies are presently used in all those scientific areas concerned with the organismic production of essential human foods. In plant breeding, molecular methods are used in combination with classical, and W. Friedt and F. Ordon look at both aspects. M. Teuber reports about genetically modified food in connection with safety assessments. Probably the gravest potential interventions in an organism are those involving human beings. However, according to M. Leipoldt, the unlimited interventions seem – and not only for ethical reasons – to be improbable at present and in the near future.

For the past 20 years the benefits and potential dangers of gene technology have been heatedly discussed. Points of view range from outright rejection to euphoristic acceptance. The benefits and dangers have been debated at various meetings in many countries since the first congress in Asilomar, California, in 1975 and were made known to the public *via* the media. Have the points of view changed? Has the consensus become more uniform? The arguments put forward at these meetings of scientists are summarized by J. Tomiuk and A. Sentker. T. Potthast's article on ethical implications brings the topic of transgenic organisms to a close. Finally, the importance of the media in the discussion of the acceptance of new technologies and progress is discussed by A. Sentker.

Transgenic Organisms – Biological and Social Implications
J. Tomiuk, K. Wöhrmann & A. Sentker (eds)
© 1996 Birkhäuser Verlag Basel/Switzerland

Inactivation of gene expression in transgenic plants

P. Meyer

Department of Biology, University of Leeds, Leeds LS2 9JT, UK

Summary. In recent years more and more reports have appeared in the literature describing instabilities of gene expression in transgenic plants. Many of these instabilities, which can be classified as transcriptional or post-transcriptional silencing, require multiple homologous sequences. Although first reported for transgenes, silencing can also affect endogenous genes. The aim of this article is to summarize recent data on homology-dependent gene silencing and to discuss the current models about the underlying mechanisms of gene silencing.

Gene inactivation in plants

The inactivation of recombinant genes that have been introduced into transgenic plants has become a widely recognized phenomenon within recent years. Early reports already mentioned the inactivation of transgenes accompanied by an increase in DNA methylation (Amasino et al., 1984) and a correlation between inactivation and the number of integrated transgene copies (Jones et al., 1987). The importance of homologous sequences as a signal for gene silencing became apparent from a publication that described the inactivation of a marker gene in a transgenic line after a second construct had been introduced into this line, carrying homology to the previously integrated marker gene (Matzke et al., 1989). One year later, two laboratories demonstrated that the introduction of sequence homology could induce the concurrent inactivation of the transgene and its endogenous homologue (Napoli et al., 1990; van der Krol et al., 1990). In transformation experiments, this effect is often exploited to generate a set of individual transformants each showing a different expression level of the gene. Depending on the efficiencies of over-expression or homology-dependent repression, distinct phenotypes for these categories can be isolated from the same transformation experiment (Boerjan et al., 1994).

Several terms have been created to categorize gene silencing events in plants. The terms "cosuppression" and "sense-inactivation" are often found in the literature. Homology-dependent gene silencing (HDGS) or repeat-induced gene silencing (RIGS) are more general terms that do not refer to a particular effect or mechanism but define the presence of homology as the common parameter among all silencing phenomena. Based on transcription and RNA turnover analysis, we can distinguish at least two different classes of homology dependent gene silencing.

Silencing at the posttranscriptional level is linked to a decay of steady-state transcript levels. Transcriptional silencing is frequently associated with an epigenetic modification of the genes, such as promoter methylation or changes in chromatin structure (Flavell, 1994). Posttranscriptional silencing can also be characterized by hypermethylation in some cases (Ingelbrecht et al., 1994) suggesting a link between transcriptional and posttranscriptional silencing.

Related phenomena in other eukaryotes

The participation of repeated sequences in gene inactivation phenomena is not limited to plants, but can be found in several other eukaryotes. A comparison of HDGS in plants with other eukaryotic silencing systems illuminates some interesting analogies that might suggest common biological functions. In the filamentous fungi *Neurospora crassa* and *Ascobulus immersus* the presence of DNA repeats triggers methylation and inactivation of the repeated regions. In *Neurospora* transformants that contain linked or unlinked duplicated sequences, a mechanism named repeat induced point mutation (RIP) induces methylation of C-residues and C to T transitions, preferentially at CA dinucleotides (Selker et al., 1987; Cambareri et al., 1989). In *Ascobulus*, gene duplication leads to *de novo* methylation and premeiotic inactivation due to a mechanism termed methylation induced premeiotically (MIP) (Goyon and Faugeron, 1989). An inverse correlation between copy number and the expression of transgenes was also shown for the asexual cycle of *Neurospora* (Pandit and Russo, 1992). After transformation with a resistance marker, vegetative and reversible inactivation of the marker occurred in multicopy transformants, which was accompanied by hypermethylation of marker genes. Treatment with 5-azacytidine induced a stable reactivation of the marker in some, but not all, transformants, which suggests a role of DNA methylation in this process. Methylation might also be involved in an unidirectional silencing event in *Neurospora*, named quelling (Romano and Macino, 1992). The expression of endogenous genes was impaired when several homologous copies were integrated at an ectopic site. Reversion of the quelling effect was correlated with a reduction in the number of ectopically integrated gene fragments and could be enhanced by 5-azacytidine treatment. Most likely *Neurospora* and *Ascobulus* contain a homology searching mechanism responsible for the specific methylation of repeated sequences. Clearly, the position of the target sequences for such a mechanism is important for its efficiency because linked repeats are better targets for methylation than unlinked repeats.

   Other well characterized examples for homology-dependent silencing include the position-effect variegation (PEV) phenomena in *Drosophila*. PEV at the *white* locus, for example, is induced after translocation of a euchromatic gene next to heterochromatin. Inactivation of the *white* gene that affects eye colour, can be monitored in individual cells by the reduction of the red eye pigment. *Cis*-inactivation of the gene, which results from the spreading of the heterochromatic state into the euchromatic neighbourhood, causes the mosaic-like phenotype of red and white cells that gave PEV its name (Spofford, 1976). PEV demonstrates the incompatibility of two kinds of chromatin: Heterochromatin, located within the pericentric regions and which remains condensed throughout the cell cycle, and euchromatin, which is located at the chromosome arms and decondenses during interphase. Interestingly, a PEV-like effect can also be triggered by repeated sequences that are closely linked to a marker gene, which suggests that pairing of repeats contributes to heterochromatin formation (Dorer and Henikoff, 1994). Homologous chromosomes of non-dividing nuclei are physically paired in somatic cells of *Drosophila*, providing the basis for the transmission of heterochromatic states to homologous alleles. One example for pairing-dependent transinactivation is dominant PEV at the *brown* locus that also affects eye colour. Heterochromatin that was imposed on a rearranged *brown* allele is transmitted to the unrearranged homologous copy, which also becomes inactivated (Dreesen et al., 1991).

In mammals, only a few reports on homology-dependent silencing events exist. Expression of a *CAT* gene in *COS1* cells was reduced after the cells were transfected with constructs expressing defective *CAT* RNA in sense orientation (Cameron and Jennings, 1991). As this effect requires transcription of sense RNA, the authors proposed various models that are based on the formation of RNA–RNA pairing or the formation of a triplex between RNA and the double stranded DNA. An RNA–DNA triplex is also favoured to explain a very similar effect that was detected in *Dictyostelium discoideum*, where transcription of a *myosin II heavy-chain* gene decreased expression of the endogenous homologue, concomitant with accumulation of the transferred sense construct (Scherczinger and Knecht, 1993).

Indirect evidence for the presence of a homology-based methylation mechanism comes from a comparative analysis of the presence of CG dinucleotides in mammalian genome (Kricker et al., 1992). CG dinucleotides provide the recognition site for mammalian DNA methyltransferase. Repeated sequences in mammalian genomes are highly depleted in CG dinucleotides, while TG dinucleotides are over-represented. This suggests that CG sites in repeated sequences are efficient targets for methylation and subsequent conversion into TG dinucleotides *via* deamination (Kricker et al., 1992). It was proposed that fragmentation of genes into introns and exons protects coding sequences against homologous interactions with their own pseudogenes, because the reduction in continuous homology prevent its recognition by a homology searching mechanism (Kricker et al., 1992).

Mechanisms of gene inactivation

There are many examples for the functional inactivation of genes by mutant polypeptides that disrupt the activity of wildtype proteins (Herskowitz, 1987). These dominant negative mutations require the production of truncated or mutated polypeptides and are often based on the titration of regulatory subunits or on the formation of nonfunctional multimeric protein complexes. Although we cannot exclude related effects to be involved in certain homology-dependent silencing effects, they clearly cannot account for most of these phenomena. Most HDGS effects described so far are accompanied by a reduction of transcription or steady-state transcript levels. Moreover, the sense constructs are preferentially transferred as full length copies and there are no indications that the transgenes generate mutated or truncated proteins at high frequencies. More likely HDGS events are mediated by changes in RNA turnover or by epigenetic changes of DNA methylation patterns or chromatin structure that influence transcription (see below). A brief overview follows on some of the mechanisms possibly involved in HDGS.

DNA methylation

Plants contain up to 30% of methylated cytosines (Adams et al., 1984), compared to only 2–8% 5 mC residues in vertebrates (Vanyushin et al., 1970). The relatively high 5 mC content is due to the fact that CG dinucleotides, the common target sequences for DNA methyltransferase in mammals and plants, are less depleted in plants. Secondly, CNG trinucleotides serve as additional methylation targets in plants. The symmetry of the target sequences CG and CNG provides the information for the faithful propagation of methylation patterns after replication. As hemimethylated DNA is the preferred substrate for DNA methyltransferase, it is assumed that the enzyme recognizes 5 mC residues in the template strand of semiconservatively replicated DNA and that it methylates C residues in the synthesized strand, if the template strand carries a 5 mC residue at a symmetrical position (Holliday and Pugh, 1975). It is not clear whether the same methyltransferases methylate both CG and CNG sequences or, whether different methyltransferases exist. In contrast to the mouse, a small multigene family in *Arabidopsis thaliana* hybridizes to methyltransferase cDNA suggesting the presence of multiple enzymes (Finnegan and Dennis, 1993).

While knockout mutants of the murine DNA methyltransferase proved the essential role of DNA methylation in mammalian gene regulation (Li et al., 1992), the function of DNA methylation in plants is less clear. A correlation between gene inactivation and DNA methylation has been shown for transgenes (Amasino et al., 1984; van Slogteren et al., 1984), transposable elements

(Chandler and Walbot, 1986; Schwarz and Dennis, 1986) and some endogenous genes (Spena et al., 1983), but in other genes no inverse correlation between DNA methylation and gene induction could be observed (Nick et al., 1986).

Usually, DNA methylation is measured using isoschizomeric restriction enzymes that recognize the same DNA sequence but different C-methylation patterns. Because the use of restriction enzymes is limited to their recognition sequences, only a fraction of a DNA region can be scanned for methylated or unmethylated cytosines. Genomic sequencing procedures that allow the control of the methylation state for every C residue within a sequence are laborious and difficult to quantify. Recently, however, an efficient PCR-based genomic sequencing procedure was developed (Frommer et al., 1992) by which C methylation patterns could be determined for individual cells. When the *A1* transgene transferred into *Petunia hybrida* was analysed by this technique, it was discovered that cytosine residues that were located outside CG or CNG sequences could also become methylated. These non-symmetrical methylation patterns were clustered in certain regions and it was not always the same C residue, but one or several C residues within a particular region that were affected (Meyer et al., 1994). These results imply that induction and conservation of non-symmetrical methylation patterns is probably not encoded in the sequence, but more likely in the secondary structure of a sequence.

So far, non-symmetrical methylation patterns have only been detected in transgenes (Ingelbrecht et al., 1994), but not in endogenous genes. It is, therefore, unclear if they are specifically imposed on transgenes or genes that have been transferred into new chromosomal environments. An indication for a specific recognition of foreign DNA comes from the comparative analysis of an *A1* transgene and its integration region in *Petunia hybrida* that showed specific *de novo* methylation within the transgene region, but no methylation in the surrounding chromosomal region (Meyer and Heidmann, 1994). These data suggest a role for DNA methylation as a protection mechanism against foreign DNA as it has been already proposed for mammalian cells (Bestor, 1990; Doerfler, 1991).

Condensation of chromatin

Changes in chromatin structure can serve as modulators in the activation and repression of gene activity. The term "condensation of chromatin structure" is used in a very broad sense for different biological systems, which most probably do not share identical molecular mechanisms. Experiments with *Drosophila*, yeast and *Xenopus* tell us that chromatin is a dynamic structure that interacts with the transcription apparatus (Lewin, 1994). The core histones that form the nucleosomes exert a regulatory function *via* modifications of the N-terminal domains, such as the acetylation of lysine residues (Loidl, 1994). *In vivo* studies of *TFIIIA* binding to *Xenopus* 5S RNA genes have shown that hyperacetylation of core histones is required to allow transcription factor binding, either because acetylation releases the N-terminal tail from the DNA, or because hyperacetylated N-termini actively support transcription factor binding (Lee et al., 1993). *In vivo* experiments performed in yeast also argue for a regulatory function of core histone N-termini. Acetylation of critical lysine residues in the N-termini of histones *H3* and *H4* released telomeric and silent mating locus repression, suggesting a function of the N-termini in the formation of heterochromatin-like complexes (Thompson et al., 1993). Besides a modification of nucleosomes, variation in the spacing between histone octamers and packing of chromatin into a condensed structure contribute to the transcriptional repression of a local genomic region. Changes in spacing between histone octamers can be influenced by histone *H1* or high mobility group proteins (Travers, 1994), while heterochromatin components contribute to the formation of a compact structure, rendering the locus inaccessible to the transcription apparatus. Specifically in *Drosophila,* a number of proteins have been identified that contribute to constitutive or tissue specific repression of genomic regions. A common protein domain has been identified that provides a link between proteins involved in heterochromatin formation and developmental silencing (Paro and Hogness, 1991). This motif, the so-called chromobox, is found in *HP1*, a major component of β-heterochromatin involved in position-effect variegation and in proteins of the polycomb group that maintains homeotic genes in a repressed state. It is also found in the *swi6* gene product, which is necessary for mating type and telomeric silencing in fission yeast, which suggests that chromatin condensation is mediated by related mechanisms in different species.

It has been proposed that chromatin condensation might be the underlying mechanism of silencing events at the transcriptional level that are based on DNA–DNA interactions (see below). An alternative hypothesis, which appears specifically applicable for posttranscriptional silencing processes proposes a role for RNA turnover.

RNA processing

The steady-state levels of individual mRNAs that finally interact with the ribosomes in the cytoplasm are not only determined by promoter activity. At the posttranscriptional level, we can consider several steps that influence the effective amounts of steady-state mRNA. This includes processing of hnRNA, the transport of precursor mRNAs to the cytoplasm and the degradation of transcripts. It is conceivable that at least certain steps within the RNA processing cascade are not independently regulated, but that they are linked by feedback control mechanisms.

Polyadenylation signals have a profound influence on the quantities of primary transcripts. When in a tobacco system the 3'-ends of different genes were placed behind an *NPTII* gene, expression levels differed up to 60-fold after the constructs had been inserted into the chromosome, while all constructs had been equally well expressed in a transient expression system (Ingelbrecht et al., 1989). These results suggest that polyadenylation efficiency and RNA stability varies for different 3'-ends.

Especially in monocot genes, introns play an important part in the regulation of expression efficiencies. In transient assays, the presence of intron 1 of the *Shrunken-1* gene enhanced chimeric gene expression in rice and maize about a 100-fold (Maas et al., 1990). Insertion of a dicot-specific intron generated 10- to 90-fold enhancements in mRNA levels of chimeric reporter constructs in transgenic rice tissue, but not in tobacco (Tanaka et al., 1991). At least for monocot transformation systems, we therefore have to consider substantial differences in steady-state transcript levels when introns are inserted into the constructs.

In recent years several, partly contradictory models have emerged on the transport of RNA from the site of transcription to the cytoplasm. In analogy to the established model of an organized movement of newly synthesized polypeptides through the cytoplasmic secretion machinery (Pryor et al., 1992), it was proposed that transcripts move to the nucleoplasm in an ordered fashion, passing through localized spots that harbour individual steps of the processing machinery. This model is supported by reports on the localization of splicing components in subnuclear domains, called speckles and foci (Spector et al., 1991) and by observations that intron-containing RNAs are targeted to speckles upon microinjection into mammalian nuclei (Spector et al., 1991). Quantitative three-dimensional analysis of nuclear components involved in precursor mRNA metabolism suggested that spliceosome assembly factors and poly-A-rich transcripts are localized at domains forming a ventrally positioned horizontal array in monolayer mammalian cells (Carter et al., 1993). A study that visualized fibronectin and neurotensin mRNAs within the mammalian interphase nucleus showed that allelic genes from homologous chromosomes are both transcriptionally active and spatially separate. Unprocessed RNA molecules accumulated at the site of transcription and the excision of introns occurs within this body of accumulated RNA, probably along a polarized track that begins at the site of transcription. In contrast to the association of RNA to localized tracks or foci, at least some excised introns diffused freely. As the foci did not generally extend to the nuclear envelope, it was unclear, whether mature mRNA is exported on another track route or, whether it diffuses freely (Xing et al., 1993).

In contrast to these reports, other studies on processing of the adenovirus type 2 (*Ad2*) major late mRNA in *HeLa* cells argue against a compartmentalization of RNA processing because splicing occurred at the sites of transcription, which were not coincident with intranuclear speckles that harbour components of the splicing machinery (Zhang et al., 1994). It is therefore conceivable that speckles serve as reservoirs for splicing factors and that not all genes have to pass processing tracks. On the other hand, it is obvious that processing of certain genes is highly localized and that individual speckles and foci serve as common processing sites for certain genes. Assuming that the position of a gene within the interphase nucleus determines the entrance of the transcript into defined processing routes, transcripts of multiple transgenes localized at different positions may enter common or separate processing routes, which should affect local concentrations of processing intermediates along individual tracks. Therefore, we not only have to consider the general quantities of steady-state RNA levels within the nucleus, but also have to account for the local concentrations of homologous RNA molecules within the processing track.

RNA stability

The effective amounts of mRNA that are available to the translation machinery depend on the efficiency of RNA processing, but they are also influenced by the stability of individual messenger molecules. It is conceivable that the susceptibility of a particular RNA to degradation is not necessarily a constant factor, but that it is influenced by the overall or local concentration of processing intermediates. Two models for quantity-dependent degradation of RNA are the production of antisense RNA *via* RNA-directed RNA polymerases and the specific degradation of mRNAs after the induction of high transcript levels.

RNA-dependent RNA polymerases are widely distributed among plants, although there has been a long dispute about the possible contamination of the material by the specific RNA polymerases of viruses (Fraenkel-Conrad, 1983). These enzymes are usually present in low amounts, which can be significantly increased upon viral infection. The enzymes in different species are clearly different in size and template specificity. When two different host species were infected with the same virus, the enzymes enhanced in each species still conserved its host specific characteristics, which supports the assumption that the enzymes are not derived from the virus but encoded by the host plant. The biological role of RNA-dependent RNA polymerases has not been finally elucidated, but it may be speculated that they create antisense molecules against specific transcripts catalysing the formation and subsequent degradation of dsRNA molecules.

It is difficult to imagine that RNA-dependent RNA polymerases have a sequence-specific target. More likely, the accumulation of threshold quantities could induce the synthesis of antisense molecules against specific RNA sequences. On the other hand, specific sequences have been found that determine mRNA stability and account for the difference in half-life of individual transcripts. The small auxin-up RNAs (SAURs), a group of auxin-induced soybean transcripts share common *DST* elements in their 3'-untranslated regions. When inserted into the 3'-untranslated region of reporter constructs, *DST* elements induced a significant decay in transcript stability (Newman et al., 1993). It is conceivable that the *DST*-specific degradation is part of a control system which adjusts high expression levels generated by auxin induction, and that related control systems exist for other inducible genes. Such control systems should also detect high amounts of transgene transcripts that carry target signals for degradation leading to a posttranscriptional degradation of transgene RNA and homologous endogenous transcripts.

Inactivation of transgenes

Transgene inactivation can comprise both transcriptional (Meyer et al., 1993) and posttranscriptional silencing (Ingelbrecht et al., 1994) of marker genes. Several inactivation events are associated with hypermethylation in promoter regions (Matzke et al., 1989), while some do not show this correlation (Goring et al., 1991). Silencing can affect single copy transgenes (Pröls and Meyer, 1992), but the integration of multiple copies appears to enhance silencing efficiencies, especially when repeated sequences are inserted in concatameric arrangements at one locus (Assad et al., 1993; Mittelsten Scheid et al., 1991). Various examples exist for transinteractions of copies, located at allelic positions (Meyer et al., 1993) or at unlinked sites (Matzke et al., 1993). Like *MIP* in *Ascobulus*, homology-based silencing is influenced by the length of homology and especially by the position of the interacting sequences. Linked copies are more efficiently silenced than unlinked copies and unlinked loci show defined differences in silencing capacity (Vaucheret, 1993) and susceptibility to become silenced (Neuhuber et al., 1994). The most efficient example of transinactivation is a tobacco line carrying a transgene with two promoters, the 19S and the 35S promoter of *CaMV*. Both genes linked to the two promoters are suppressed and this locus transinactivates newly introduced constructs that provide only 90 bp of common homology (Vaucheret, 1993).

It has been suggested that at least certain transgene inactivation events are mediated by DNA–DNA interactions between homologous sequences (Matzke and Matzke, 1990; Meyer et al., 1993), which would account for the position dependence of efficient silencing. In a DNA–DNA pairing model the interaction between two homologous sequences would be determined by the probability with which the two loci associate in interphase nuclei. This probability should be higher for tandemly linked copies, compared to unlinked, ectopic copies. To account for the enhancement of silencing by DNA–DNA pairing, there are at least two possible mechanisms, which might act alternatively or synergistically. Either paired homologous sequences provide a signal for a plant-specific homology detection mechanism, or pairing facilitates the transfer of cellular components that control silencing from one homologous locus to the other. The second possibility implies that silenced loci that are hypermethylated or carry a condensed chromatin structure are more efficient than active and unmodified loci in silencing homologous sequences. A DNA–DNA pairing model would certainly not account for all transgene silencing phenomena reported, but it provides an attractive working hypothesis, which links transgene silencing to inactivation mechanisms in *Drosophila*.

Silencing of transgenes and endogenous genes

Homology-based gene silencing is not a transgene-specific event, but endogenous genes can also be inactivated when homologous sequences are transferred into the genome. This very surprising effect became noticeable when researchers shifted their attention towards the transfer of endogenous genes rather than marker genes. Many such experiments dealt with the introduction of antisense constructs to inactivate gene expression of particular genes. As a putative negative control for antisense experiments, sense-constructs were expressed in parallel. In 1990, the first reports appeared about the coordinated inactivation of a sense construct and the corresponding endogenous sequences in *Petunia hybrida* (Napoli et al., 1990; van der Krol et al., 1990). The detection of this effect, which was termed "cosuppression", was facilitated because both groups had transferred a sense construct of chalcone synthase (*CHS*), a key enzyme in anthocyanin pigment synthesis. Up to 50% of the transformant showed white sectors or completely white flowers due to the extinction of anthocyanin production. Cosuppressed plants had reduced steady-state levels of *CHS* mRNA, but RNA transcription was not affected (Kooter and Mol, 1993; Flavell, 1994). The cosuppression phenotype can be inherited in the next generation, but can also revert to other states, which are somatically inherited. Transformants with white flowers can develop branches with purple flowers or floral sectors. All flowers of such branches have a similar phenotype, which suggests that the underlying epigenetic change had occurred early during meristem formation of the branch. From these data, it was suggested that the capacity of the transgene to cause cosuppression is subject to developmental influences (Flavell, 1994).

Many cosuppression phenomena have been described today, many of which show similarities to cosuppression of *CHS* genes in petunia. Nevertheless, we still lack a consistent picture about the underlying mechanisms, mainly because we are dealing with a variety of plant and plant virus genes and because there is not always sufficient information about the individual step in gene expression that is affected by cosuppression. In most cases, cosuppression occurs posttranscriptionally (de Carvalho et al., 1992; van Blokland et al., 1994; Mueller et al., 1995), but suppression at the transcriptional level has also been reported (Brussian et al., 1993). The efficiency of cosuppression varies among individual genes. The most remarkable example is the nuclear encoded cytosolic *GS2* gene that is cosuppressed in all transformants tested (Brears et al., in press). An unexpected link between cosuppression and transgene mediated resistance against plant viruses was observed from the analysis of several transgenic plants resistant against different members of the potyvirus group. Contrary to the model of coat protein-mediated resistance, resistance was not dependent on the accumulation of viral coat proteins, but it was sufficient to express untranslatable viral sequences. Resistant lines showed low steady-state levels of the transgene constructs due to posttranscriptional suppression (Lindbo et al., 1993; Smith et al., 1994; Mueller et al., 1995).

An important feature of many cosuppression phenomena is their developmental control, which needs to be considered in models that aim to explain the molecular mechanism of cosuppression. Several such models have been elaborated, and they are not mutually exclusive. The *antisense model* (Grierson et al., 1991) suggests that silencing is based on the formation of RNA duplex molecules after pairing of homologous sense and antisense molecules. Due to the activity of transgene promoters or plant promoters located near the integrated transgene, antisense molecules may be produced constitutively or by particular states of the transgene that change during development. The *biochemical switch model* (Meins and Kunz, 1994) suggests that transcription involves the production of a diffusible activator molecule which maintains stability of the transcript. Assuming that the activator is degraded in a first-order process, the final transcript levels would depend on activator concentrations that are regulated by its synthesis and degradation levels. The *threshold model* (Dougherty et al., 1994) also proposes a feedback mechanism that is dependent on transcription levels. Accumulation of critical threshold levels could have a feedback inhibitory effect, which also could be developmentally regulated. The threshold theory corresponds with the dependency of some cosuppression events on gene dosage (de Carvalho et al., 1992; Dehio and Schell, 1994) and with an observation made for cosuppression of the polygalacturonase gene in tomato, where silencing only became active in fruits when the endogenous *PG* gene was transcribed (Dehio and Schell, 1994). Threshold levels may induce specific degradation mechanisms or they may trigger the activity of RNA-dependent RNA polymerases which produce quantitative or catalytic amounts of antisense RNA. Another model proposes that the interacting genes can change their states that mediate differentiation and development (Jørgenson, 1994). Different states might affect processing or transport efficiencies of the transcript, thus inducing threshold levels when processing intermediates accumulate locally. In this context, cosuppression would serve as a monitor system for endogenous mechanisms that regulate the developmental activity of certain genes (Jørgenson, 1994).

Concluding remarks

The most astonishing fact about homology-dependent gene silencing is the relatively long period until this phenomenon was recognized by the scientific community. The first report about homology-based silencing appeared in 1989, when literally hundreds of transformation experiments had already been evaluated and published. In the following years, additional reports arose, but transgene instability was usually classified as enigmatic, exceptional and irrelevant. The neglect and repression of silencing in published work shows our bias against results that deviate from our expectations.

There is no doubt that stable activity of transgenes can be selected (Dehio and Schell, 1993) and that transgene technology has been extremely successful in the production of new traits and culture lines (Stone, 1994). Compared to classical breeding approaches that are based on mutagenesis and selection, the underlying genetic mechanisms are clearly much better understood in transgenic plants. This makes transgene technology not only a faster and more efficient technology than classical breeding, but probably also decreases the risk of unintended secondary effects. Nevertheless, we should realize and even appreciate that we cannot consider a plant to be a simple test-tube and that our understanding of genome organisation and gene expression is still limited. The comparative analysis of the various silencing systems and their underlying molecular characteristics will certainly contribute to an improvement of transgene technology. More importantly, however, it will broaden our understanding of the complexity and the regulation of gene expression in plants.

Acknowledgements
I would like to thank Sarah Grant for her helpful suggestions and critical reading of the manuscript. This work was supported by a Heisenberg scholarship of the German Research Foundation (DFG).

References

Adams, R.L.P., Davis, T., Fulton, J., Kirk, D., Qureshi, M. and Burdon, R.H. (1984) Eukaryotic DNA methylase properties and action on native DNA and chromatin. *Curr. Top. Microbiol. Immun.* 108: 143–156.

Amasino, R.M., Powell, A.L.T. and Gordon, M.P. (1984) Changes in T-DNA methylation and expression are associated with phenotypic variation and plant regeneration in a crown gall tumor line. *Mol. Gen. Genet.* 197: 437–446.

Assad, F.F., Tucker, K.L. and Signer, E.R. (1993) Epigenetic repeat-induced gene silencing (RIGS) in *Arabidopsis*. *Plant Mol. Biol.* 22: 1067–1085.

Bestor, T.H. (1990) DNA methylation: Evolution of a bacterial immune function into a regulator of gene expression and genome structure in higher eukaryotes. *Phil. Trans. R. Soc. Lond. B* 326: 179–187.

Boerjan, W., Bauw, G., van Montagu, M. and Inze, D. (1994) Distinct phenotypes generated by overexpression and suppression of S-adenosyl-L-methionine synthetase reveal developmental patterns of gene silencing in tobacco. *The Plant Cell* 6: 1401–1414.

Brears, T., Liu, C., Knight, T.J. and Coruzzi, G.M. (1996) Ectopic overexpression of glutamine synthetase genes in transgenic tobacco causes an increase in plant growth; *in press*.

Brussian, J.A., Karlin-Neumann, G.A., Huang, L. and Tobin, E.M. (1993) An *Arabidopsis* mutant with a reduced level of cab140 RNA is a result of cosuppression. *The Plant Cell* 5: 667–677.

Cambareri, E.B., Jensen, B.C., Schabtach, E. and Selker, E.U. (1989) Repeat-induced G-C to A-T mutations in *Neurospora. Science* 244: 1571–1775.

Cameron, F.H. and Jennings, P.A. (1991) Inhibition of gene expression by a short sense fragment. *Nucleic Acids Res.* 19: 469–475.

Carter, K.C., Bowman, D., Carrington, W., Fogarty, K., McNeil, J.A., Fay, F.S. and Lawrence, J.B. (1993) A three-dimensional view of precursor messenger RNA metabolism within the mammalian nucleus. *Science* 259: 1330–1335.

Chandler, V.L. and Walbot, V. (1986) DNA modification of a maize transposable element correlates with loss of activity. *Proc. Natl. Acad. Sci. USA* 83: 1767–1771.

de Carvalho, F., Gheysen, G., Kushnir, S., van Montagu, M., Inze, D. and Castresana, C. (1992) Suppression of beta-1.3-glucanase transgene expression in homozygous plants. *EMBO J.* 11: 2595–2602.

Dehio, C. and Schell, J. (1993) Stable expression of a single-copy *rolA* gene in transgenic *Arabidopsis thaliana* allows an exhaustive mutagenesis analysis of the transgene-associated phenotype. *Mol. Gen. Genet.* 241: 359–366.

Dehio, C. and Schell, J. (1994) Identification of plant genetic loci involved in posttranscriptional mechansim for meiotically reversible transgene silencing. *Proc. Natl. Acad. Sci. USA* 91: 5538–5542.

Doerfler, W. (1991) Patters of DNA methylation – Evolutionary vestiges of foreign DNA inactivation as a host defense mechanism. *Biol. Chem. Hoppe-Seyler* 372: 557–564.

Dorer, R.D. and Henikoff, S. (1994) Expansion of transgene repeats cause heterochromatin formation and gene silencing in *Drosophila. Cell* 77: 993–1002.

Dougherty, W.G., Lindbo, J.A., Smith, H.A., Parks, T.D., Swaney, S. and Proebsting, W.M. (1994) RNA-mediated virus resistance in transgenic plants: Exploitation of a cellular pathway possibly involved in RNA degradation. *Mol. Plant-Microbe Interact.* 7: 544–552.

Dreesen, T.D., Henikoff, S. and Loughney, K. (1991) A pairing-sensitive element that mediates transinactivation is associated with the *Drosophila brown* gene. *Genes Develop.* 5: 331–340.

Finnegan, E.J. and Dennis, E.S. (1993) Isolation and identification by sequence homology of a putative cytosine methyltransferase from *Arabidopsis thaliana. Nucleic Acids Res.* 21: 2383–2388.

Flavell, R.B. (1994) Inactivation of gene expression in plants as a consequence of specific sequence duplication. *Proc. Natl. Acad. Sci. USA* 91: 3490–3496.

Fraenkel-Conrad, H. (1983) RNA-dependent RNA polymerases of plants. *Proc. Natl. Acad. Sci. USA* 80: 422–424.

Frommer, M., McDonald, L.E., Millar, D.S., Collis, C.M., Watt, F., Grigg, G.W., Molloy, P.L. and Paul, C.L. (1992) A genomic sequencing protocol that yields a positive display of 5-methylcytosine residues in individual DNA strands. *Proc. Natl. Acad. Sci. USA* 89: 1827–1831.

Goring, D.R., Thomson, L. and Rothstein, S.J. (1991) Transformation of a partial nopaline synthase gene into tobacco suppresses the expression of a resident wildtype gene. *Proc. Natl. Acad. Sci. USA* 88: 1770–1774.

Goyon, C. and Faugeron, G. (1989) Targeted transformation of *Ascobulus immersus* and *de novo* methylation of the resulting duplicated DNA sequences. *Mol. Cell. Biol.* 9: 2818–2827.

Grierson, D., Fray, R.G., Hamilton, A.J., Smith, C.J.S. and Watson, C.F. (1991) Does co-suppression of sense genes in transgenic plants involve antisense RNA? *TIBTECH* 9: 122–123.

Herskowitz, I. (1987) Functional inactivation of genes by dominant negative mutations. *Nature* 329: 219–222.

Holliday, R. and Pugh, J.E. (1975) DNA modification mechanisms and gene activity during development. *Science* 187: 226–232.

Ingelbrecht, I.L.W., Herman, L.M.F., Dekeyser, R.A., van Montagu, M.C. and Depicker, A.G. (1989) Different 3'-end regions strongly influence the level of gene expression in plant cells. *The Plant Cell* 1: 671–680.

Ingelbrecht, I., van Houdt, H., van Montague, M. and Depicker, A. (1994) Posttranscriptional silencing of reporter transgenes in tobacco correlates with specific methylation. *Proc. Natl. Acad. Sci. USA* 91: 10502–10506.

Jones, J.D.G., Gilbert, D.E., Grady, K.L. and Jørgensen, R.A. (1987) T-DNA structure and gene expression in petunia plants transformed by *Agrobacterium tumefaciens* C58 derivatives. *Mol. Gen. Genet.* 207: 478–485.

Jørgenson, R. (1994) Developmental significance of epigenetic imposition on the plant genome: A paragenetic function for chromosomes. *Develop. Genet.* 15: 523–532.

Kooter, J.M. and Mol, J.N.M. (1993) Transinactivation of gene expression in plants. *Plant Biotech.* 4: 166–171.

Kricker, M.C., Drake, J.W. and Radman, M. (1992) Duplication-targeted DNA methylation and mutagenesis in the evolution of eukaryotic chromosomes. *Proc. Natl. Acad. Sci. USA* 89: 1075–1079.

Lee, D.Y., Hayes, J.J., Pruss, D. and Wolffe, A.P. (1993) A positive role for histone acetylation in transcription factor access to nucleosomes. *Cell* 72: 73–84.

Lewin, B. (1994) Chromatin and gene expression: Constant questions, but changing answers. *Cell* 79: 397–406.

Li, E., Bestor, T.H. and Jaenisch, R. (1992) Targeted mutation of the DNA methyltransferase gene results in embryonic lethality. *Cell* 69: 915–926.

Lindbo, J.A., Silva-Rosales, L., Proebsting, W.M. and Dougherty, W.G. (1993) Induction of a highly specific antiviral state in transgenic plants: Implications for regulation of gene expression and virus resistance. *The Plant Cell* 5: 1749–1759.

Loidl, P. (1994) Histone acetylation: Facts and questions. *Chromosoma* 103: 441–449.

Maas, C., Laufs, J., Grant, S., Korfhage, C. and Werr, W. (1990) The combination of a novel stimulatory element in the first exon of the maize *Shrunken-1* gene with the following intron 1 enhances reporter gene expression up to 1000-fold. *Plant Mol. Biol.* 16: 199–207.

Matzke, M.A., Priming, M., Trnovsky, J. and Matzke, A.J.M. (1989) Reversible methylation and inactivation of marker genes in sequentially transformed tobacco plants. *EMBO J.* 8: 643–649.

Matzke, M.A. and Matzke, A.J.M. (1990) Gene interactions and epigenetic variation in transgenic plants. *Develop. Genet.* 11: 214–223.

Matzke, M.A., Neuhuber, F. and Matzke, A.J.M. (1993) A variety of epistatic interactions can occur between partially homologous transgene loci brought together by sexual crossing. *Mol. Gen. Genet.* 236: 379–386.

Meins, J.F. and Kunz, C. (1994) Silencing of chitinase expression in transgenic plants: An autoregulatory model. *In*: J. Paszkowski (ed.): *Gene Inactivation and Homologous Recombination in Plants*. Kluwer Academic Publishers, Dordrecht, pp 335–348.

Meyer, P., Heidmann, I. and Niedenhof, I. (1993) Differences in DNA-methylation are associated with a paramutation phenomenon in transgenic petunia. *The Plant Journal* 4: 86–100.

Meyer, P. and Heidmann, I. (1994) Epigenetic variants of a transgenic petunia line show hypermethylation in transgene DNA: An indication for specific recognition of foreign DNA in transgenic plants. *Mol. Gen. Genet.* 243: 390–399.

Meyer, P., Niedenhof, I. and Ten Lohuis, M. (1994) Evidence for cytosine methylation of non-symmetrical sequences in transgenic *Petunia hybrida*. *EMBO J.* 13: 2084–2088.

Mittelsten Scheid, O., Paszkowski, J. and Potrykus, I. (1991) Reversible inactivation of a transgene in *Arabidopsis thaliana*. *Mol. Gen. Genet.* 228: 104–112.

Mueller, E., Gilbert, J., Davenport, G., Brigneti, G. and Baulcombe, D.C. (1995) Homology-dependent resistance: Transgenic virus resistance in plants related to homology-dependent gene silencing. *The Plant Journal; in press*.

Napoli, C., Lemieux, C. and Jørgensen, R. (1990) Introduction of a chimeric chalcone synthese gene into petunia results in reversible cosuppression of homologous genes in trans. *The Plant Cell* 2: 279–289.

Neuhuber, F., Park, Y.D., Matzke, A.J.M. and Matzke, M.A. (1994) Susceptibility of transgene loci to homology-dependent gene silencing. *Mol. Gen. Genet.* 244: 230–241.

Newman, T.C., Ohme-Takagi, M.O., Taylor, C.B. and Green, P.J. (1993) *DST* sequences, highly conserved among plant SAUR genes, traget reporter transcripts for rapid decay in tobacco. *Plant Cell* 5: 701–714.

Nick, H., Bowen, B., Ferl, R.J. and Gilbert, W. (1986) Detection of cytosine methylation in the maize alcohol dehydrogenase gene by genomic sequencing. *Nature* 319: 243–246.

Pandit, N.N. and Russo, V.E.A. (1992) Reversible inactivation of a foreign gene, *hph*, during the sexual cycle of *Neurospora crassa* transformants. *Mol. Gen. Genet.* 234: 412–422.

Paro, R. and Hogness, D. (1991) The polycomb protein shares a homologous domain with a heterochromatin-associated protein in *Drosophila*. *Proc. Natl. Acad. Sci. USA* 88: 263–267.

Pröls, F. and Meyer, P. (1992) The methylation patterns of chromosomal integration regions influence gene activity of transferred DNA in *Petunia hybrida*. *The Plant Journal* 2: 465–475.

Pryor, K.N., Wuestehube, L.J. and Schekman, R. (1992) Vesicle-mediated protein sorting. *Annu. Rev. Biochem.* 61: 471–516.

Romano, N. and Macino, G. (1992) Quelling: Transient inactivation of gene expression in *Neurospora crassa* by transformation with homologous sequences. *Mol. Microbiol.* 6: 3343–3353.

Scherczinger, C.A. and Knecht, D.A. (1993) Co-Suppression of *Dictyostelium discoideum* myosin II heavy-chain gene expression by a sense orientation transcript. *Antisense Res. Develop.* 3: 207–217.

Schwarz, D. and Dennis, E. (1986) Transposase activity of the *Ac* controlling element in maize is regulated by its degree of methylation. *Mol. Gen. Genet.* 205: 476–482.

Selker, E.U., Cambareri, E.B., Jensen, B.C. and Haack, K.R. (1987) Rearrangement of duplicated DNA in specialized cells of *Neurospora*. *Cell* 51: 741–752.

Smith, C.J.S., Watson, C.R., Ray, J., Schuch, W. and Grierson, D. (1990) Expression of a truncated tomato polygalacturonase gene inhibits expression of the endogenous gene in transgenic plants. *Mol. Gen. Genet.* 244: 447–481.

Smith, H.A., Swaney, S.L., Parks, T.D., Wernsman, E.A. and Dougherty, W.G. (1994) Transgenic plant virus resistance mediated by untranslatable sense RNAs: Expression, regulation and fate of nonessential RNAs. *The Plant Cell* 6: 1441–1453.

Spector, D.L., Fu, X.-D. and Maniatis, T. (1991) Associations between distinct pre-mRNA splicing components and the cell nucleus. *EMBO J.* 10: 3467–3481.

Spena, A., Viotti, A. and Pirrotta, V. (1983) Two adjacent genomic zein sequences: Structure, organization and tissue specific restriction pattern. *J. Mol. Biol.* 169: 799–811.

Spofford, J.B. (1976) Position effect variegation in Drosophila. *In*: M. Ashburner and E. Novitski (eds): *Genetics and Biology of Drosophila*. Academic Press, London. pp 955–1019.

Stone, R. (1994) Large plots are next test for transgenic crop safety. *Science* 266: 1472–1473.

Tanaka, A., Mita, S., Ohta, S., Kyozuka, J., Shimamoto, K. and Nakamura, K. (1991) Enhancement of foreign gene expression by a dicot intron in rice but not in tobacco is correlated with an increased level of mRNA and an efficient splicing of the intron. *Nucleic Acids Res.* 18: 6767–6770.

Thompson, J.S., Ling, X. and Grunstein, M. (1993) Histone *H3* amino terminus is required for telomeric and silent mating locus repression in yeast. *Nature* 369: 245–247.

Travers, A.A. (1994) Chromatin structure and dynamics. *BioEssays* 16: 657–662.

van Blokland, R., van der Geest, N., Mol, J.N.M. and Kooter, J.M. (1994) Transgene-mediated suppression of chalcone synthase expression in *Petunia hybrida* results from an increase in RNA turnover. *The Plant Journal* 6: 861–877.

van der Krol, A.R., Mur, L.A., Beld, M., Mol, J. and Stuitje, A.R. (1990) Flavonoid genes in petunia: Addition of a limiting number of copies may lead to a suppression of gene expression. *The Plant Cell* 2: 291–299.

van Slogteren, G.M.S., Hooykaas, P.J.J. and Schilperoot, R.A. (1984) Silent T-DNA genes in plant lines transformed by *Agrobacterium tumefaciens* are activated by grafting and 5-azacytidine treatment. *Plant Mol. Biol.* 3: 333–336.

Vanyushin, B.F., Tkacheva, S.G. and Belozersky, A.N. (1970) Rare bases in animal DNA. *Nature* 225: 948–949.

Vaucheret, H. (1993) Identification of a general silencer for 19S and 35S promoters in a transgenic tobacco plant: 90 bp of homology in the promotersequence are sufficient for trans-inactivation. *C. R. Acad. Sci. Paris* 316: 1471–1483.

Xing, Y., Johnson, C.V., Dobner, P.R. and Lawrence, J.B. (1993) Higher level organization of individual gene transcription and RNA splicing. *Science* 259: 1326–1330.

Zhang, G., Taneja, K.L., Singer, R.H. and Green, M.R. (1994) Localization of pre-mRNA splicing in mammalian nuclei. *Science* 372: 809–812.

Transgenic Organisms – Biological and Social Implications
J. Tomiuk, K. Wöhrmann & A. Sentker (eds)
© 1996 Birkhäuser Verlag Basel/Switzerland

The impact of transposable elements on genome evolution in animals and plants

W.J. Miller, L. Kruckenhauser and W. Pinsker

Institut für Allgemeine Biologie, AG Genetik, Universität Wien, Medizinische Fakultät, Währingerstraße 17, A-1090 Wien, Austria

Summary. Barbara McClintock's discovery of mobile DNA sequences in maize more than 40 years ago has changed our understanding of genome stability. Since that time many of these jumping genes have been characterized in bacteria, fungi, plants and animals. It is now generally recognized that the majority of all spontaneous mutations having significant phenotypic effects are caused by the mobility of transposable elements (TEs). The biological consequences for the host genome range from point mutations to gross chromosomal rearrangements. Although most TE-induced mutations seem to exert a negative effect on the fitness of their host, a growing body of evidence indicates that in the course of evolution at least some TE-mediated changes have become established features of the host genomes. Examples of how TEs contribute to the evolution of their host are presented in our report: (i) TEs can supply adjacent genes with specific *cis*-regulatory domains thus altering their expression pattern; (ii) they can provide their protein coding sections for the formation of novel host genes; (iii) they may take over basic cellular functions.

Evolutionary life cycle of transposable elements

Eukaryotic genomes harbour a considerable fraction of mobile DNA. In *Drosophila*, one of the most thoroughly studied experimental systems, 10–12% of the genome (Spradling and Rubin, 1981) consists of transposable elements (TEs). TEs are rather short DNA sequences (1.3–10 kb) with limited coding capacity, usually for one up to a few proteins. The genes coded by TEs are mainly required for self-propagation in the genome, e.g., for transposition. Eventually TEs may pick up sequences which serve other functions, e.g., genetic factors which increase the survival of the host thus indirectly supporting the maintenance of the TE in the gene pool (Ajioka and Hartl, 1989). According to their structure and the mechanism of transposition three general classes can be distinguished: DNA transposons, LTR retrotransposons and poly-A retrotransposons. In the genome TEs behave as rather independent entities. They are able to replicate autonomously, inserting copies of themselves at different chromosomal locations. The increase of the copy number brought about by the transpositional activity leads to a propagation of the sequence in the host genome (Hickey, 1992). Since the genetic information carried by TEs is primarily used for self-multiplication without direct benefits for the host, they act like molecular parasites exploiting the environment provided inside living cells (Doolittle and Sapienza, 1980; Orgel and Crick, 1980).

The insertion of TEs at new sites has mainly negative consequences, as it is the case with any randomly occurring mutation. The effects span the entire spectrum from the disruption of single genes (Lambert et al., 1988) to gross chromosomal rearrangements (Goldberg et al., 1983; Lim and Simmons, 1994). Although TE-induced mutations can under certain circumstances provide some advantage to the host, e.g., the increase of adaptability by additional variation or stabilization of coadapted gene arrangements through formation of chromosomal inversions (Lyttle and Haymer, 1992), the detrimental effects predominate. In spite of the negative consequences the replicative mechanism of transposition enables newly activated transposons to invade the gene pool of the host population. The initial transpositional burst, however, cannot proceed forever. Natural selection tends to minimize the harmful mutagenic effects, eliminating the TE from the gene pool through the reduced fitness of the afflicted individuals. On the other hand, selection favours mechanisms that act as negative regulators of transpositional activity. One source of transpositional suppressors is provided by the TEs themselves: Defective copies, which are generated by the inaccurate mode of replication. These incomplete copies lack essential functions of active transposons but they are able to inhibit the transposition process through interactions with the active members of their TE family at the DNA or protein level. Another possibility is the modification of host encoded factors required for transposition. Favoured by natural selection these mechanisms will eventually get control over tranpositional activity and finally the propagation of the TE in the genome will come to a halt.

Once inactivated, TEs are still transmitted vertically like any other genomic sequence. As long as the functionally important sections remain intact, resurrection of the silent TE is possible. Outcrossing of individuals carrying inactivated TEs, e.g., through migration to an uninfected population lacking specific suppressors in the genome, may initiate a new transpositional burst. Likewise, external stress factors (chemical and physical mutagens, temperature) can temporarily lead to a breakdown of the suppressing mechanism and the transposition activity can be resumed (Junakovic et al., 1986). In the long run, however, unexpressed TEs are doomed to genomic extinction. As a result of recombination and random genetic drift the copy number will decrease by "stochastic loss" and finally the immobile TEs will be eliminated from the gene pool. But even those copies which are transmitted over many generations will suffer degradation of their information content. Without selective constraints maintaining the functionality of the sequence, an immobilized TE will degrade by accumulating mutations, a process finally leading to "vertical inactivation" (Lohe et al., 1995). The upper temporal limit for successful reactivation of a silenced gene has been estimated as 6 million years (Marshall et al., 1994). Thus, in a phylogenetic lineage undergoing fast adaptive radiation, successive periods of inactivation and reactivation of a TE family along different branches may bring about a "spotty" distribution pattern among closely related species. In fact, discontinuous distribution of TEs which is not in accordance with the phylogeny of the host species appears as a widespread phenomenon observed with several TE families (Kidwell, 1992, 1993).

One way for TEs to escape the fate of stochastic loss or vertical inactivation is horizontal transmission into the gene pool of another species. Evidence for the transfer of TE families across species barriers has been provided for several TE families of *Drosophila* (Abad et al., 1989; Daniels et al., 1990; Mizrokhi and Mazo, 1990; Hagemann et al., 1992; Maruyama and Hartl, 1991; Lawrence and Hartl, 1992; Simmons, 1992; Clark et al., 1994; Lohe et al., 1995) and other species (Calvi et al., 1991; Heinemann, 1991; Smith et al., 1992; Kidwell, 1993; Robertson, 1993). In contrast to non-mobile genes, TEs have a much greater chance to become integrated into a novel genomic environment. Due to their mobility they are likely to disperse through the gene pool of the new host, causing all the side effects connected with a transpositional burst.

Up to this point the role of TEs in the evolution of genomes has been described more or less as an additional source of variation. The evolutionary life cycle of the TE families has been considered as a succession of three phases: Dynamic replication, inactivation and degradation (see Fig. 1). To start a new cycle, TEs can either remain dormant for a longer period or invade a new gene pool. There is, however, still another possibility how TE sequences can be maintained in the genome. Getting captured by control mechanisms of the host, they may acquire novel functions and thus become stably integrated into the genome in a process of coadaptation (von Sternberg et al., 1992; McDonald, 1993). Although this alternative to the regular cycle probably occurs only sporadically, comparable to the rareness of advantageous point mutations, the impact on the evolution may be more important than the additional random mutations caused by the transpositional activity. Therefore it is the goal of this chapter to compile cases where former TEs have apparently become "domesticated" (Miller et al., 1992) by the host, serving as stable and useful components of the genome.

Regulation of host gene expression

When transposable elements were first discovered more than forty years ago, it was suggested that these sequences might play a role in the control of the gene expression (McClintock, 1951). Later it was proposed (Fincham and Sastry, 1974) that on an evolutionary time scale TE-induced mutations could also be responsible for lasting changes of expression patterns, modifying, e.g., the tissue specificity of genes or their temporal expression during development. This assumption is confirmed by recent findings. In general, TEs have a preference to integrate in the 5'-non-coding regions of genes thus causing mutations in regulatory sequences (Voelker et al., 1990). Moreover, TEs carry signals that are not only necessary for various aspects of their

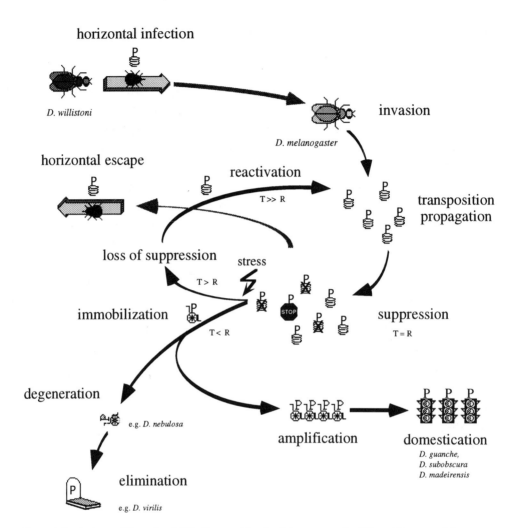

Figure 1. Evolutionary life cycle of TEs illustrated by the *P* transposons in different lineages of the genus *Drosophila*. A new gene pool is invaded through horizontal infection. The best example for this phenomenon is the recent transfer from *D. willistoni* to *D. melanogaster*. The invasion is followed by a phase of active transposition and rapid propagation of the TE in the new genomic environment. In order to suppress the detrimental side effects of TE-induced mutations, regulating mechanisms develop, leading to the immobilization of the TE. As long as the functionally important sequences are still intact, loss of suppression can lead to the reactivation of the dormant TE thus triggering a new phase of multiplication and genomic mobility. After a longer period of inactivity, however, the TE will accumulate deleterious mutations. Damage in functionally important sections of the sequence will cause permanent immobilization (e.g., *D. nebulosa*) and finally the elimination of the TE from the genome (e.g., *D. virilis*). This fate can be averted either by the horizontal escape into an uninfected gene pool, starting there with a new cycle, or by molecular domestication, the stable integration into the genome combined with functional modification (e.g., the tandemly amplified *P* homologues in *D. guanche* and its relatives).

own expression, but can also act on adjacent genes or interfere with the transcriptional machinery of these genes. The regulatory functions of TEs have been extensively studied in cultured cell lines (for review see, Arkhipova and Ilyin, 1992), but little is known about expression patterns in living multicellular organisms. Recently, Ding and Lipshitz (1994) have examined the expression patterns of 15 different families of LTR-containing retrotransposons during the normal development of *D. melanogaster*. This investigation included the well known *copia*-like elements *17.6*, *297*, *412*, *blood*, *copia*, *gypsy*, *mdg1*, *mdg3*, and *roo*. For each type of TE the analyses were carried out with independent wild-type strains differing in the chromosomal insertion sites of the respective transposon. Nevertheless, the results demonstrate that each of the 15 TE families exhibits a typical pattern of spatial and temporal expression during embryogenesis. These specific patterns are strictly conserved although the strains differed widely with respect to the insertion sites. It can be concluded from these data that each TE harbours internal *cis*-regulatory sequences which interact with specific host-encoded transcription factors. A good example is given by the *D. melanogaster* retrotransposon *412*. Here the *cis*-regulating sequences, which are located within the LTRs, are recognized by the homeobox protein coded by the *ubx* gene, a homeotic gene otherwise involved in the determination of segment identity along the anterior–posterior axis. The direct DNA–protein interaction between the LTR-specific regulatory sequence and the UBX protein has been proved by *in vitro* as well as *in vivo* experiments.

A similar situation is found with the transcriptional regulation of the TE *copia* (Cavarec and Heidmann, 1993). This retrotransposon of *D. melanogaster* possesses homeoprotein binding sites located inside its left LTR, close to a previously identified enhancer element described by Mount and Rubin (1985). In addition, ectopic expression of the *fushi tarazu* protein, which is normally involved in the formation of body segments, results in a 2- to 3-fold increase of endogenous *copia* transcripts in transgenic embryos (Cavarec and Heidmann, 1993). Interestingly, the structural organization of the *copia* LTR and the adjacent homeoprotein binding site shows striking similarities to corresponding sequences in the *Drosophila* retrotransposons *412*, *297*, and *gypsy* as well as in the yeast retroelement *Ty*. In addition, Ding and Lipshitz (1994) have demonstrated that the tissue-specific expression patterns are shared among several retrotransposon families. For example, the TEs *297*, *mdg3* and *copia* are expressed in the central nervous system late in embryogenesis, whereas *blood*, *roo*, and *gypsy* are expressed in yolk nuclei of early embryos.

These findings suggest that the retrotransposons in the *Drosophila* genome contain a rich repertoire of transcriptional regulatory elements with the potential to confer novel spatial and temporal patterns on flanking host genes. Therefore TE-induced mutations in the *cis*-regulatory regions of duplicated genes may be more important for their functional diversification than mutations in the coding sequences (Fincham and Sastry, 1974; King and Wilson, 1975).

Li and Noll (1994) provided the first experimental evidence for this hypothesis. The three *Drosophila* genes, the pair-rule gene *paired* (*prd*), the segment-polarity gene *gooseberry* (*gsb*), and the neural specific gene *gooseberry neuro* (*gsbn*), have distinct developmental functions during *Drosophila* embryogenesis. The three corresponding proteins PRD, GSB and GSBN act as transcription factors. Whereas their N-terminal halves are highly conserved, their C-terminal regions as well as their *cis*-regulatory regions, which determine their specific spatial and temporal expression, show a high degree of variation. Li and Noll (1994) could demonstrate that the three proteins are interchangeable with respect to their regulatory function. For example, they were able to rescue a *gsb⁻* phenotype by using a chimeric transgene consisting of the *gsb* control region fused to the entire protein coding region of the *prd* gene. Thus the PRD protein can serve as a substitute for the GSB protein, provided that its expression is governed by a *gsb* control region. In addition, in footprinting analyses carried out *in vitro* all three proteins bind precisely to the same (A+T)-rich DNA target sites. The sequence similarities between the three proteins as well as the conservation of their DNA binding function indicate that they evolved from an ancestral single-copy gene by repeated duplication. The duplicated coding sequences of the original gene might have been juxtaposed to new *cis*-regulatory elements which changed their expression patterns and thus gave them novel stage and tissue specific functions.

These data should be discussed in the context of recent findings obtained with TEs in plants (Bureau and Wessler, 1992, 1994a, 1994b). In these studies evidence is provided that retroelements are involved in the evolution of gene structure and expression by supplying duplicated genes with new regulatory sequences. In contrast to *Drosophila*, only a few complete retrotransposons are known to be transcriptionally active in plants grown under normal conditions. Consequently, TE induced mutations are rare. On the other hand, the plant genomes have huge amounts of interspersed repetitive components derived from *copia*-like elements which are no longer able to transpose. Retrotransposon-like sequences are ubiquitous and diverse in plants, suggesting that these mobile elements are an ancient component of plant genomes.

Recently, White et al. (1994) have isolated a full-sized member of a novel family of *copia*-like elements causing a mutation in the *waxy* gene of maize. This retrotransposon, called *Hopscotch*, is 4.8 kb long, has identical 231 bp LTRs and one single ORF with the coding capacity for a 1440 aa protein. It contains all the amino acid domains known to be essential for autonomously active retro-elements (nucleic acid binding, protease, integrase, reverse transcriptase, and RNase H). Database searches with the *Hopscotch* amino acid sequence revealed that 21 previously described normal plant genes harbour remnants of functional domains of this element family in their flanking regions. Many of the retrotransposon-like sequences in the flanks of these normal genes represent quite ancient insertions. In three cases retro-like sequences have been found in several members of one gene family at homologous positions, indicating that the insertions predated the origin of the gene family by successive duplications. For example, five out of seven sequenced members of the 19 kDa *zein* family of maize (*pms1*, *pms2*, *zel9ba*, *zeil9* and *ze25*) have retrotransposon-like sequences at homologous positions of their 5'-flanking regions. The LTRs of these insertion sequences provide an additional promoter which may influence the expression of the adjacent genes. Another example for modification of the expression pattern is provided by three members of the ribulose-bisphosphate carboxylase (*rbcS*) family in maize, namely *rbcS-E9*, *rbcS-8.0* and *rbcS-3.6*. These genes share a common TE insertion in their upstream flanking regions and are coordinately expressed. In contrast, the *rbc-3 A* gene, which is also a member of this gene family but lacks the TE insertion, is expressed in a different manner. The data suggest that the upstream insertion of a *copia*-like TE shared by *rbcS-E9*, *rbcS-8.0*, and *rbcS-3.6* may be responsible for the lower expression of these genes in leaves and seeds compared to *rbc-3 A*. Therefore, in this case the insertions apparently serve as negative regulators of transcription.

   The important role of TEs as modifiers of the expression of normal plant genes is further confirmed by investigations on the mobile sequence *Tourist* (Bureau and Wessler, 1994a). This small TE belongs to the group of SINE-like mobile elements which accounts for a significant fraction of the repetitive DNA in plants (Bureau and Wessler, 1992). It has inverted repeats at the termini and a total length of only 128 bp. The widespread presence in different plant species, e.g., maize, grasses, rice, and sorghum, indicates that *Tourist* may be a phylogenetically old component of the plant genomes (>50 million years). Nevertheless, it seems to have retained its mobility as has been proved by the observation of recent transpositional activity detected in maize, sorghum, and rice. Using *Tourist* as a query sequence in a GenBank search, all *Tourist* elements identified so far turned out to be components of normal non-mutant genomic sequences, located either in their upstream regions or within their intronic sequences. In maize, the *Tourist*-like elements are found to be associated with 23 normal genes, a number representing one-third of the genomic maize gene sequences available in the data base. Similarily, 6 out of 35 genomic rice gene sequences, 5 out of 8 sorghum gene sequences, and 2 out of 40 barley sequences in the GenBank data base contain *Tourist*-like elements in their control regions.

   Stowaway, another member of the inverted repeat type family of TEs, shares many structural features with the *Tourist* element but no significant similarity at the sequence level (Bureau and Wessler, 1994b). This new family of TEs has been identified within a *Tourist* insertion located at the extreme 5'-end of the sorghum phosphoenolpyruvate carboxylase *CP21* gene sequence. The GenBank search revealed 47 plant gene sequences harbouring *Stowaway* elements with an overall sequence similarity among elements of more than 60%. Unlike *Tourist* elements which were found only in a limited number of cereal grasses, the *Stowaway* family has a much wider distribution, with members present in both monocotyledonous and dicotyledonous plants. Interestingly, these TE-derived elements that are mostly located in the 5'-flanking regions of normal plant genes harbour domains previously identified as *cis*-acting regulatory motifs. For example, two putative embryogenesis specific nuclear factors bind within the internal sequence of the *Stowaway* element located at the 5'-region of the carrot gene *DC59*.

So far, more than 100 normal plant genes were found to contain mobile elements or their remnants (*Hopscotch*, *Tourist* and *Stowaway*). The correspondence of many of their locations to previously identified *cis*-regulatory domains provide strong evidence that these elements have influenced the evolution of the expression patterns of normal plant genes. The conservation of their insertion site in different members of one gene family indicates that TE-insertions often occurred prior to gene duplication of the ancestor gene. The *cis*-regulatory sequences derived from insertions of former mobile elements have been sequestrated by the plant genes in the course of evolution and are now involved in normal gene regulation.

In contrast to plants, data bank searches in *D. melanogaster* have failed to detect any *copia*-like insertions located in the flanks of normal genes (White et al., 1994), although these elements are highly expressed and acted as the causative agent of many spontaneous mutations in *Drosophila* (for review see, Bingham and Zachar, 1989). In plants, *copia*-like TEs are transcribed at low levels under normal conditions and have been found to be responsible for only a few mutations. This disparity between plant and *Drosophila* genes can be interpreted by assuming a different historic pattern in the two lineages. The *copia*-like elements of plants are thought to be quite old components of the genomes and thus probably had a longer time frame to become inserted into the flanking regions of genes. Among the many insertions that occurred, only those became finally fixed which exerted selectively favourable effects on the adjacent coding sequences. In contrast, *copia*-like elements of *Drosophila* are suspected to have entered the lineage more recently by horizontal transmission as indicated by the analysis of both the codon usage of *copia* and its sequence relationships to other retrotransposons (Mount and Rubin, 1985). Thus for the *copia*-like TEs of *Drosophila* the period of time spent in the lineage may have been simply too short to insert at adequate locations where stable integration led to a selective advantage of the host.

In species other than plants, only a few examples have been reported where TEs are thought to have contributed to the evolution of host gene expression. In humans the amylases are encoded by a gene family comprising five sequences. Two subfamilies, named *amy-1* and *amy-2*, exist which differ in tissue specifity. Whereas the two *amy-2* genes are expressed in the pancreas, the three *amy-1* genes are active only in the parotid salivary gland. All members of the amylase gene family share a common insertion consisting of a gamma-actin pseudogene 0.2 kb upstream of the first exon. Robins and Samuelson (1992) could show that the tissue-specific expression pattern of the *amy-1* genes has been caused by a retroviral-like insertion in the pseudogene, obviously by a reactivation of a cryptic promoter within the pseudogene. Salivary amylase is produced, e.g., in primates, rodents, lagomorphs but not in all species within these orders. The inconsistency of salivary expression suggests that the tissue specifity has been acquired independently in different lineages. The human amylase genes have derived from a common ancestor as all members harbour the gamma-actin pseudogene. This insertion ocurred approximately 40 million years ago, after the separation of the mammalian lineages. There are no gamma-actin or retroviral inserts in the mouse amylase cluster, which also have salivary-specific genes.

An other example provides evidence that the specific expression of the mouse sex-limited protein (SlP) has been caused by the ancient insertion of a retrovirus-like sequence (Stavenhagen and Robins, 1988; Robins and Samuelson, 1992). This androgen-responsive gene is a member of the murine histocompatibility complex which arose by tandem duplication of the fourth component of the complement *C4* gene. Sequence comparisons between the genes for *C4* and *Slp* revealed that their major regulatory difference in mice is caused by an ancient retroviral-like insertion, which alters the expression of the gene located 2 kb downstream *via* LTR regulatory sequences that act as hormone responsive enhancers. The specifity to androgen appears to depend on accessory factors that bind within the LTR and cooperate with bound hormone receptors (Adler et al., 1991).

Formation of novel protein coding genes

As shown above, TEs can contribute to the functional diversification of genes by supplying *cis*-regulatory domains altering the expression pattern. In the following, we will describe the first example for the conversion of TE-derived coding sequences into stably integrated host genes in *Drosophila* (Miller et al., 1992, 1995). *P* elements comprise families of TEs which are typically found as dispersed repetitive sequences in several *Drosophila* species (for review see, Kidwell, 1994). In contrast, the genomes of the closely related species *D. guanche, D. subobscura*, and *D. madeirensis* harbour only immobile *P* element derivatives which are tandemly clustered at a single chromosomal location (Paricio et al., 1991; Miller et al., 1992). Although these defective transposon derivatives have lost their termini and one of their exons, they have still retained the coding capacity for a "*P*-repressor-like" 66 kDa protein. The conserved genomic position in this triad of species indicates that the loss of mobility has taken place in the common ancestor at least 3 Myr ago (Pinsker et al., 1993). No complete *P* elements could be detected in these species, only in *D. subobscura* eroded relics of *P* element derived sequence fragments were found in the β-heterochromatin (Paricio et al., 1994).

How these *P* elements became immobilized has been described in a plausible model (Pinsker et al., 1993). But which selective forces were responsible for the conservation of the clustered

P element-derived protein coding sequences? Northern blot analyses show that transcripts of the expected size of 2.1 kb are produced in adults of both *D. guanche* and *D. subobscura* (Miller et al., 1995). Moreover, the *P* homologues of *D. guanche* are even transcribed in a stage and tissue specific pattern (W.J. Miller, unpublished data). Both the conservation of the ORFs and the transcriptional activity can be taken as an evidence that these sequences have adopted new functions. This requires that the coding sequences as well as the regulatory units have become modified accordingly.

Since these *P* homologues have lost both termini including the original promotor sequences (Rio, 1990), their control regions have been destroyed. The spacer region located between adjacent copies of the tandem array has no similarity to the missing segments of mobile *P* transposons. Sequence analysis of the noncoding upstream region from independent *P* homologues of *D. guanche* and *D. subobscura* revealed the existence of two different types of cluster units, designated A- and G-type. Although both types harbour typical eukaryotic motifs of transcription initiation (GC-box, octamer motif, CAAT-box and TATA-box) not present in the canonical *P* element, they have different evolutionary origins. The A-type promoter is composed of footprints left behind by at least two different insertions of TEs. One sequence shows strong similarity to *ISY3*, a *LINE*-type insertion sequence of *D. miranda* (Steinemann and Steinemann, 1993), the other resembles the inverted repeat of a still unnamed dispersed repetitive element (*UDRE-812*) of *D. subobscura* (Marfany and Gonzàlez-Duarte, 1992). The 192 bp section similar to a subregion of *ISY3* includes putative binding sites for transcription factors: A GC-box, a CAAT-box, and the promoter-associated octamer motif ATTTGCAT (Steinemann and Steinemann, 1993). A TATA-box is provided by the section homologous to a 179 bp stretch from the *UDRE-812* inverted repeat (Miller et al., 1995). Activation by the captured promoter has probably resulted in a different expression pattern initiating a novel function of the protein.

In contrast to the situation found in plants and mammals, where the acquisition of novel control regions seems to be a more widespread phenomenon (see above), this is so far the only example of a TE-derived promotor in wild-type genes of *Drosophila*. Moreover, in this case even the coding region does not stem from a preexisting host gene but from still another TE, the *P* element. To elucidate the potential function of these *P*-derived repressor-like proteins, the amino acid sequences coded by several cluster units from *D. guanche* and *D. subobscura* were compared. The 66 kDa proteins are strongly conserved within and between the related species, with the exception of the N-terminal region, a hypervariable section comprising 62 amino acids. Although being highly variable, these amino-terminal sections of the proteins have apparently preserved structural features of functional importance. Each protein has retained zinc finger-like motifs which were also found in the active transposase-coding *P* elements of other species. In addition, the proteins harbour a coiled-coil domain and three leucine zippers. Although this assumption has to be confirmed by experimental data, the sequence analysis suggests that the proteins encoded by the *P* homologues might function as DNA-binding factors probably regulating the expression of still unknown genomic target genes (Miller et al., 1995). The formation of a gene cluster made up of diverged repeat units, each composed of three different TEs, suggests the functional transition of former mobile elements into controlling elements integrated by the host genome into a complex regulatory network.

Acquisition of novel cellular functions

In the examples described above the TEs adopted by the host genome lost their mobility and became stably integrated at fixed locations. A third possiblity, where TEs adopt cellular functions without losing their ability to transpose is illustrated by recent results on the organisation of telomere sequences in *Drosophila*. The DNA isolated from the chromosomal termini of most eukaryotes contains tandem repeats of a short sequence (2 – 8 bp), commonly referred to as terminal or telomeric repeats (for review see, Blackburn, 1991; Biessmann and Mason, 1992). These sequences have been detected in protozoans, fungi, algae, slime molds, nematodes, humans and flowering plants. These terminal blocks are heterogeneous in size (hundreds to thousands of repeats). Their length is thought to be determined by a dynamic equilibrium between loss caused by incomplete replication of the termini or terminal degradation, and the addition of terminal repeats mediated by telomerase activity. Although telomeric sequences made up of short repeats are typical for eukaryotes, the telomeres of *Drosophila* have apparently developed a different mechanism. Several lines of evidence indicate that here the loss of terminal sequences is compensated by the addition of repetitive elements generated through transposition (Traverse and Pardue, 1988; Biessmann et al., 1992b; Levis et al., 1993). The first TEs discovered as spontaneous additions to broken chromosome ends were the *HeT-A* sequences (Traverse and Pardue, 1988). The *HeT-A* element is a member of the LINE family, non-LTR retrotransposons with a typical oligo(A) tract at the 3'-end. These LINE-like elements possess two open reading frames. One codes for a *gag*-like protein, the second shows similarity to a reverse transcriptase. Another telomere specific LINE-like element, named *TART*, has been discovered in native *Drosophila* telomeres (Levis et al., 1993). Both, *HeT-A* and *TART* share sequence similarity only in ORF 2 and belong to the same class of LINE-like retrotransposons (Sheen and Levis, 1994). At the proximal side of the *TART* element there were three other TE insertions of two *HeT-A* and one additional *TART* sequence. All these elements were truncated at their 5'-ends but organized in the same 5'– 3' orientation. It has been postulated that this tandem arrangement of two different types of LINE-like elements at the *Drosophila* chromosome ends has resulted from successive transposition of the four elements into the terminus. Additional evidence comes from *in situ* hybridizations on polytene chromosomes. When probed with *TART* elements, signals were obtained only from the telomeric tips indicating that this TE family resides exclusively at the chromosomal tips.

The isolation of a full-sized *TART* element provided direct evidence that these elements retrotranspose preferentially into the termini of chromosomes. This site-directed transposition appears therefore as a part of the essential process by which *Drosophila* telomeres are maintained (Sheen and Levis, 1994). With a length of 5.1 kb the 3'-UTR of *TART* elements has an exceptional size compared to other LINE elements (<0.6 kb for all other LINE*s* of *Drosophila* known so far). The average size of both the *HeT-A* (Biessmann and Mason, 1992) and the *TART* elements (Sheen and Levis, 1994) is 7 kb, a length quite atypical for this class of elements. Biessmann et al. (1992a) measured the transposition rate for *HeT-A* elements on the X chromosome as 0.7– 1.0% per generation. Progressive telomeric DNA loss due to incomplete replication occurs at the chromosome end at a rate of 70–80 bp per fly generation (Levis, 1989). The average size of the telomere addition by one of the two LINE-like elements (7 kb) multiplied by their rate of transposition (0.7–1.0%) results in a net addition of 50–70 bp per chromosome end. This stochastic telomere lengthening caused by *HeT-A* and *TART* transpositions should be sufficient to counterbalance the progressive loss of chromosome ends. The accumulation of these sequences in the telomeric region thus provides a buffer that prevents loss of single-copy genes from the receding ends.

This system found in *Drosophila* represents a balanced equilibrium between terminal loss and sequence addition by transposition. After loss of the primordial short eukaryotic telomeres in *Drosophila*, a fine-tuned system between the host genome and mobile elements has developed to take over a basic cellular function. The insertion frequency of a specific class of transposable elements must have been adapted to match the average rate of telomere loss in order to ensure constant chromosome size. These results present a fascinating example how natural selection has managed to force a "selfish" TE into the framework of basic cellular mechanisms without curtailing the most characteristic feature of this class of DNA sequences: Genomic mobility.

Conclusions

In biotechnology mobile DNA sequences are used as potent tools and gene vector systems. One of the applications is the generation of transgenic organisms. Although the functional aspects of TEs are quite well understood, little is known about their evolutionary fate and their implications on host genome evolution. It is generally known that insertions of TEs can induce new mutations, a side effect that may counter any advantage provided by the transferred gene. Another problem is posed by the instability of the transferred gene carried on a piece of mobile DNA as a vector and its interaction with endogenous jumping genes. In our contribution we provide evidence that in the course of evolution TEs have become stable elements of the genome. This, however, happened in evolutionary time scales and under the control of natural selection. We cannot expect that a balance between the transferred gene, the TE vector, and the host genome is established hastily, i.e., within the life time of a scientist. Although TEs have the potential to exert positive effects on the host genome, it is not certain whether these effects will actually be brought about and whether they will be considered positive from the viewpoint of the experimenter.

References

Abad, P., Vaury, C., Pèlisson, A., Chaboissier, M.-C., Busseau, I. and Bucheton, A. (1989) A long interspersed repetitive element – the *I* factor of *Drosophila teissieri* – is able to transpose in different *Drosophila* species. *Proc. Natl. Acad. Sci. USA* 86: 8887–8891.

Adler, A.J., Scheller, A., Hoffman, Y. and Robins, D.M. (1991) Multiple components of a complex androgen-dependent enhancer. *Mol. Endocrinol.* 5: 1587–1596.

Ajioka, J.W. and Hartl, D.L. (1989) Population dynamics of transposable elements. *In*: D.E. Berg and M.M. Howe (eds): *Mobile DNA*. American Society of Microbiology, Washington, pp 939–959.

Arkhipova, I.R. and Ilyin, Y.V. (1992) Control of transcription of *Drosophila* retrotransposons. *BioEssays* 14: 161–168.

Bingham, B.M. and Zachar, Z. (1989) Retrotransposons and *FB* transposons from *Drosophila melanogaster*. *In*: D.E. Berg and M.M. Howe (eds): *Mobile DNA*. American Society of Microbiology, Washington, pp 485–502.

Biessmann, H. and Mason, J.M. (1992) Genetics and molecular biology of telomeres. *Adv. Genet.* 30: 185–249.

Biessmann, H., Champion, L.E., O'Hair, M., Ikenaga, K., Kasravi, B. and Mason, J.M. (1992a) Frequent transposition of *Drosophila melanogaster HeT-A* transposable elements to receding chromosome ends. *EMBO J.* 11: 4459–4469.

Biessmann, H., Valgiersdottir, A., Lofsky, A., Chin, C., Ginther, B., Levis, R. and Pardue, M.-L. (1992b) *Het-A*, a transposable element specifically involved in "healing" broken chromosome ends in *Drosophila*. *Mol. Cell. Biol.* 12: 3910–3918.

Blackburn, E.H. (1991) Structure and function of telomeres. *Nature* 350: 569–573.

Bureau, T.E. and Wessler, S.R. (1992) *Tourist*: A large family of small inverted repeat elements frequently associated with maize genes. *Plant Cell* 4: 1283–1294.

Bureau, T.E. and Wessler, S.R. (1994a) Mobile inverted-repeat elements of the *Tourist* family are associated with the genes of many cereal grasses. *Proc. Natl. Acad. Sci. USA* 91: 1411–1415.

Bureau, T.E. and Wessler, S.R. (1994b) *Stowaway*: A new family of inverted repeat elements associated with the genes of both monocotyledonous and dicotyledonous plants. *Plant Cell* 6: 907–916.

Calvi, B., Hong, T.J., Findley, S.D. and Gelbart, W.M. (1991) Evidence for a common evolutionary origin of inverted repeat transposons in *Drosophila* and plants: *hobo*, *Activator*, and *Tam3*. *Cell* 66: 465–471.

Cavarec, L. and Heidmann, T. (1993) The *Drosophila copia* retrotransposon contains binding sites for transcriptional regulation by homeoproteins. *Nucleic Acids Res.* 21: 5041–5049.

Clark, J.B., Maddison, W.P. and Kidwell, M.G. (1994) Phylogenetic analysis supports horizontal transfer of *P* transposable elements. *Mol. Biol. Evol.* 11: 40–50.

Daniels, S.B., Peterson, K.R., Strausbaugh, L.D., Kidwell, M.G. and Chovnick, A. (1990) Evidence for horizontal transmission of the *P* transposable element between *Drosophila* species. *Genetics* 124: 339–355.

Ding, D. and Lipshitz, H.D. (1994) Spatially regulated expression of retrovirus-like transposons during *Drosophila melanogaster* embryogenesis. *Genet. Res. Camb.* 64: 167–181.

Doolittle, W.F. and Sapienza, C. (1980) Selfish genes, the phenotype paradigm and genome evolution. *Nature* 284: 601–603.

Fincham, J.R.S. and Sastry, G.R.K. (1974) Controlling elements in maize. *Annu. Rev. Genet.* 8: 15–50.

Goldberg, M.L., Sheen, J.Y., Gehring, W.J. and Green, M.M. (1983) Unequal crossing-over associated with asymmetrical synapsis between nomadic elements in the *Drosophila melanogaster* genome. *Proc. Natl. Acad. Sci. USA* 80: 5017–5021.

Hagemann, S., Miller, W.J. and Pinsker, W. (1992) Identification of a complete *P* element in the genome of *Drosophila bifasciata*. *Nucleic Acids Res.* 20: 409–413.

Heinemann, J.A. (1991) Genetics of gene transfer between species. *Trends Genet.* 7: 181–185.

Hickey, D.A. (1992) Evolutionary dynamics of transposable elements in prokaryotes and eukaryotes. *Genetica* 86: 269–274.

Junakovic, N., di Franco, C., Barsanti, P. and Palumbo, G. (1986) Transposition of *copia*-like nomadic elements can be induced by heat-shock. *J. Mol. Evol.* 24: 89–93.

Kidwell, M.G. (1992) Horizontal transfer. *Curr. Opin. Genet. Dev.* 2: 868–873.

Kidwell, M.G. (1993) Lateral transfer in natural populations of eukaryotes. *Annu. Rev. Genet.* 27: 235–256.

Kidwell, M.G. (1994) The evolutionary history of the *P* family of transposable elements. *J. Hered.* 85: 339–346.

King, M.C. and Wilson, A.C. (1975) Evolution at two levels in human and chimpanzees. *Science* 188: 107–116.

Lambert, M.E., McDonald, J.F. and Weinstein, I.B. (1988) *Eukaryotic Transposable Elements as Mutagenic Agents.* Cold Spring Harbor Press, New York.

Lawrence, J.G. and Hartl, D.L. (1992) Interference of horizontal genetic transfer from molecular data: An approach using the bootstrap. *Genetics* 131: 753–760.

Levis, R.W. (1989) Viable deletions of a telomere from a *Drosophila* chromosome. *Cell* 58: 791–801.

Levis, R.W., Ganesan, R., Houtchens, K., Tolar, L.A. and Sheen, F. (1993) Transposons in place of telomeric repeats at a *Drosophila* telomere. *Cell* 75: 1083–1093.

Li, X. and Noll, M. (1994) Evolution of distinct developmental functions of three *Drosophila* genes by acquisition of different *cis*-regulatory regions. *Nature* 367: 83–87.

Lim, J.K. and Simmons, M.J. (1994) Gross chromosome rearrangements mediated by transposable elements in *Drosophila melanogaster*. *BioEssays* 16: 269–275.

Lohe, A.R., Moriyama, E.N., Lidholm, D.-A. and Hartl, D. (1995) Horizontal transmission, vertical inactivation, and stochastic loss of *mariner*-like transposable elements. *Mol. Biol. Evol.* 12: 62–72.

Lyttle, T.W. and Haymer D.S. (1992) The role of transposable element *hobo* in the origin of endemic inversions in wild populations of *Drosophila melanogaster*. *Genetica* 86: 113–126.

Marfany, G. and Gonzàlez-Duarte, R. (1992) Evidence for retrotranscription of protein-coding genes in the *Drosophila subobscura* genome. *J. Mol. Evol.* 35: 492–501.

Marshall, C.R., Raff, E.C. and Raff, R.A. (1994) Dollo's law and the death and resurrection of genes. *Proc. Natl. Acad. Sci. USA* 91: 12283–12287.

Maruyama, K. and Hartl, D. (1991) Evidence for interspecific transfer of the transposable element *mariner* between *Drosophila* and *Zaprionus*. *J. Mol. Evol.* 33: 514–524.

McClintock, B. (1951) Chromosome organization and genic expression. *Cold Spring Harbor Symposia on Quantitative Biology* 16: 13–47.

McDonald, J.F. (1993) Evolution and consequences of transposable elements. *Curr. Opin. Genet. Dev.* 3: 855–864.

Miller, W.J., Hagemann, S., Reiter, E. and Pinsker, W. (1992) *P* homologous sequences are tandemly repeated in the genome of *Drosophila guanche*. *Proc. Natl. Acad. Sci. USA* 89: 4018–4022.

Miller, W.J., Paricio, N., Hagemann, S., Martinez-Sebastián, M.J., Pinsker, W. and DeFrutos, R. (1995) Structure and expression of the clustered *P* element homologues in *Drosophila subobscura* and *D. guanche*. *Gene* 156: 167–174.

Mizrokhi, L.J. and Mazo, A.M. (1990) Evidence for horizontal transmission of the mobile element *jockey* between distant *Drosophila* species. *Proc. Natl. Acad. Sci. USA* 87: 9216–9220.

Mount, S.M. and Rubin, G.M. (1985) Complete nucleotide sequence of the *Drosophila* transposable element *copia*: Homology between *copia* and retroviral proteins. *Mol. Cell. Biol.* 5: 1630–1638.

Orgel, L.E. and Crick, F.H.C. (1980) Selfish DNA: The ultimate parasite. *Nature* 284: 604–607.

Paricio, N., Pérez-Alonso, M., Martínez-Sebastian, M.J. and DeFrutos, R. (1991) *P* sequences of *Drosophila subobscura* lack exon 3 and may encode a 66kd repressor-like protein. *Nucleic Acids Res.* 19: 6713–6718.

Paricio, N., Martínez-Sebastian, M.J. and DeFrutos, R. (1994) A heterochromatic *P* sequence in the *D. subobscura* genome. *Genetica* 92: 177–186.

Pinsker, W., Miller, W.J. and Hagemann, S. (1993) *P* elements of *Drosophila*: Genomic parasites as genetic tools. *In:* K. Wöhrmann and J. Tomiuk (eds): *Transgenic Organisms: Risk Assessment of Deliberate Release.* Birkhäuser Verlag, Basel, pp 25–42.

Rio, D.C. (1990) Molecular mechanisms regulating *Drosophila P* element transposition. *Annu. Rev. Genet.* 24: 543–578.

Robertson, H.M. (1993) The *mariner* transposable element is widespread in insects. *Nature* 362: 241–245.

Robins, D.M. and Samuelson, L.C. (1992) Retrotransposons and the evolution of mammalian gene expression. *Genetica* 86: 191–201.

Sheen, F. and Levis, R.W. (1994) Transposition of the LINE-like retrotransposon *TART* to *Drosophila* chromosome termini. *Proc. Natl. Acad. Sci. USA* 91: 12510–12514.

Simmons, G.M. (1992) Horizontal transfer of *hobo* transposable elements within the *Drosophila melanogaster* species complex: Evidence from DNA sequencing. *Mol. Biol. Evol.* 9: 1050–1060.

Smith, M.W., Feng, D.-F. and Doolittle, R.F. (1992) Evolution by acquisition: The case for horizontal gene transfers. *Trends Biochem. Sci.* 17: 489–493.

Spradling, A.C. and Rubin, G.M. (1981) *Drosophila* genome organization: Conserved and dynamic aspects. *Annu. Rev. Genet.* 15: 219–264.

Stavenhagen, J.B. and Robins, D.M. (1988) An ancient provirus has imposed androgen regulation on the adjacent mouse sex-limited protein gene. *Cell* 55: 247–254.

Steinemann, M. and Steinemann, S. (1993) A duplication including the *Y* allele of *Lcp2* and the *TRIM* retrotransposon at the *Lcp* locus on the degenerating neo-Y chromosome of *Drosophila miranda*: Molecular structure and mechanism by which it may have arisen. *Genetics* 134: 497–505.

Traverse, K.L. and Pardue, M.L. (1988) A spontaneously open ring chromosome of *Drosophila melanogaster* has acquired *He-T* DNA at both new telomeres. *Proc. Natl. Acad. Sci. USA* 85: 8116–8120.

von Sternberg, R.M., Novick, G.E., Gao, G.-P. and Herrera, R.J. (1992) Genome canalization: The coevolution of transposable and interspersed repetitive elements with single copy DNA. *Genetica* 86: 215–246.

Voelker, R.A., Graves, J., Gibson, W. and Eisenberg, M. (1990) Mobile element insertions causing mutations in the *Drosophila suppressor* of *sable* locus occur in DNase I hypersensitive subregions of 5'-transcribed nontranslated sequences. *Genetics* 126: 1071–1082.

White, S.E., Habera, L.F. and Wessler, S.R. (1994) Retrotransposons in the flanking regions of normal plant genes: A role for *copia*-like elements in the evolution of gene structure and expression. *Proc. Natl. Acad. Sci. USA* 91: 11792–11796.

Transgenic Organisms – Biological and Social Implications
J. Tomiuk, K. Wöhrmann & A. Sentker (eds)

Evolutionary changes of the structure of mobile genetic elements in *Drosophila*

L. Bachmann

Department of Population Genetics, University of Tübingen, Auf der Morgenstelle 28, D-72076 Tübingen, Germany

Summary. Mobile genetic elements have been intensively studied during the last decades and, especially from *Drosophila* species, many transposons have been characterized. These species may be model organisms for the understanding of the evolutionary processes of mobile elements. Heterochromatic regions of chromosomes that exhibit a low recombination rate are believed to be a "graveyard" for transposons, i.e., regions where they degenerate. Recent studies indicate that mobile elements can also be activated from heterochromatic locations. Furthermore, the combination of sequence motifs obviously plays an important role in the evolution of mobile elements.

Introduction

Mobile genetic elements make up a significant portion of eukaryotic genomes; for example it is thought that they contribute about 10–15% to the *Drosophila melanogaster* genome, which is in turn four times more than was estimated for its close relative *D. simulans* (Dowsett and Young, 1982). It is not surprising therefore that mobile genetic elements have received great attention during the last few years and that a large number of these elements have been detected and characterized. This is especially true for *Drosophila* species. However, for geneticists mobile elements are not only interesting from a theoretical point of view. The genetic mobility of these elements can also make them suitable for the construction of vectors to create transgenic organisms, e.g., the *P* elements of *Drosophila*. If so, the question of the genetic stability of the vectors introduced to a foreign genome should be addressed. Although no general conclusions can be drawn, a view on the evolutionary dynamics of mobile elements might be helpful in this context.

Comparative analyses of several different families of mobile elements within large species groups reveal that the distribution of a certain family is usually limited to species of the same taxonomic group (Eickbush, 1994). Consequently, it can be concluded that families of mobile elements "continuously" arise and get lost with time. In respect of the evolutionary changes of the structure of transposons, it is first of all not important whether the mobile elements arise within a species or invade a genome by horizontal transfer. It seems to be more important where the transposons are located in the genome.

Once present in a genome, most mutations caused by transposons, such as insertions into coding regions, are thought to be deleterious, although beneficial mutations may also occur. Therefore, selection should eliminate them rapidly. This is confirmed by restriction mapping analyses of well known genomic regions of *D. melanogaster*. Mobile elements inserting into non-coding regions, on the other hand, should only have minor effects and the role of selection is reduced to nearly zero. Thus, it seems not surprising that transposable elements are frequently overabundant in chromosomal regions where the rate of recombination is relatively low, e.g., in heterochromatic regions (Charlesworth and Langley, 1989; Charlesworth et al., 1994).

Degeneration of mobile elements located in heterochromatin

Transposons which insert into heterochromatic regions are believed to be inactivated and to be able to persist over relatively long evolutionary periods as truncated elements. This can be illustrated by several examples:

I elements

Crozatier et al. (1988) and Vaury et al. (1990) characterized several *I* elements from a reactive *R* strain of *D. melanogaster*. These *I* elements are retrotransposons without long terminal repeats. Although *R* strains lack transposable *I* factors, degenerated *I* elements are still found in various unrelated strains in the pericentric heterochromatin at constant locations. The authors, therefore, assume that these defective *I* elements had already been present in the pericentric heterochromatin before *D. melanogaster* was invaded by active *I* factors between 1930 and 1970 (Bucheton et al., 1984). The analyses showed that defective *I* factors mainly degenerate by accumulating nucleotide substitutions and insertions/deletions. Most of the insertions/deletions affect the ORF2 of functional *I* elements, which encodes for a protein with significant homology to viral reverse transcriptases; they vary in size from 1 bp to several hundred base pairs. However, the overall similarity of the degenerated *I* factors still exceeds 90%, which is of the same order of magnitude as their similarity to active *I* elements.

SCLR repeats

Recently, Nurminsky et al. (1994) characterized various heterochromatic *SCLR* repeats from the distal X chromosome of *D. melanogaster*. These scrambled repeats are 60 kb long. The majority of the repeats consists of the copia-like retrotransposons *aurora* (Shevelyov, 1993), *mdgI*^{-het} (Nurminsky, 1993) and *GATE* (DiNocera et al., 1986) and the LINE element *G* (DiNocera et al., 1986) as well as type I ribosomal insertions (Jakubczak et al., 1990); the minority consists of heterochromatin specific variants of the *Stellate* genes and rDNA fragments. The *SCLR* repeats are interpreted as being the result of a series of events: (a) Mobile elements first insert into *Stellate* and rDNA genes; (b) internal duplications lead to tandem arrays; (c) subsequent recombination mixes the *Stellate* and rDNA repeats; (d) unequal sister chromatid exchanges generate copy number variability of *SCLR* repeats between different strains. Cloned *SCLR* repeats differ from each other by insertions, deletions and point mutations and the herein integrated mobile elements have lost their functionality.

Neo-Y chromosome

A similar scenario as that for the *SCLR*-repeats was described for the neo-Y chromosome of *D. miranda* (Steinemann and Steinemann, 1993). A comparison of the chromosomal regions of the *Lcp1-Lcp4* genes (Larval cuticle proteins) between the X2- and the neo-Y chromosome reveals numerous significant rearrangements (insertions, deletions and duplications) within the neo-Y chromosome. Several insertion sequences were described and some of them were identified as retrotransposons (*TRIM* and *TRAM*) or as homologues of domesticated *P* elements (*ISY3*). Additionally, *in situ* hybridization data showed a massive accumulation of *ISY3* elements on the neo-Y chromosome (Steinemann and Steinemann, 1992).

P elements

Paricio et al. (1994) identified and sequenced a heterochromatic *P* element in *D. subobscura*. *P* elements are DNA transposons and received greater attention as they have invaded the genome of *D. melanogaster* within the last few decades. Curiously, *P* homologues that lack the exon 3 and both termini of a mobile *P* element were described as tandemly repeated genes in the closely related sibling species *D. subobscura*, *D. madeirensis* and *D. guanche* (Paricio et al., 1991; Miller et al., 1992). The detailed analysis of the heterochromatic *P* sequence shows that this sequence is highly degenerated in comparison to the euchromatic *P* homologues of *D.*

subobscura. Sequence similarity exists only in exon 1 and intron 2 within regions of 425 bp and 37 bp, respectively. Beside numerous base substitutions several insertions and deletions were observed; some of them are only 1 bp, 2 bp or 5 bp long but there are also longer ones that span 92 bp or 100 bp. Furthermore, the clone containing the heterochromatic *P* sequence seems to be a mosaic of *P* and non *P*-related elements. The observed similarity of only 85–87% between the heterochromatic and euchromatic *P* homologues of *D. subobscura* is significantly lower than that between respective euchromatic copies of *D. subobscura* and its close relative *D. guanche*. This leads to the assumption that the heterochromatic *P* sequences are phylogenetically older than the active euchromatic *P* homologues of these species (see Miller et al., this volume). It should be added that heterochromatic *P* sequences are not a peculiarity of *D. subobscura* but are also found in other *Drosophila* species, e.g., *D. nebulosa* (Lansman et al., 1987).

Activation of mobile elements from heterochromatic locations

The examples listed above may support the idea that heterochromatin serves as a "trap" or "graveyard" for mobile elements, where they – once inserted – can only degenerate. However, there are a few other examples which illustrate the clear oversimplification of such a view and indicate that active transposons might also originate in heterochromatic regions. They may lead to a revision of our understanding of the evolution of transposable elements.

HeT-A elements

The *HeT-A* elements (Biessmann et al., 1992, 1993) and the recently identified *TART* family (Levis et al., 1993) are heterochromatin specific and functionally involved in the maintenance of the telomeric structure of *D. melanogaster*. Once transposed to the end of a chromosome *HeT-A* elements degenerate by terminal loss before attachment of the next element. This leads to arrays of truncated *HeT-A* elements of unequal length (Mason and Biessmann, 1995).

Bari-1 elements

Caizzi et al. (1993) discovered the *Bari-1* elements in *D. melanogaster* of which most map within the h39 region of the centromeric heterochromatin of chromosome 2 and exhibit a clear tandem organization that is otherwise typical for non-coding satellite DNA. A few *Bari-1* copies were also detected at euchromatic sites, however, vary in different strains. On the basis of these observations *Bari-1* elements are regarded as mobile elements. Euchromatic *Bari-1* elements encode a 339 amino-acid protein similar to the *Tc1* transposase of *Caenorhabdites elegans* transposons. The required ORFs for the transposase of the heterochromatic elements show no degeneration and retain coding potential. Although clear experimental data are missing, the authors also provide further evidence for functionality of the heterochromatic *Bari-1* transposons.

pDv elements

The *pDv* transposons of *D. virilis* (Evgen'ev et al., 1982; Zelentsova et al., 1986) are admittedly not well-defined mobile elements. They basically consist of two components, the long direct terminal repeats and the internal, tandemly repeated 36 bp motifs that are transcribed and translated. Both components seem also to be present in the genome of *D. littoralis*, which is a species of the *D. virilis*-group but is placed in a different phylad. However, the terminal repeats of *pDv* elements show a significant sequence similarity to a specific *pvB370* satellite-DNA of the centromeric heterochromatin that can be detected in a number of species of the *D. virilis*-group (Heikkinen et al., 1995). It is argued that the *pDv* transposons result from recombination events in the heterochromatin of *D. virilis* which led to a combination of the two components (*pvB370* sequences and 36 bp repeats) and to the mobility of the *pDv* elements.

Combination of sequence motifs in mobile elements

The examples of the *pDv* transposon of *D. virilis* and of the *SCLR repeat*s may illustrate another aspect of the possible evolutionary changes in structure of mobile elements, namely the combination of formerly unrelated sequence motifs and a possible subsequent acquisition of new properties. Despite the mentioned evolutionary relationship of the *pDv* transposons and the *pvB370* satellite DNA family, a gene bank search also reveals significant sequence similarity of the 5' upstream region of the *pDv* element described by Zelentsova et al. (1986) to other mobile elements (Fig. 1), i.e., to the *TRIM* retrotransposon of *D. miranda* (Steinemann and Steinemann, 1993), to the retrosequence *812* of *D. subobscura*, *D. madeirensis* and *D. pseudoobscura* (Marfany and Gonzalez-Duarte, 1992) and to the upstream regions of the *P* homologues of *D. subobscura*, *D. madeirensis* and *D. guanche* (see Miller et al., this volume). Furthermore, sequence similarity exists with the genes *sevenless* of *D. virilis* (Michael et al., 1990) and *spalt* of *D. orena* (Reuter et al., 1989). However, motifs of the mobile elements similar to the *pDv* transposon are the main components of the highly repetitive *SGM* satellite DNA sequences of *D. subobscura*, *D. madeirensis* and *D. guanche* (Bachmann, 1990; Gutknecht, 1993).

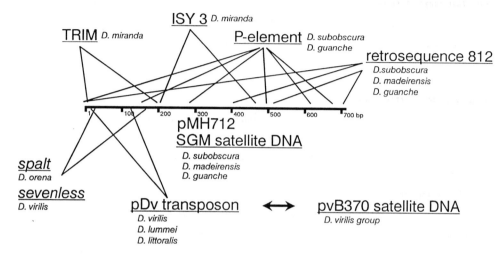

Figure 1. Schematical illustration of shared sequence motives of various mobile elements and specific satellite DNA families from *Drosophila* species. The nucleotide sequence of the *SGM* clone *pMH712* of *D. madeirensis* is given on the top. The central position of the *SGM* satellite DNA family was only chosen for graphical reasons and does not reflect any central phylogenetic position.

Finally, one more example should just be mentioned in order to avoid the impression that structural changes of mobile elements always involve heterochromatic variants. Csink and McDonald (1995) sequenced the 5' Long Terminal Repeats (LTRs) and adjacent untranslated leader sequences of 27 full-size *copia* elements from natural populations of *D. melanogaster, D. mauritiana* and *D. simulans*, all of which are of euchromatic origin. Some structural differences such as a characteristic 39 bp deletion specific for *D. simulans* and *D. mauritiana* were observed. Nevertheless, the authors stress that structural aberrations were not detected in those regions being critical for *copia* expression.

Of course, the complicated relationships among specific sequence motifs and sequence families hinder a clear interpretation of the evolution of mobile elements. The growing body of data on this topic indicates that we are only at the beginning of understanding the role of mobile elements in evolution. It seems therefore to be impossible to predict the genetic fate of transposons or of sequence motives of mobile elements introduced into foreign genomes. The recent spread of *P* elements through natural populations of *D. melanogaster* may be given here to illustrate possible consequences. *P* elements are believed to be recently introduced into the genome of *D. melanogaster* by horizontal gene transfer from *D. willistoni via* the mite *Proctolaelaps regalis* (Houck et al., 1991). As could be demonstrated by a worldwide survey of natural populations of *D. melanogaster*, the *P* elements most probably originated in America and spread bidirectionally through the European and Asian populations (Anxolabéhère et al., 1988). Nevertheless, the spread of *P* elements was not accompanied by significant negative ecological effects for the species.

References

Anxolabéhère, D., Kidwell, M.G. and Periquet, G. (1988) Molecular characteristics of diverse populations are consistent with the hypothesis of a recent invasion of *Drosophila melanogaster* by mobile *P* elements. *Mol. Biol. Evol.* 5: 252–269.

Bachmann, L. (1990) *Vergleichende Untersuchungen über die Satelliten-DNA verschiedener Arten der Drosophila-obscura Gruppe*. Ph.D. thesis, University of Tübingen.

Biessmann, H., Valgeirsdottir, K., Lofsky, A., Chin, C., Ginther, B., Levis, R.W. and Pardue, M.L. (1992) *Het-A*, a transposable element specifically involved in "healing" broken chromosome ends in *D. melanogaster. Mol. Cell. Biol.* 12: 3910–3918.

Biessmann, H., Kasravi, B., Jakes, K., Bui, T., Ikenaga, K. and Mason, J.M. (1993) The genomic organization of *HeT-A* retroposons in *D. melanogaster. Chromosoma* 102: 297–305.

Bucheton, A., Paro, R., Sang, H.M., Pelisson, A. and Finnegan, D.J. (1984) The molecular basis of *I-R* hybrid dysgenesis in *Drosophila melanogaster*: Identification, cloning and properties of the *I* factor. *Cell* 38: 153–163.

Caizzi, R., Caggese, C. and Pimpinelli, S. (1993) *Bari-1* a new transposon-like family in *D. melanogaster* with a unique heterochromatic organization. *Genetics* 37: 675–689.

Charlesworth, B. and Langley, C.H. (1989) The population genetics of *Drosophila* transposable elements. *Annu. Rev. Genet.* 23: 251–287.

Charlesworth, B., Sniegowski, P. and Stephan, W. (1994) The evolutionary dynamics of repetitive DNA in eukaryotes. *Nature* 371: 215–220.

Crozatier, M., Vaury, C., Busseau, I., Pelisson, A. and Bucheton A. (1988) Structure and genomic organization of *I* elements involved in *I-R* hybrid dysgenesis in *Drosophila melanogaster. Nucleic Acids Res.* 16: 9199–9213.

Csink, A.K. and McDonald, J.F. (1995) Analysis of *copia* sequence variation within and between *Drosophila* species. *Mol. Biol. Evol.* 12: 83–93.

DiNocera, P.P., Graziani, F. and Lavorgna, G. (1986) Genomic and structural organization of *Drosophila melanogaster G* elements. *Nucleic Acids Res.* 14: 675–691.

Dowsett, A.P. and Young, M.W. (1982) Differing levels of dispersed repetitive DNA among closely related species of *Drosophila. Proc. Natl. Acad. Sci. USA* 79: 4570–4574.

Eickbush, T.H. (1994) Origin and evolutionary relationships of retroelements. *In*: S.S. Morse (ed.): *The Evolutionary Biology of Viruses*. Raven Press Ltd., New York, pp 121–157.

Evgen'ev, M.B., Yenikolopov, G.N., Peunova, N.I. and Ilyin, Y.V. (1982) Transposition of mobile genetic elements in interspecific hybrids of *Drosophila. Chromosoma* 85: 375–386.

Gutknecht, J. (1993) *Isolierung und Charakterisierung von hochrepetitiver DNA aus Drosophila subsilvestris, D. subobscura, D. guanche und D. madeirensis.* Diploma thesis, University of Tübingen.

Heikkinen, E., Launonen, V., Müller, E. and Bachmann, L. (1995) The *pvB370* satellite DNA family of the *Drosophila virilis* group and its evolutionary relation to mobile dispersed genetic *pDv* elements. *J. Mol. Evol.* 41: 604–614.

Houck, M.A., Clark, J.B., Peterson, K.R. and Kidwell M.G. (1991) Possible horizontal transfer of *Drosophila* genes by the mite *Proctolaelaps regalis. Science* 253: 1125–1129.

Jakubczak, J.L., Xiong, Y. and Eickbush, T.H. (1990) Type I (*R1*) and type II (*R2*) ribosomal DNA insertions of *Drosophila melanogaster* are retrotransposable elements closely related to those of *Bombyx mori. J. Mol. Biol.* 212: 37–52.

Lansman, R.A., Shade, R.O., Grigliatti, T.A. and Brock, H.W. (1987) Evolution of *P* transposable elements: Sequences of *D. nebulosa* elements. *Proc. Natl. Acad. Sci. USA* 84: 6491–6495.

Levis, R.W., Ganesan, R., Houtchens, K., Tolar, A. and Sheen, F.-M. (1993) Transposon in place of telomeric repeats at a *Drosophila* telomere. *Cell* 75: 1083–1093.

Marfany, G. and Gonzalez-Duarte, R. (1992) Evidence for retrotranscription of protein-coding genes in the *Drosophila subobscura* genome. *J. Mol. Evol.* 35: 492–501.

Mason, J.M. and Biessmann, H. (1995) The unusual telomeres of *Drosophila. Trends Genet.* 11: 58–62.

Michael, W.M., Bowtell, D.D.L. and Rubin, G.M. (1990) Comparison of the *sevenless* genes of *Drosophila virilis* and *Drosophila melanogaster. Proc. Natl. Acad. Sci. USA* 87: 5351–5353.

Miller, W.J., Hagemann, S., Reiter, E. and Pinsker, W. (1992) *P* element homologous sequences are tandemly repeated in the genome of *D. guanche. Proc. Natl. Acad. Sci. USA* 89: 4018–4022.

Nurminsky, D.I. (1993) Two subfamilies of *mdgl* retrotransposons with different evolutionary fates in *Drosophila melanogaster. J. Mol. Evol.* 37: 496–503.

Nurminsky, D.I., Shevelyov, Y.Y., Nuzhdin, S.V. and Gvozdez, V.A. (1994) Structure, molecular evolution and maintenance of copy number of extended repeated structures in the X-heterochromatin of *Drosophila melanogaster. Chromosoma* 103: 277–285.

Paricio, N., Perez-Alonso, M., Martinez-Sebastian, M.J. and de Frutos, R. (1991) *P* sequences of *D. subobscura* lack exon 3 and may encode a 66 kd repressor-like protein. *Nucleic Acids Res.* 19: 6713–6718.

Paricio, N., Martinez-Sebastian, M.J. and de Frutos, R. (1994) A heterochromatic *P* sequence in the *D. subobscura* genome. *Genetica* 92: 177–186.

Reuter, D., Schuh, R. and Jäckle, H. (1989) The homeotic gene *spalt* (*sal*) evolved during *Drosophila* speciation. *Proc. Natl. Acad. Sci. USA* 86: 5483–5486.

Shevelyov, Y.Y. (1993) Non-mobile retrotransposons *aurora* in *D. melanogaster* heterochromatin. *Mol. Gen. Genet.* 239: 205–208.

Steinemann, M. and Steinemann, S. (1992) Degenerating Y-chromosome of *Drosophila miranda*: A trap for retrotransposons. *Proc. Natl. Acad. Sci. USA* 89: 7591–7595.

Steinemann, M. and Steinemann, S. (1993) A duplication including the *Y* allele of *Lcp2* and the *TRIM* retrotransposon at the *Lcp* locus on the degenerating neo-Y-chromosome of *Drosophila miranda*: Molecular structure and mechanisms by which it may have arisen. *Genetics* 134: 497–505.

Vaury, C., Abad, P., Pelisson, A., Lenoir, A. and Bucheton, A. (1990) Molecular characteristics of the heterochromatic *I* elements from a reactive strain of *Drosophila melanogaster. J. Mol. Evol.* 31: 424–431.

Zelentsova, E.S., Vashakidze, R.P., Krayev, A.S. and Evgen'ev, M.B. (1986) Dispersed repeats in *Drosophila virilis*: Elements mobilized by interspecific hybridization. *Chromosoma* 93: 469–476.

Transgenic Organisms – Biological and Social Implications
J. Tomiuk, K. Wöhrmann & A. Sentker (eds)
© 1996 Birkhäuser Verlag Basel/Switzerland

Mechanisms and consequences of horizontal gene transfer in natural bacterial populations

M.G. Lorenz and W. Wackernagel

Department of Genetics, Faculty of Biology, University of Oldenburg, D-26111 Oldenburg, Germany

Summary. Gene transfer mechanisms among bacteria such as conjugation, transformation and transduction have been known for decades, but only in recent years is their impact on the life of bacteria in the environment beginning to be fully recognized. An increasing body of molecular ecological studies performed in environment-simulating microcosms or in natural populations have suggested that there is a high potential for horizontal gene transfers to occur in bacterial habitats including soil, sediments and water. Also, the footprints of recombination events have been identified in genes and genomes of bacteria isolated from natural populations. Evidence is accumulating that sex (horizontal gene transfer and recombination) is a vital part of the life of many if not all bacteria in their natural habitats. In risk assessments on the release of recombinant organisms a perpetual gene flux between bacteria and perhaps occasionally between bacteria and eukaryotes must be considered.

Introduction

The use of genetically engineered microorganisms (GEMs) for applications in the environment can serve a variety of important commercial and public health purposes (Wilson and Lindow, 1993). These include the clean-up of chemically polluted sites in terrestrial and aquatic systems (bioremediation), the production of metals from low grade ores (biomining), replacement of mineral fertilizers (biofertilization), and the control of weeds and protection of plants against insects in agriculture. These applications require large quantities of GEMs to be released into the environment. Such releases have been considered a risk because recombinant nucleotide sequences may be spread by horizontal gene transfer processes from GEMs to other organisms in the environment. This could result in new genotypes with unforeseen properties. The higher the potential for gene transfer in natural populations and communities the more risk management would be necessary. Extensive efforts were made to study the various bacterial horizontal gene transfer processes known from laboratory experiments directly in the environment or under habitat-simulating conditions. Some of the data will be summarized here. Also, various aspects contributing to genetic exchange among bacteria in their habitats and the genetic outcome of such exchanges will be discussed. An increasing body of observations suggests that in bacteria, as in higher organisms, genetic variability is produced not only by mutation but also by gene exchange coupled with recombination.

Mechanisms of gene transfer and their potential in bacterial habitats

Textbooks on genetics and microbiology provide ample information on the basic bacterial mechanisms of horizontal gene transfer. These are conjugation (mediated by plasmids during cell-to-cell contact), transduction (mediated by bacteriophages) and natural genetic transformation (active uptake of free DNA by the cell). The latter is considered the genuine bacterial gene exchange mechanism, because the uptake and heritable integration of DNA into the chromosome require proteins and regulatory functions which are encoded by the bacterial genome. In contrast, conjugation and transduction depend on genetic elements which appear to be parasitic or com-mensalic to bacteria. Selective forces may, however, have led to the acquisition of genes beneficial to the host cell by extrachromosomal genetic elements, either as permanent part of the element (e.g., antibiotic resistance determinants on plasmids or transposons) or as in the form of random-ly contained DNA (e.g., as in transduction, where only a subpopulation of the phages contains non-phage DNA). This may have changed parasitic elements into commensalic or even symbiotic elements. Natural transformation has been detected so far in at least 43 bacterial species, of which roughly two thirds live in terrestrial and aquatic environments (Lorenz and Wackernagel, 1994 and unpublished data). Recent reviews have focused on factors which are considered important for these three basic mechanisms to be active in natural bacterial habitats in the environment (Levy and Miller, 1989; Lorenz and Wackernagel, 1993). For instance, the efficiency of genetic trans-formation of several soil bacteria (*Acetobacter vinelandii, Bacillus subtilis, Pseudomonas stutzeri*) depends on the quality of the carbon and energy source and is enhanced upon the onset of starvation for a specific nutrient like the carbon source or iron (Lorenz and Wackernagel, 1994). Evidence of environmental sensing coupled to the development of DNA uptake ability (competence) was also found in *Streptococcus pneumoniae* in which a mutation in a gene of the signal transducing cascade prevented competence development (Guenzi et al., 1994).

Further reviews on bacterial horizontal gene transfer have recently highlighted the general inter-est in this field. Genetic transformation has received increasing attention as documented by specific reviews on the molecular mechanisms involved (Dubnau, 1991a,b; Lorenz and Wackernagel, 1994), on its potential in bacterial habitats of the environment (Lorenz and Wackernagel, 1994), and its energetics (Palmen et al., 1994). The DNA translocation across bacterial cell walls occurring during phage infection (transduction), conjugation and transforma-tion has been extensively reviewed (Dreiseikelmann, 1994). Similarly, the trans-kingdom DNA transfer from *Agrobacterium tumefaciens* to cells of higher plants has recently been summarized (Zambryski, 1992; Citovsky and Zambryski, 1993). Further, there are several recent reports on the transfer of plasmids from bacteria to yeast (Heinemann and Sprague, 1989; Heinemann, 1991; Hayman and Bolen, 1993; Nishikawa et al., 1992). Gene transfer processes have been observed which do not fit clearly into one of the three categories mentioned, such as intra- and interspecies cell-to-cell contact transformation (Lorenz and Wackernagel, 1994) or conjugative transposon transfer (Clewell and Gawron-Burke, 1986; Salyers and Shoemaker, 1994).

A remarkable case of horizontal gene transfer by recombinational exchange of genes between a phage and the chromosome of its host which has led to the evolution of a new phage has recently been detected (Moineau et al., 1994). Apparently, under selective pressure the phage has acquired genes from the host which change its morphology and, more important, allow the phage to overcome the abortive phage-defense mechanism of its host. The possibility that phages change or increase their host range by mutational alteration of their adsorption apparatus has been known for a long time. Some phages have *per se* a wide host range, e.g., *P1* was shown to transduce 16 different Gram-negative species and a *Myxobacterium* (Murooka and Harada, 1979; Kaiser and Dworkin, 1975). In addition, it must be considered that recombination between genes determining parts of the adsorption apparatus from different phages (Sandmeier, 1994) could alter the host range of phages and thereby could effect the gene transfer betweeen bacterial species previously not linked by a common transducing phage.

In general, studies under environment-simulating conditions (i.e., in artificial habitats or environmental microcosms) have not identified the factors that would in principle preclude the mechanisms and processes of bacterial gene transfer initially detected in laboratory experiments from occurring in the natural environment.

Processes incorporating new genetic information in bacterial cells

Once DNA has entered a cell there are two routes by which the DNA can become a permanent part of the cell: (i) Establishment as an autonomously replicating element and (ii) recombinational integration into a preexisting replicon of the cell (chromosome, plasmid or phage).

Establishment of a replicon following transfer mostly requires circularization of the DNA. In plasmid *R1162*, which can be mobilized into diverse bacterial species, an enzyme has been detected which cleaves the plasmid strand before its transfer and fuses it by a ligation-like process in the recipient cell (Bhattacharjee and Meyer, 1991). The protein remains covalently bound to the 5'-end of the transferred strand. Synthesis of the complementary strand produces a circular duplex DNA molecule. Replication functions including one or more origin(s) of replication on the DNA are necessary for establishment in the new cell. Successful establishment further requires absence of other replicons belonging to the same incompatibility group from the cell (Novick, 1987). Plasmids can also establish when only linear single strands or fragments of single strands originating from a plasmid are taken up by competent cells in the course of natural transformation

(Canosi et al., 1981). Reconstitution of a complete plasmid in the cytoplasm by hybridization and repair synthesis of DNA mostly requires (overlapping) fragments from two or even three plasmid molecules. This is probably the reason for the second or third order dependence of transformation frequency on monomer plasmid DNA concentration (Canosi et al., 1981; Chamier et al., 1993). Finally, also phages can transduce plasmid DNA, which is probably packaged in the phage particle as a linear concatemer and circularizes in the cell by homologous recombination of redundant duplex segments (Saye et al., 1987). The experimental identification of replicon establishment following DNA transfer is facilitated by the expression of genes with a selective phenotype like antibiotic resistance determinants. It is also probable that in natural habitats selective pressure will act in favour of the establishment of replicons which contain genes giving increased fitness, such as genes with new metabolic functions (van der Meer et al., 1992).

Recombinative integration of foreign DNA can occur by a variety of mechanisms. Many integration processes depend on extensive homology between the interacting DNA molecules (homologous recombination). When a single strand enters the cell during genetic transformation, homologous recombination to resident DNA proceeds by a strand displacement process (strand assimilation) driven by *RecA* protein. Natural genetic transformation in bacteria is blocked by *recA* mutations (*Streptococcus pneumoniae,* Martin et al., 1995; *Acinetobacter calcoaceticus,* Gregg-Jolly and Ornston, 1994; *Haemophilus influenzae,* Smith et al., 1981; *Bacillus subtilis,* de Vos and Venema, 1982; *Pseudomonas stutzeri,* Vosman and Hellingwerf, 1991). Double-stranded DNA fragments transported into the cell by transducing phages would require two crossover events for integration into the chromosome or into a plasmid. Such homologous recombination events depend on *RecA* and *RecBCD* proteins and are stimulated by specific nucleotide sequences which are hot spots of recombination in *Escherichia coli* and possibly other bacteria (for review see, Myers and Stahl, 1994). Interestingly, the observed barrier to general recombination between DNA molecules which share only limited homology (i.e., DNA from different species) is disrupted by mutations eliminating mismatch repair (Rayssiguier et al., 1989; Matic et al., 1995). Thus, mismatch repair processes (or the mismatch repair enzymes themselves) prevent recombination between DNA molecules with similar, but not identical nucleotide sequences. On the other hand, the *SOS* response of bacteria triggered by genotoxic stress (Walker, 1985) appears to increase the chance of incorporation of only partially homologous sequences by the machinery for homologous recombination (Matic et al., 1995). It seems as if damage to the genome relieves the stringency of nonhomologous discrimination. Previously it was shown that the DNA damage-induced *SOS* response also relieves DNA restriction (Day, 1977; Thoms and Wackernagel, 1984). DNA restriction normally allows cells to identify (by lack of the specific DNA methylation pattern) and to degrade by endonucleolytic cleavage foreign duplex DNA entering the cell (Bickle and Krüger, 1993).

Without the need for extensive homology, foreign DNA can also be integrated by site-specific recombination. In these cases specific integrases catalyze crossover events at short identical nucleotide sequences. Examples are transposons carrying a gene for transposase and one or several other genes coding for example, antibiotic resistance determinants (for review see, Berg and Howe, 1989). Similarly, integrons and mobile gene cassettes depend on specific sites on the DNA and integrases for their translocation (Hall and Collis, 1995). Recombination events leading to the evolution of bacteriophage tail fibre genes composed of segments of different origins (Sandmeier, 1994) are examples for this type of recombination. It should be noted that circular DNA structures require only a single recombination event for integration (Campbell model). Lysogenization of host cells by phages cointegrates formation between plasmids and transposition of *Tn916*, and similar transposons rely on site-specific *recA*-independent recombination events. In some instances target sites in DNA are duplicated during integration, in others they are not (Salyers and Shoemaker, 1994). It is possible that many of the so-called illegitimate recombination events (Ehrlich, 1989) do not represent recombinational accidents leading to mutations (e.g., deletions, duplications) but result from rare events of rather site-specific recombination processes involving small regions of little or no homology.

Reservoirs of genetic material outside living bacteria

Sources of genetic material from which bacteria could derive genes, parts of genes or other nucleotide sequences are manifold. These sources can be extracellular DNA, DNA in non-dividing or dead cells or phage particles. The sources are presumably present in most habitats, although probably with changing levels of concentrations and availability.

Extracellular DNA has been found in abundant quantities in aquatic and terrestrial habitats (Lorenz and Wackernagel, 1994). The molecular size of extracellular DNA, when determined, indicated that it would be appropriate for genetic transformation of bacteria. The DNA can be of bacterial or eukaryotic origin and would include chromosomal fragments as well as plasmids. The possible excretion of DNA by bacteria as well as the release during cell lysis has been examined in a variety of bacterial species (Lorenz and Wackernagel, 1994) including soil bacteria (Lorenz et al., 1991). The previous assumption that naked DNA in soils, sediments and waters is immediately and completely degraded by extracellular microbial DNases has been challenged by studies following the fate of DNA introduced in aquatic and terrestrial environments. It was found that the persistence of free DNA in non-sterile waters exceeded several hours and days and in non-sterile soils even two months (Lorenz and Wackernagel, 1994; Wackernagel, in press). DNA protection was found to be due to adsorption of DNA to particulate material, both in studies employing purified mineral fractions (Lorenz and Wackernagel, 1994) or different non-sterile soils (Blum et al., in press). Adsorbed DNA was shown to be available for uptake by bacteria during transformation (Lorenz and Wackernagel, 1990; Chamier et al., 1993; Romanowski et al., 1993; Wackernagel, in press). DNA present in non-dividing bacterial cells either dormant or dead may also persist and, upon eventual release during lysis of the cells or excretion, would contribute to an extracellular bacterial gene pool. In deep sea sediments the half life of DNA molecules in dead cells was several days (Novitzky, 1986). It must be emphasized that free DNA may persist even under conditions which would be deleterious for cells, such as dryness, high temperature or certain conditions including low pH-values or the presence of bactericidal chemicals like detergents and phenolic compounds. Consequently, DNA may be available to (transformable) bacterial cells long after the donor of the DNA has lost its reproductive potential.

Bacterial genes may also be stored in transducing phage particles. About 50 transducing phage particles of bacteriophage *P1* represent an amount of DNA equivalent to a complete large bacterial genome (5000 kb). Many bacterial genomes have only one fifth the size (Cole and Girons, 1994). Some phages are rather sensitive to environmental conditions, whereas others can persist for long times. Association of phages with mineral surfaces has been shown to increase the stability of the phages (Stotzky et al., 1981). It is known that phages persist with their host cells in the aquatic and terrestrial environment in a dynamic interaction (Ogunseitan et al., 1990; Pantastico-Caldas et al., 1992).

Another source of genes may be cells which have lost their reproductive potential, e.g., by lethal damage of their chromosome or total chromosomal loss. Under such conditions intact plasmid DNA can still be present and may be mobilized in the course of plasmid retrotransfer processes. It was shown that cells without a chromosome could be potent donors of plasmid DNA during conjugation when the transfer functions were provided (Heinemann and Ankenbauer, 1993). Thus, there are ubiquitous sources of DNA for bacterial gene transfer processes and in these sources DNA may be preserved for extended periods. Future studies are necessary to examine whether these sources are exploited by bacterial populations.

Sexual population structure and genetic molecular mosaics in bacteria

The examination of bacterial populations with population genetic methods such as multilocus enzyme electrophoresis (Selander et al., 1986) was initiated several years ago. Vegetative reproduction of bacteria (mostly by binary fission) results in a clonal population structure in which new genotypes would arise only by mutation. In fact, clonal population structures have been found in *Escherichia coli, Salmonella typhimurium, Bordetella bronchiseptica, Legionella pneumophila, Haemophilus influenzae* and *Pseudomonas syringae* (Whittam et al., 1983; Maynard Smith et al., 1993). However, several species were shown to have sexual (panmictic) population structures, including the pathogenic organisms *Neisseria gonorrhoeae* and *N. meningitidis* (Maynard Smith et al., 1993) and the soil organisms *Bacillus subtilis* (Istock et al., 1992) and *B. licheniformis* (Duncan et al., in press) from desert soil, a *Bacillus* species (Lorenz and Wackernagel, in press) from podsol, several *Rhizobium* species (Maynard Smith et al., 1993) and *Burkholderia* (*Pseudomonas*) *cepacia* (Wise et al., 1995). The sexual population structure is characterized by a random association of alleles (linkage equilibrium) which reflects extensive horizontal gene transfer and recombination in the population.

Nucleotide sequence comparison of specific genes from members of a panmictic population have confirmed the mosaic structure of the genome at the gene level. For instance, the gene determining low resistance to penicillin in the pathogens *Neisseria meningitidis* and *N. gonorrhoeae* turned out to contain segments derived from commensals *N. flavescens* and *N. cinerea* (Maynard Smith et al., 1993; Bowler et al., 1994). From the analyses of penicillin-resistant *Streptococcus pneumoniae* species it was similarly concluded that mosaic genes were present with parts from related species (Sibold et al., 1994). These examples demonstrate the occurrence of interspecies gene exchange, which may rely on genetic transformation of these naturally competent species. Hybrid genes have been identified in natural populations of many bacteria (Gill et al., 1994; Lorenz and Wackernagel, 1995).

However, in members of clonal organisms, such as *E. coli* and *Salmonella typhimurium,* a mosaic structure of genes or of small genome segments has been identified by molecular methods including DNA sequencing and DNA restriction pattern analysis (Groisman et al., 1992, 1993; Milkman and McKane Bridges, 1993). Such mosaics have been attributed to "localized sex" and would probably not be detected by methods like multilocus enzyme electrophoresis encompassing whole genomes. Recently, in *E. coli* such localized sex was related to transduction (McKane and Milkman, 1995), a gene transfer process that transports only small portions of a genome. Moreover, interspecies transfer (or transfer between subspecies) by transduction may be strongly affected by DNA restriction, because duplex DNA is subject to restriction/modification (Bickle and Krüger, 1993). In fact, the size pattern of mosaic structures produced by transduction was shown to consist of smaller segments in transductions between different *E. coli* strains suggesting a possible role for restriction/modification systems in the production of sequence patterns in nature (McKane and Milkman, 1995). High selective pressure conferred upon a population by the environment (such as by the presence of antibiotics or the immune response of the host organism) may support the establishment of genetic mosaics produced in the course of gene transfer processes in a population, particularly of mosaics in genes which are important for the colonization of the habitat. Examples are the (transformation-enhanced) variability of pilin and adhesin genes of *N. gonorrhoeae* (Meyer et al., 1990) or the flagellin genes of *Campylobacter jejuni* (Wassenaar et al., 1995). This situation may, in turn, lead to genetic polymorphism of genes which, although not subject to strong selection, are located close to loci under selection and which diversify as the result of frequent intraspecies and interspecies gene exchanges (Nelson and Selander, 1994).

Certain genes of bacteria and plants or animals show a high degree of sequence similarity which suggests that these genes represent cases of interkingdom gene exchange in the past and that these genes have diverged independently thereafter (Smith et al., 1992; Syvanen, 1994). In conclusion, there is accumulating evidence of horizontal gene fluxes within species, between species and even between kingdoms.

Genetic variability due to gene transfer-independent mechanisms

Besides intraspecies and interspecies gene transfer events there are other mechanisms based on the appearance of mutations, which promote divergence in populations and, under natural selection, genetic adaptation of organisms to a changing environment. *In vitro* experiments show that in an unstructured environment (a glucose-limited continuous culture) a growing clone of *E. coli* acquires a certain degree of polymorphism, which may reflect incipient processes of diversification of asexual organisms (Helling et al., 1987; Rosenzweig et al., 1994). The evolved coexistence after 773 generations of three subpopulations with altered modes of carbon and energy source metabolism was considered to be the evolution of resource partitioning. A structured environment (surface of a solid agar medium) appears to accelerate such evolutionary diversification by promoting high morphological polymorphism as found in the soil bacterium *Comamonas* sp. (Korona et al., 1994).

The primary production of genetic diversity may be caused by complex events besides spontaneous point mutations. DNA rearrangements leading to phenotypic switches have been observed under conditions of transformative DNA transfer (see above), but also in a number of cases by flip-flop mechanisms and transpositions (Dybvig, 1993). For instance, in *Mycoplasma pulmonis* the DNA element encoding the enzymes for restriction and modification of DNA is invertible leading to switch on/off of DNA restriction. When the system is switched on, double strand breaks generated by restriction may be the cause of rearrangements seen in the surviving clones (Dybvig and Yu, 1994). In resting bacteria the generation of genetic diversity has been directly linked to DNA rearrangements and transposition events (Arber et al., 1994).

Microbial genetic ecology and new safety systems

Only a few years ago the various bacterial gene transfer mechanisms were considered to play a negligible, or no role in the life of free-living bacteria. The results of extensive work spent on the search of gene transfer processes in natural bacterial habitats or on simulating such events in laboratory experiments suggested that these transfers may occur permanently in aquatic and terrestrial habitats as well as in the hosts of parasitic or commensalic microorganisms. Moreover, gene transfer may not be confined to a sexually active population, but may extend to members of populations of different species within the community. In conclusion, genetic flux, besides mutation, appears to be a basic principle of diversity production and thereby of genetic adaptation of bacteria in their natural habitats. Hence, research in microbial ecology (including microbial diversity) has to include the examination of gene transfer potentials of populations and communities. These considerations are particularly important in the context of releases of GEMs (Kellenberger, 1994). Because we are now aware of the relatively high potential for horizontal gene transfer in the environment, there is a general convention that potentially dangerous genes (e.g., pathogenic determinants) should not be contained in GEMs to be released.

For biosafety of the use of GEMs, differing biological containment systems have been developed during the past years (Schweder, this volume). The most advanced principle is to have a controlled suicide system which triggers endogenous killing of GEMs either stochastically or at a certain time point as the result of physical or chemical change of the milieu, e.g., by increase of the temperature or depletion of a nutrient (Molin et al., 1993). Several suicide systems have been established in which the killing gene produces a small protein that destroys the cell membrane (Molin et al., 1993). However, the cellular DNA of a GEM may constitute a separate risk, because the DNA can be transferred by transformation to other organisms even after death and lysis of the GEM. These considerations recently led to the establishment of a suicide system in which the DNA itself is the target of the killing gene (Ahrenholtz et al., 1994). Here the killing function is encoded in a nuclease gene (the extracellular nuclease of *Serratia marcescens*) which was deleted for the signal peptide-coding region resulting in intracellular RNase and DNase activity of the nuclease upon derepression of the killing gene. Thus, suicide of the cell is brought about by degradation of its own DNA, thereby preventing transfer of genes from GEMs to other organisms. This type of safety device under appropriate control and incorporated into GEMs can help to limit their unwanted survival and the dissemination of their recombinant DNA.

Conclusions

Several biological phenomena and processes which are important for the mobility of bacterial genes have been analyzed by experiments and are summarized here. These included the gene transfer mechanisms of bacteria, the ways by which new genetic material can be incorporated into bacterial cells and the various sources of genetic material available to bacteria for retrieval of genes or other nucleotide sequences. Mechanisms that create new alleles and genomic rearrangements in cells are briefly summarized. Further, evidence was presented that the consequences of recombination following horizontal gene transfer can be traced in the genomes of bacteria. Such traces have been identified in bacteria with an apparent sexual population structure and also in bacteria having clonal populations. It is becoming clear that gene exchange by horizontal gene transfer and recombination is a source of genetic diversity in bacteria, as well as mutation. The diversity derives from intergenic recombination (resulting in new combinations of alleles), intragenic recombination (resulting in new alleles) and integrative recombination (resulting in addition of genetic material). Thus, among bacteria, sex including interspecies gene transfers probably provides a major contribution to the genetic variability of populations. Gene transfer, therefore, appears to constitute a vital part of the life of bacterial populations and communities and is probably fundamental to their genetic adaptation and speciation.

Acknowledgements
The help of B. Thoms in the preparation of this chapter is gratefully acknowledged. The work of the authors was supported by the Bundesministerium für Bildung, Wissenschaft, Forschung und Technologie and by the Fonds der Chemischen Industrie.

References

Ahrenholtz, I., Lorenz, M.G. and Wackernagel, W. (1994) A conditional suicide system in *Escherichia coli* based on the intracellular degradation of DNA. *Appl. Environ. Microbiol.* 60: 3746–3751.
Arber, W., Naas, T. and Blot, M. (1994) Generation of genetic diversity by DNA rearrangements in resting bacteria. *FEMS Microbiol. Ecol.* 15: 5–14.
Berg, D.E. and Howe, M.M. (1989) *Mobile DNA*. American Society for Microbiologie, Washington.
Bhattacharjee, M.K. and Meyer, J. (1991) A segment of a plasmid gene required for conjugal transfer encodes a site-specific, single-strand DNA endonuclease and ligase. *Nucleic Acids Res.* 19: 1129–1137.
Bickle, T.A. and Krüger, D.H. (1993) Biology of DNA restriction. *Microbiol. Rev.* 57: 434–450.
Blum, S.A.E., Lorenz, M.G. and Wackernagel, W. Persistence of DNA in natural soils: Adsorption to particulate material provides protection against nucleolytic degradation. *In*: E.R. Schmidt and T. Hankeln (eds): *Transgenic Organisms and Biosafety. Horizontal Gene Transfer, Stability of DNA and Expression of Transgenes.* Springer-Verlag, Heidelberg; *in press.*
Bowler, L.D., Zhang, Q.-Y., Riou, J.-Y. and Spratt, B.G. (1994) Interspecies recombination between the *penA* genes of *Neisseria meningitidis* and commensal *Neisseria* species during the emergence of penicillin resistance in *N. meningitidis*: Natural events and laboratory simulation. *J. Bacteriol.* 176: 333–337.
Canosi, U., Iglesias, A. and Trautner, T.A. (1981) Plasmid transformation in *Bacillus subtilis*: Effects of insertion of *Bacillus subtilis* DNA into plasmid *pC194*. *Mol. Gen. Genet.* 181: 434–440.

Chamier, B., Lorenz, M.G. and Wackernagel, W. (1993) Natural transformation of *Acinetobacter calcoaceticus* by plasmid DNA adsorbed on sand and groundwater aquifer material. *Appl. Environ. Microbiol.* 59: 1662–1667.

Citovsky, V. and Zambryski, P. (1993) Transport of nucleic acids through membrane channels: Snaking through small holes. *Annu. Rev. Microbiol.* 47: 167–197.

Clewell, D.B. and Gawron-Burke, C. (1986) Conjugative transposons and the dissemination of antibiotic resistance in streptococci. *Annu. Rev. Microbiol.* 40: 635–659.

Cole, S.T. and Girons, I.S. (1994) Bacterial genomics. *FEMS Microbiol. Rev.* 14: 139–160.

Day, R.S. (1977) UV-induced alleviation of K-specific restriction of bacteriophage l. *J. Virol.* 21: 1249–1251.

de Vos, W.M. and Venema, G. (1982) Transformation of *Bacillus subtilis* competent cells: Identification of a protein involved in recombination. *Mol. Gen. Genet.* 187: 439–445.

Dreiseikelmann, B. (1994) Translocation of DNA across bacterial membranes. *Microbiol. Rev.* 58: 293–316.

Dubnau, D. (1991a) Genetic competence in *Bacillus subtilis. Microbiol. Rev.* 55: 395–424.

Dubnau, D. (1991b) The regulation of genetic competence in *Bacillus subtilis. Mol. Microbiol.* 5: 11–18.

Duncan, K.E., Ferguson, N., Kimura, K., Zhou, X. and Istock, C.A. Fine scale genetic and phenotypic structure in natural populations of *Bacillus subtilis* and *Bacillus licheniformis*: Implications for bacterial evolution and speciation. *Evolution; in press.*

Dybvig, K. (1993) DNA rearrangements and phenotypic switching in prokaryotes. *Mol. Microbiol.* 10: 465–471.

Dybvig, K. and Yu, H. (1994) Regulation of a restriction and modification system *via* DNA inversion in *Mycoplasma pulmonis. Mol. Microbiol.* 12: 547–560.

Ehrlich, D. (1989) Illegitimate recombination in bacteria. *In*: D.E. Berg and M.M. Howe (eds): *Mobile DNA.* American Society for Microbiology, Washington, pp 799–832.

Gill, M.J., Jayamohan, J., Lessing, M.P.A. and Ison, C.A. (1994) Naturally occurring PIA/PIB hybrids of *Neisseria gonorrhoeae. FEMS Microbiol. Letters* 119: 161–166.

Gregg-Jolly, L.A. and Ornston, L.N. (1994) Properties of *Acinetobacter calcoaceticus recA* and its contribution to intracellular gene conversion. *Mol. Microbiol.* 12: 985–992.

Groisman, E.A., Saier, M.H., Jr. and Ochman, H. (1992) Horizontal transfer of a phosphatase gene as evidence for mosaic structure of the *Salmonella* genome. *EMBO J.* 11: 1309–1316.

Groisman, E.A., Sturmoski, M.A., Solomon, F.R., Lin, R. and Ochman, H. (1993) Molecular, functional, and evolutionary analysis of sequences specific to *Salmonella. Proc. Natl. Acad. Sci. USA* 90: 1033–1037.

Guenzi, E., Gasc, A.-M., Sicard, M.A. and Hakenbeck, R. (1994) A two-component signal-transducing system is involved in competence and penicillin susceptibility in laboratory mutants of *Streptococcus pneumoniae. Mol. Microbiol.* 12: 505–515.

Hall, R.M. and Collis, C.M. (1995) Mobile gene cassettes and integrons: Capture and spread of genes by site-specific recombination. *Mol. Microbiol.* 15: 593–600.

Hayman, G.T. and Bolen, P.L. (1993) Movement of shuttle plasmids from *Escherichia coli* into yeasts other than *Saccharomyces cerevisiae* using trans-kingdom conjugation. *Plasmid* 30: 251–257.

Heinemann, J.A. and Sprague, G.F., Jr. (1989) Bacterial conjugative plasmids mobilize DNA transfer between bacteria and yeast. *Nature* 340: 205–209.

Heinemann, J.A. (1991) Genetics of gene transfer between species. *Trends Genet.* 7: 181–185.

Heinemann, J.A. and Ankenbauer, R.G. (1993) Retrotransfer of the *Inc*P plasmid *R751* from *Escherichia coli* maxicells: Evidence for the genetic sufficiency of self-transferable plasmids of bacterial conjugation. *Mol. Microbiol.* 10: 57–62.

Helling, R.B., Vargas, C. and Adams, J. (1987) Evolution of *Escherichia coli* during growth in a constant environment. *Genetics* 116: 349–358.

Istock, C.A., Duncan, K.E., Ferguson, N. and Zhou, X. (1992) Sexuality in a natural population of bacteria: *Bacillus subtilis* challenges the clonal paradigm. *Molec. Ecol.* 1: 95–103.

Kaiser, D. and Dworkin, M. (1975) Gene transfer to a *Myxobacterium* by *Escherichia coli* phage *P1. Science* 187: 653–654.

Kellenberger, E. (1994) Genetic ecology: A new interdisciplinary science, fundamental for evolution, biodiversity and biosafety evaluations. *Experientia* 50: 429–437.

Korona, R., Nakatsu, C.H., Forney, L.J. and Lenski, R.E. (1994) Evidence for multiple adaptive peaks from populations of bacteria evolving in a structured habitat. *Proc. Natl. Acad. Sci. USA* 91: 9037–9041.

Levy, S.B. and Miller, R.V. (1989) *Gene Transfer in the Environment.* McGraw-Hill, New York.

Lorenz, M.G. and Wackernagel, W. (1990) Natural genetic transformation of *Pseudomonas stutzeri* by sand-adsorbed DNA. *Arch. Microbiol.* 154: 380–385.

Lorenz, M.G., Gerjets, D. and Wackernagel, W. (1991) Release of transforming plasmid and chromosomal DNA from two cultured soil bacteria. *Arch. Microbiol.* 156: 319–326.

Lorenz, M.G. and Wackernagel, W. (1993) Bacterial gene transfer in the environment. *In*: K. Wöhrmann and J. Tomiuk (eds): *Transgenic Organisms: Risk Assessment of Deliberate Release.* Birkhäuser Verlag, Basel, pp 43–64.

Lorenz, M.G. and Wackernagel, W. (1994) Bacterial gene transfer by natural genetic transformation in the environment. *Microbiol. Rev.* 58: 563–602.

Lorenz, M.G. and Wackernagel, W. (1995) Die Rolle der Sexualität in der Evolution von Prokaryoten. *BioSpektrum* 1: 28–33.

Martin, B., Garcia, P., Castanié, M.-P. and Claverys, J.-P. (1995) The *recA* gene of *Streptococcus pneumoniae* is part of a competence-induced operon and controls lysogenic induction. *Mol. Microbiol.* 15: 367–379.

Matic, I., Rayssiguier, C. and Radman, M. (1995) Interspecies gene exchange in bacteria: The role of *SOS* and mismatch repair systems in evolution of species. *Cell* 80: 507–515.

Maynard Smith, J., Smith, N.H., O'Rourke, M. and Spratt, B.G. (1993) How clonal are bacteria? *Proc. Natl. Acad. Sci. USA* 90: 4384–4388.

McKane, M. and Milkman, R. (1995) Transduction, restriction and recombination patterns in *Escherichia coli*. *Genetics* 139: 35–43.

Meyer, T.F., Gibbs, C.P. and Haas, R. (1990) Variation and control of protein expression in *Neisseria*. *Annu. Rev. Microbiol.* 44: 451–477.

Milkman, R. and McKane Bridges, M. (1993) Molecular evolution of the *E. coli* chromosome. IV. Sequence comparisons. *Genetics* 133: 455–468.

Moineau, S., Pandian, S. and Klaenhammer, T.R. (1994) Evolution of a lytic bacteriophage *via* DNA acquisition from the *Lactococcus lactis* chromosome. *Appl. Environ. Microbiol.* 60: 1832–1841.

Molin, S., Boe, L., Jensen, L.B., Kristensen, C.S., Givskov, M., Ramos, J.L. and Bej, A.K. (1993) Suicidal genetic elements and their use in biological containment of bacteria. *Annu. Rev. Microbiol.* 47: 139–166.

Murooka, Y. and Harada, T. (1979) Expansion of the host range of coliphage *P1* and gene transfer from enteric bacteria to other gram-negative bacteria. *Appl. Environ. Microbiol.* 38: 754–757.

Myers, R.S. and Stahl, F.W. (1994) *C* and the *RecBCD* enzyme of *Escherichia coli*. *Annu. Rev. Genet.* 28: 49–70.

Nelson, K. and Selander, R.K. (1994) Intergeneric transfer and recombination of the 6-phosphogluconate dehydrogenase gene (*gnd*) in enteric bacteria. *Proc. Natl. Acad. Sci. USA* 91: 10227–10231.

Nishikawa, M., Suzuki, K. and Yoshida, K. (1992) DNA integration into recipient yeast chromosomes by transkingdom conjugation between *Escherichia coli* and *Saccharomyces cerevisiae*. *Curr. Genet.* 21: 101–108.

Novick, R.P. (1987) Plasmid incompatibility. *Microbiol. Rev.* 51: 381–395.

Novitzky, J.A. (1986) Degradation of dead microbial biomass in a marine sediment. *Appl. Environ. Microbiol.* 52: 504–509.

Ogunseitan, O.A., Sayler, G.S. and Miller, R.V. (1990) Dynamic interaction of *Pseudomonas aeruginosa* and bacteriophages in lake water. *Microb. Ecol.* 19: 171–185.

Palmen, R., Driessen, A.J.M. and Hellingwerf, K.J. (1994) Bioenergetic aspects of the translocation of macromolecules across bacterial membranes. *Biochem. Biophys. Acta* 1183: 417–451.

Pantastico-Caldas, M., Duncan, K.E. and Istock, C.A. (1992) Population dynamics of bacteriophage and *Bacillus subtilis* in soil. *Ecology* 73: 1888–1902.

Rayssiguier, C., Thaler, D.S. and Radman, M. (1989) The barrier to recombination between *Escherichia coli* and *Salmonella typhimurium* is disrupted in mismatch-repair mutants. *Nature* 342: 396–401.

Romanowski, G., Lorenz, M.G. and Wackernagel, W. (1993) Plasmid DNA in a groundwater aquifer microcosm – Adsorption, DNase resistance and natural genetic transformation of *Bacillus subtilis*. *Molec. Ecol.* 2: 171–181.

Rosenzweig, R.F., Sharp, R.R., Treves, D.S. and Adams, J. (1994) Microbial evolution in a simple unstructured environment: Genetic differentiation in *Escherichia coli*. *Genetics* 137: 903–917.

Salyers, A.A. and Shoemaker, N.B. (1994) Broad host range gene transfer: Plasmids and conjugative transposons. *FEMS Microbiol. Ecol.* 15: 15–22.

Sandmeier, H. (1994) Acquisition and rearrangement of sequence motifs in the evolution of bacteriophage tail fibres. *Mol. Microbiol.* 12: 343–350.

Saye, D.J., Ogunseitan, O., Sayler, G.S. and Miller, R.V. (1987) Potential for transduction of plasmids in a natural freshwater environment: Effect of plasmid donor concentration and a natural microbial community on transduction in *Pseudomonas aeruginosa*. *Appl. Environ. Microbiol.* 53: 987–995.

Selander, R.K., Caugant, D.A., Ochman, H., Musser, J.M., Gilmour, M.N. and Whittam, T.S. (1986) Methods of multilocus enzyme electrophoresis for bacterial population genetics and systematics. *Appl. Environ. Microbiol.* 51: 873–884.

Sibold, C., Henrichsen, J., König, A., Martin, C., Chalkley, L. and Hakenbeck, R. (1994) Mosaic *pbpX* genes of major clones of penicillin-resistant *Streptococcus pneumoniae* have evolved from *pbpX* genes of a penicillin-sensitive *Streptococcus oralis*. *Mol. Microbiol.* 12: 1013–1023.

Smith, H.O., Danner, D.B. and Deich, R.A. (1981) Genetic transformation. *Annu. Rev. Biochem.* 50: 41–68.

Smith, M.W., Feng, D.F. and Doolittle, R.F. (1992) Evolution by acquisition: The case for horizontal gene transfer. *Trends Biochem. Sci.* 17: 489–493.

Stotzky, G., Schiffenbauer, M., Lipson, S.M. and Yu, B.H. (1981) Surface interactions between viruses and clay minerals and microbes: Mechanisms and implications. *In*: M. Goddard and M. Butler (eds): *Viruses and Wastewater Treatment*. Pergamon Press, Oxford, pp 199–204.

Syvanen, M. (1994) Horizontal gene transfer: Evidence and possible consequences. *Annu. Rev. Genet.* 28: 237–261.

Thoms, B. and Wackernagel, W. (1984) Genetic control of damage-inducible restriction alleviation in *Escherichia coli K12*: An *SOS* function not repressed by *lexA*. *Mol. Gen. Genet.* 197: 297–303.

van der Meer, J.R., de Vos, W.M., Harayama, S. and Zehnder, A.J.B. (1992) Molecular mechanisms of genetic adaptation to xenobiotic compounds. *Microbiol. Rev.* 56: 677–694.

Vosman, B. and Hellingwerf, K.J. (1991) Molecular cloning and functional characterization of a *recA* analog from *Pseudomonas stutzeri* and construction of a *P. stutzeri recA* mutant. *Antonie van Leeuwenhoek J. Microbiol.* 59: 115–123.

Wackernagel, W. The persistence of DNA in the environment and its potential for bacterial genetic transformation. *In*: E.R. Schmidt and T. Hankeln (eds): *Transgenic Organisms and Biosafety. Horizontal Gene Transfer, Stability of DNA and Expression of Transgenes.* Springer-Verlag, Germany; *in press.*

Walker, G.C. (1985) Inducible DNA repair systems. *Annu. Rev. Biochem.* 54: 425–457.

Wassenaar, T.M., Fry, B.N. and van der Zeijst, B.A.M. (1995) Variation of the flagellin gene locus of *Campylobacter jejuni* by recombination and horizontal gene transfer. *Microbiol.* 141: 95–101.

Whittam, T.S., Ochman, H. and Selander, R.K. (1983) Multilocus genetic structure in natural populations of *Escherichia coli. Proc. Natl. Acad. Sci. USA* 80: 1751–1755.

Wilson, M. and Lindow, S.E. (1993) Release of recombinant microorganisms. *Annu. Rev. Microbiol.* 47: 913–944.

Wise, M.G., Shimkets, L.J. and McArthur, J.V. (1995) Genetic structure of a lotic population of *Burkholderia (Pseudomonas) cepacia. Appl. Environ. Microbiol.* 61: 1791–1798.

Zambryski, P. (1992) Chronicles from the *Agrobacterium*-plant cell DNA transfer story. *Annu. Rev. Plant Physiol. Plant Mol. Biol.* 43: 465–490.

Transgenic Organisms – Biological and Social Implications
J. Tomiuk, K. Wöhrmann & A. Sentker (eds)
© 1996 Birkhäuser Verlag Basel/Switzerland

Evolutionary genetic considerations on the goals and risks in releasing transgenic crops

K.D. Adam and W.H. Köhler[1]

University of Waikato, Dept. of Biological Sciences, Private Bag 3105, Hamilton, New Zealand. Current address: Rayonier New Zealand Ltd., P.O.Box 9283, Newmarket, Auckland, New Zealand
[1]*Justus-Liebig-Universität, Biometrie und Populationsgenetik, Ludwigstraße 27, D-35390 Giessen, Germany*

Summary. Concerns have been raised that transgenic crop plants released into the environment might either change their ecological role and turn into weeds, or, more importantly, that the transgenic contructs they contain might escape to related wild species through introgressive hybridization, increasing the weed potential of those relatives. This chapter deals with the possible consequences of such an escape and discusses microevolutionary processes involved in the potential spread of a transgene from an initial hybrid into the recipient population of a wild relative. Using the hypothetical example of a transgene being transferred from *Brassica napus* to wild populations of *B. campestris*, we describe the potential role of mutation, selection, random genetic drift and different mating systems. We also discuss the involvement of migration, competition and coevolution. Starting from the assumption that in the wild hybrids between crops and their wild relatives *per se* will always have a reduced overall fitness compared to pure wild relatives, we first use population genetic models involving selection and migration, and selection, mutation and drift, to predict the outcome of a hypothetical escape, depending on different fitness levels for hybrids carrying the transgene. We show that a transgene awarding a net selective advantage will almost always be able to persist and spread through the recipient population. Even a slightly disadvantageous transgene, however, could persist, either as a result of frequent hybridization, or due to random genetic drift in small populations. A second approach enlists some simple models developed by plant breeders for optimising introgressive hybridization from wild relatives into crops. This approach has the advantage of accomodating multiple loci and different breeding systems. Based on the worst-case scenario of a "recombinant" hybrid, a plant containing the transgenic construct, but alleles from the wild relative species at all other loci, we predict that such a genotype is far less likely to occur in self-fertile species, where the initial hybrid develops into a series of inbred lines, than in outcrossing species, where the inital hybrid can quickly shed deleterious crop alleles through repeated backcrossing.

Introduction

Transgenic crops are cultivars genetically modified by the introduction of foreign DNA into their genome by molecular genetic methods. The ability to produce transgenic plants became a reality more than ten years ago (Fraley at al., 1988), and due to technical improvements in plant tissue culture and transformation and the growing involvement of the biotechnology industry, agricultural projects now rank as the third most common activity by biotechnology firms (Dibner, 1991; Seidler and Levin, 1994). The improvement of crops through genetic transformation has two goals: One is a more efficient production of a sufficient amount of food of good quality for human nutrition. This can be achieved for example by introducing transgenes which confer resistance against pests and diseases, or the ability to grow in less favourable environments. In a similar way weeds in herbicide-tolerant transgenic crops are easier to control. The second goal is the breeding of "tailor-made" cultivars designed to produce improved raw materials for the industry, such as potatoes with altered starch biosynthesis pathways in favour of amylopectin. This starch dissolves easily and is more stable at high temperatures, making it easier to handle in food production processes (see Teuber, this volume). For some plant breeders and farmers, the goal is the manipulation of all agriculturally and industrially relevant traits of a crop by gene technological methods (DLG, 1995).

The debate over the field release of genetically engineered organisms concentrates on the potential risks associated with such releases. Critics have cited potential social, cultural and ethical implications of an "escape" of transgenes into the environment, i.e., the uncontrolled transfer of transgenes from target species into others, altering the ecological role of the recipient species and causing disturbances of our ecosystems. In addition, there would be an economic risk in having to repair damage to the environment or to adjust to ecological changes. There is, however, no agreement on how big these risks are. Many molecular geneticists and biotechnology companies see the risks associated with releasing transgenic plants as merely hypothetical (Schuster, 1991), whereas critics maintain that because of the novel genetic modifications introduced by genetic engineering, the risk of negative consequences from this technology cannot be predicted (Breyer, 1991). For example, transgenes could turn crops into weeds, or the transfer of transgenes from crops to their wild relatives through spontaneous hybridization could increase the weed potential of such wild relatives and cause a reduction of ecological diversity through the displacement of other species.

In order to identify and measure the potential ecological and evolutionary risks of releasing transgenic plants into the environment, many research projects have been and are being carried out in laboratories, greenhouses, and (small scale) field experiments (Wöhrmann et al., 1993). These experiments should reveal mechanisms by which transgenic crops could turn into weeds or transgenes could escape to non-target species. They should also allow an assessment of the frequency of such escapes and of the detrimental effects (if any) resulting from such accidents. In summary, such research should lead to estimates of accident probabilities for all accident mechanisms and components. The final step of risk assessment then consists of a description and evaluation of possible accidents in terms of the probability of their occurrence and the size of the resulting damage (probabilistic safety assessment). To estimate these probabilities, expert opinions and expert panels are often being used (Naimon, 1991). The relevant areas of expertise for a probabilistic safety assessment for field release of transgenic plants are: (i) Studies on plant reproductive systems, in particular incompatibility systems and pollen dispersal mechanisms and dynamics; (ii) studies on introgressive hybridization in crop breeding; (iii) studies on plant colonization (transgenic crop or hybrids with its wild relatives as colonizer); (iv) studies on sympatric speciation; (v) "classical" population genetics literature.

However, no data are available so far to support the extrapolation of results from these experiments and the probabilistic safety assessment to large scale application in commercial plant production. Therefore, it has been suggested that monitored field experiments be conducted over several years to test for the effects of transgenic plant cultivars under field conditions, as it is done with conventional cultivars (Dietz, 1993; Nöh, 1994).

Recently, the Environmental Protection Agency (EPA, Washington, D.C.) approved the first pesticidal transgenic plants for "limited" commercialization in the United States, relying upon years of study of the toxins engineered into plants without reports of toxicity to animals, humans, or the environment. The agency declared that it "does not believe that there will be adverse effects to humans, non-target organisms, or the environment from the limited use of the products". The EPA concluded that "there is no unreasonable risk of unplanned pesticide production through gene capture and expression of *Bacillus thuringiensis* (*B.t.*) in wild relatives of the transformed plants". A limited pre-market registration for corn, potatoes, and cotton plants carrying the *B.t.* gene was granted to three companies, all of which agreed to a careful post-commercialization monitoring of pesticidal crops for the development of *B.t.* resistance (Hoyle, 1995; see also Braun, this volume; Parker and Bartsch, this volume).

The direct effects of transgenes and the behaviour of the genetically modified organism can be predicted with some certainty, but difficulties arise in assessing effects on the interrelations within ecosystems or in predicting the effects of long-term and large-scale industrial deployment of transgenic cultivars. This brought Gabriel (1993) to the conclusion that "biologists should honestly state that the risk of technologically modified genes cannot be assessed with our present knowledge".

A lack of any extensive theoretical studies combining population genetics and ecology means that current theories of evolutionary and ecological genetics cannot be used for general statements on the risks involved with field-releasing transgenic plants (Gabriel, 1993). Evolutionary genetic considerations of the goals and risks in releasing transgenic plants can therefore only be very specific and should be done on a case-by-case basis.

Evolutionary mechanisms

In population genetics evolutionary forces are studied through their effects on gene and genotype frequencies (Köhler and Braun, 1995). Looking at a single Hardy–Weinberg population (Sperlich, 1988; Hartl and Clark, 1989), the basic forces of change in such a population are: (i)

Mutation, which will increase genetic variability; (ii) selection, which allows individuals of a favourable genotype to reproduce in increased numbers at the cost of other, less favourable genotypes; (iii) changes in the mating system, which result in isolation and population subdivision and (iv) random genetic drift, which changes gene frequencies in small populations subject to sampling errors.

Considering a subdivided population, or the coexistence of more than one population (or species), respectively, one has to consider other evolutionary mechanisms such as: (i) Migration, which is the exchange of genes between (sub)populations; (ii) competition, which describes the differential ability of populations or species to persist in an ecosystem; (iii) coevolution, which is the evolutionary response of interacting populations.

For each of these factors we have to ask ourselves whether there are experimental or theoretical results and insights that can contribute to the risk assessment for the release of transgenic organisms into our environment.

Mutation

Mutation in its broadest sense means the origin of new hereditary types. It is the ultimate origin of all genetic variation and provides the raw material on which selection can act. At the DNA level most mutations are single nucleotide substitutions, insertions or deletions, resulting in changes of single amino acids in proteins, the primary gene products. Other mutations involve the relocation of larger pieces of DNA (transposable elements). At the level of chromosomes, structural changes (deletion/duplication, inversion, and translocation) can involve substantial segments of chromosomes, and numerical changes affect entire (sets of) chromosomes (aneuploidy or polyploidy).

Today, additional sources of new genetic variation are provided by the artifical introduction of entire genes or a group of genes ("designer" genes) from another, usually unrelated species by genetic engineering. Past attempts to artificially change the genetic make-up of domesticated species through mutagenic radiation or chemicals have mostly proved unsuccessful, because they were not able to target specific genes, and because the changes they brought about usually decreased rather than improved the functionality of the genes affected. Genetic transformation, the vector-mediated introduction of a "foreign" gene, is a form of "directed" mutation insofar as a defined gene is added to the target genome. It is not directed, though, with respect to the placement of the introduced gene(s) in the target genome, and the number of copies introduced, but post-transformation selection usually results in single (or low copy number) introductions into regions of the genome where other genes are not affected. Viewing the introduction of designer genes as recurrent mutations might help us to improve our understanding of risk-determining factors (Gabriel, 1993).

Selection

Simply speaking, natural selection is the mechanism by which populations adapt to their environment. It is the differential reproduction of genotypes and can be due to differential fertility and/or mortality. Darwinian fitness of a genotype is its average number of progeny relative to that of another genotype taken as a standard. The selection coefficient is defined as the deviation of relative fitness from unity, and the rate of change of gene frequencies under natural selection depends of the magnitude of the selection coefficients. If the fitnesses of all genotypes are known, these changes may be described and calculated by simple models (e.g., Bodmer and Cavalli-Sforza, 1976). In reality, though, natural selection is perhaps the least understood process in evolutionary biology (Hartl, 1985), because: (i) The "favorability" of a gene does not depend solely on the gene itself, but involves interactions between different genes and their products, and between the overall phenotype and the environment; (ii) most mutations will be fitness-neutral and subject only to random genetic drift; (iii) favorable mutations are not constantly arising and becoming fixed; the occurrence and fixation of favorable mutants is usually not observed on an ecological time scale except in cases of extreme selection pressures such as antibiotic resistance in bacteria, dynamic host-parasite systems (e.g., plant/fungus), pesticide resistance in insects, or heavy metal tolerance in plants; (iv) major changes in phenotype will normally reduce rather than increase fitness, because the genotype of an organism is an integrated system, and the introduction of major changes will result in disturbance. Most mutations which result in visible effects on the phenotype are harmful to the organism, and there are no examples of "hopeful monsters".

In terms of risk assessment for transgenic plants, what we are interested in is the fate of a particular mutant in a population. This will depend on the effect of the mutation on the overall phenotype and involves both the interaction between the gene affected by the mutation and other genes, as well the environment. Since it is difficult and often impossible to analyse such interactions because of their complexity, or to accurately measure the fitness of a mutant, we cannot assume that we will be able to *predict* the fate of a particular mutation. Besides, the fate of a new mutation or introduced gene is also affected by evolutionary forces other than selection, such as isolation, random genetic drift or a specific mating system.

The ecological success of an organism such as a transgenic plant depends on its specific context, the nature of the introduced gene(s), and the environment. Genetically modified plants are unlikely to be superior competitors in their natural environment because it is probable that the introduced gene(s) will disrupt a balanced system of gene interactions. The fundamental subspace (fundamental niche) in which they can survive, defined by their genetic make-up, should on average be smaller than that of their non-transgenic relative. The same should be true for the realized niche – the subspace of the fundamental niche which the species is actually utilizing, defined by the ecological opportunity. The introduction of "designer genes" could add new dimensions to the fundamental niche of transgenic plants, but whether they will be able to realize these new dimensions cannot be predicted.

Mating system

Isolation is the preference or avoidance of specific partners during their reproductive phase. Apart from geographic barriers it depends on a variety of pre- and postzygotic isolation mechanisms. Isolation allows the coexistence of sympatric populations with separate genetic identities. Within the panmictic unit of a Hardy–Weinberg population, isolation is ineffective as all individuals should freely cross-breed. Changes in the reproductive system of a panmictic population leading to a stronger isolation could hinder the spread and survival of a transgene with lower fitness. Breaking isolation, on the other hand, could increase the fitness of transgenes. Any changes in the isolation status of a population will change its structure, resulting in different subpopulations with different selection schemes. Because of a smaller population size, the influence of genetic drift will increase and the chance of survival (or extinction) of a transgene may therefore be enhanced.

Random genetic drift

Random genetic drift is the change of allele frequencies by sampling error during reproduction, and as such it solely depends on population size. If a designer gene is selectively neutral in a given environment, then random genetic drift will randomly alter its frequencies up and down, eventually leading either to fixation, the situation where all plants in a recipient population carry a copy of the designer gene, or its elimination from that population. Random genetic drift can override selection in sufficiently small populations, so that even if the fitness of the carriers of a transgene should be lower (or higher) than that of its non-transgenic relatives, such a transgene could become fixed (or eliminated) in a small population.

Migration

In the preceding sections we discussed processes involving single populations completely isolated from others. Gene frequencies in a system of (sub)populations may be changed, however, by the migration of individuals between populations (physical movement of individuals) or by pollen dispersal (physical movement of gametes), given that migrants are able to mix and reproduce in the new population. Migration in a broad sense includes the exchange of genetic elements within and among species (horizontal gene transfer), but reproductive isolation between species usually prevents the exchange of genes between them.

Competition

The persistence of a transgenic plant in the field depends on its ability to compete in the natural ecosystem. There are two approaches to studying the competitive ability of transgenic crops, the Traditional Breeding Model and the Introduced Species Model (see also Friedt and Ordon, this volume).

Decades of experience with traditional plant breeding demonstrate that there is little or no hazard from the release of improved cultivars developed that way. The same could be true for transgenic plants, but these plants typically have properties that the traditional techniques could not have produced. Yet it seems likely that "at least 99 out of 100 genetically engineered organisms from the combined classes used in research and industry will present no health or environmental problems" (Regal, 1994). The risk of ecological damage may increase, though, if a fitness-enhancing trait, e.g., resistance to a common disease, is engineered into a population of plants which as such is viable in the wild, rather than into highly selected or otherwise crippled cultivars or lines.

Introduced species can have severe and far-reaching ecological effects on the ecological communities which they enter. The same could be true for genetically modified plants, but it is argued that these organisms are not likely to have high competitive abilities. Yet they could survive in small numbers, making them hard to trace and requiring long-term monitoring.

Changes in the flora and fauna of countries subjected to (European) colonisation in the last millenium provide many examples for the ill effects of introduced species, both plant and animal, and the decimation of New Zealand's native fauna by invaders provides a prominent case in point (King, 1984; Adam et al., 1993). Among others, about fifty-five species and subspecies of native birds have been made extinct or nearly so in New Zealand since the first human colonisation more than one thousand years ago.

Coevolution

Potential evolutionary responses of native organisms to genetically modified plants could be important for the risk assessment of transgenic plants, and we need to consider the impact of short-term (up to 100 years or so) coevolutionary processes on population dynamics and community-level interactions, e.g., in plant-herbivore or plant-pathogen systems. These aspects will be discussed in detail in the chapter of Paul Braun in this volume.

Some case studies

Spontaneous hybridization in Brassicaceae

Oilseed rape (*Brassica napus*, canola, Brassicaceae) is the most important oil crop in Europe, Canada, and Japan. It is now generally accepted that *B. napus* has arisen from the spontaneous hybridization of *B. campestris* (synonym *B. rapa*) and *B. oleracea*. It is uncertain whether oilseed rape exists in a truly wild form, but it is found as a weed outside arable land areas in habitats close to fields, roadsides, and ruderals. Genes providing herbicide tolerance or insect or fungal resistance are introduced into rapeseed with relative ease by transformation, and field trials of transgenic oilseed rape in North America and Europe indicate that certified seed of transgenic cultivars could be marketed within a few years (Jørgensen and Andersen, 1994).

One of its parental species, *B. campestris*, is a common weed in cultivated areas, mostly in oilseed rape fields and in its wild form is spread over large parts of the world. The data for central Europe are doubtful, though, and it is possible that *B. campestris* is a strictly atlantic species (Rufener Al Mazyad and Ammann, 1995). Some cultivated forms of *B. campestris* (syn. *B. rapa*, canola) are used for oilseeds and together with the spring form of *B. napus* are common oil crops in parts of Northern Europe and North America, while other forms are grown as vegetables in Asia.

Cultivated *B. napus* is said to hybridize spontaneously with cultivated and wild forms of *B. campestris* and other closely related species *(B. adpressa, B. juncea, B. oleracea, Raphanus raphanistrum, Sinapis arvensis*; see, e.g., Kapteijns, 1993; Sentker et al., 1994), but the incidence of such spontaneous hybridization is under debate (Downey et al., 1980; Jørgensen and Andersen, 1994). Several studies reported difficulties using oilseed rape to artificially cross-pollinate its parental species, other closely related *Brassica* plants, or closely related taxa (Miller, 1991). Estimates of spontaneous hybridization between cultivated varieties vary between 1 and

3%, and in crosses of *B. napus* with wild plants of *B. campestris* ssp. *rapifera* of unknown origin, rates range from almost zero to 88%.

In a detailed study Jørgensen and Andersen (1994) and Jørgensen et al. (in press) estimated the frequencies of spontaneous hybridization between *B. napus* and weedy *B. campestris* ssp. *campestris* in agricultural fields by various methods (isozyme and RAPD genetic markers, cyto-genetic analysis, morphology, and pollen fertility). In a 1:1 mixture in the field *B. campestris* produced 13% hybrid seeds and *B. napus* 9%. In two other experiments with single plants of the self-incompatible *B. campestris* widely spaced within fields of oilseed rape, however, the percentage of hybrid seeds of the mother plants increased to 56% and 93%, respectively. Furthermore, the analysis of a weedy population of *B. campestris* in fields of oilseed rape revealed an incidence of 60% hybrid seeds, and backcrossing of the F_1 hybrids to the weedy *B. campestris* seems to occur.

The results of studies of introgression from oilseed rape into wild *B. campestris* detected in natural populations were ambiguous. Some molecular markers show distributions with larger similarity between *B. napus* and Danish accessions of wild and *cultivated B. campestris*, others do not. There appeared a sequence difference between one marker in *B. campestris* and in *B. napus* and *oleracea*, making introgression an unlikely explanation for the observed distribution (Jørgensen et al., 1994).

In a first attempt to compare the competitive ability of transgenic lines of oilseed rape to that of commercial cultivars by studying intra-specific yield density relationships, Poulsen and Fredshavn (1995) found no significant differences between the two. This indicates that the trans-genic plants behave like the non-transformed ones, and that at least in the short run the transgenic lines are not likely to become more weedy. The authors hold, however, that it is not possible at this stage to predict the ecological consequences of a field release of transgenic *B. napus*. In a parallel study still under way, Hauser et al. (1995) are analysing the fitness components of impor-tant life stages of the F_1 hybrids between *B. napus* and *B. campestris*, and their backcrosses.

Linder and Schmidt (1994) have investigated the possible escape of a transgene *via* the seed bank. In their field experiments seeds from crosses between non-transgenic *B. campestris* cv "Tobin" and wild *B. campestris* displayed intermediate levels of dormancy relative to their parents, with each of the reciprocal hybrids more closely resembling its maternal parent. They concluded that hybrid seeds will remain in the seed bank under established pure cultivar or wild populations, an important condition for persistence, and that hybrid seeds sired by transgenic pollen may be a particular cause for concern, especially in wild populations outside agricultural fields.

Others (Eber et al., 1995; Chèvre et al., 1995) report interspecific crosses under optimal conditi-ons between oilseed rape and three widespread weeds: Wild radish (*Raphanus raphanistrum*), hoary (*B. adpressa*), and wild mustard (*Sinapis arvensis*), all of which flower at the same time as oilseed rape. Their results indicate that for *B. adpressa* and *S. arvensis* the risk of gene exchange seems to be very low either because of the low incidence of interspecific hybrids (*S. alba*), or because of the low frequency of chromosome pairing (*B. adpressa*). As far as *B. napus* x *R. raphanistrum* is concerned, further experiments are required before any firm conclusions on the incidence of spontaneous hybridization can be drawn.

In conclusion, oilseed rape (*B. napus*) appears to be a highly "leaky" crop revealing several possibilities for spontaneous hybridization with its wild relatives, especially *B. campestris*, which could provide an escape route for transgenes.

Quantitative models for the potential escape of designer genes – population genetic models

The escape of a designer gene can be modeled using a combination of evolutionary forces. We need to consider selection, because it is a premise of risk assessment that transgenes are not selectively neutral. Additionally, we can look at migration, or random genetic drift, or a combination of drift and mutation (and migration). In each case a number of simplifying assumptions have to be made: (i) Only the transgene is considered, ignoring genetic changes at all other loci (which undoubtedly occur during hybridization). We can try to accomodate this situation by assuming that the selective (dis)advantage we award to individuals carrying the transgene is a net effect, accounting for the selective (dis)advantages of hybridization at all other loci (see also section on plant breeding models below). (ii) It is assumed that the transgene is dominant over a recessive allele for which the recipient species, e.g., *B. campestris*, is fixed, whereas in reality hybrid transgenics would be "hemizygous", i.e., the transgenes such as a virus-introduced transgenic construct has no allelic equivalent in the *B. campestris* genome. Escaped coat-protein genes which confer resistance to plant viruses, for example, would fit well into such a framework and can quite appropriately be treated like a resistance gene migrating into a uniformly susceptible population.

Selection and migration

The escape of a designer gene can be modeled using the continent–island model of migration, where such a gene would migrate from the "continent" of a transgenic crop (e.g., *B. napus*) into the "island" of another species, say wild *B. campestris*. This would actually happen if a female *B. campestris* was fertilized by transgenic *B. napus* pollen and produced fertile offspring. While such an event may possibly be described as rare, it still is sufficiently regular to be treated like a

Table 1. Equilibrium frequencies q^* for a dominant transgene in the island population

Selective (dis)advantage	$|s| \gg m$	$|s| \ll m$
$s > 0$ (A favourable)	1	1
$s < 0$ (A deleterious)	$m/(-s)$	1

migration process. Under the general continent–island migration model with selection the change in allele frequency is

$$\Delta q = [s_1 + (s_2 - 2s_1) q] q (1 - q) - m(q - Q) \qquad [1]$$

(Wright, 1969), where q and Q are the frequencies of allele A in the recipient (island) population and the migrants, respectively. Let m be the proportion of migrants in each generation, and s_1 and s_2 are the selective (dis)advantages of genotypes AA and Aa compared to aa, i.e., $w_{AA} = 1 + s_2$, $w_{Aa} = 1 + s_1$, and $w_{aa} = 1$. For the purpose of this exercise, we can assume that all migrants are homozygous for the transgene A, i.e., $Q = 1$, and assign a to its "non-existent" wild-type allele. Furthermore, we have already stated that A is completely dominant over a, so that $s_2 = s_1 = s$ and $w_{AA} = w_{Aa} = 1 + s$ and $w_{aa} = 1$. Under these circumstances, [1] changes to

$$\Delta q = (1 + q) sq (1 - q) - m (q - 1). \qquad [2]$$

Determining q^*, the equilibrium frequencies for A, involves a cubic equation in q for which a general solution is not readily available. Wright (1969) provides approximate solutions for the cases of both the dominant transgene A being favourable, and A being deleterious, and for the situations where selection is the predominant force $(|s| \gg m)$, and where migration is the predominant force $(|s| \ll m)$. As can be expected for a situation where the "continental" population is fixed and hence migrants are always AA, the transgene becomes fixed in the "island" population in all situations bar one, where A is deleterious and migration rates are low, which is a realistic scenario in the context of risk assessment for transgenics (see Tab. 1 and Fig. 1 for details).

These results show that under this model the only situation where the transgene would *not* completely penetrate the recipient population is where it confers no net fitness advantage and hybridization rates are low.

Selection and mutation

Apart from migration, the escape of a transgene from a crop such as *B. napus* into a wild relative such as *B. campestris* could be described as a (recurrent) mutation event in a finite population. Again the same simplifying assumptions as for the migration model are made. In this situation the fate of the transgene will depend on three factors: "mutation" rates, i.e., rates of escape by spontaneous hybridization; effective population size; and the overall selective (dis)advantage conferred by the transgene.

Again, let *A* be a dominant transgene with frequency *q* and *a* its "non-existent" wild-type allele, with genotype fitnesses $w_{AA} = w_{Aa} = 1$ and $w_{aa} = 1 - s$. Furthermore, let *n* be the "mutation rate" for *a* to *A*, i.e., the rate at which escaped transgenes *A* enter the population of the wild relative *B. campestris*. Under this scenario, we can ignore "reverse mutations" (*A* to *a*) and set $m = 0$.

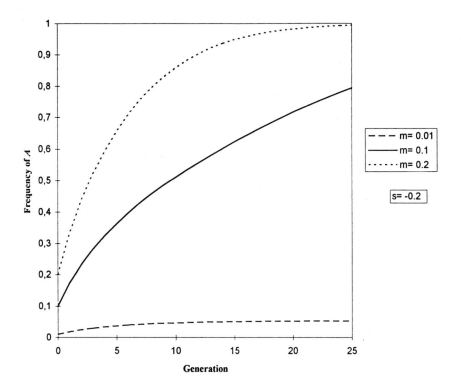

Figure 1. Frequency of deleterious transgene under the migration model.

In finite populations random genetic drift is one of the evolutionary forces, and it is useful to describe the outcome of evolutionary processes in terms of the probability density function of allele frequencies, thereby summarizing the outcome of a large number of experiments or observations on a large number of (sub)populations, all of which were subject to sampling error. Wright (1969) provides expressions for both the rates of change and the steady-state distribution of q, the frequency of a favourable transgene A in finite populations of size N as

$$\Phi(q) = Ce^{-2Ns(1-q)^2} q^{4Nv-1} (1-q)^{4N\mu-1} \qquad [3]$$

which, in the absence of reverse mutations, can be simplified to

$$\Phi(q) = Ce^{-2Ns(1-q)^2} q^{4Nv-1} (1-q)^{-1} \qquad [4]$$

For smaller populations ($4Ns < 3$) this distribution would show a typical U-shape or J-shape (Li, 1976), depending on the size of s, indicating the predominant influence of drift, with most (sub)populations being fixed for A if $s > 0$, and some for a, but only a few still containing both alleles. If populations are larger ($4Ns > 5$) and the influence of random genetic drift reduced

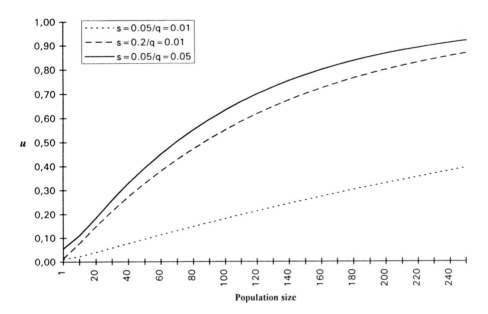

Figure 2. Fixation probability u for a favourable transgene for different values of selective advantage (s) and initial allele frequency (q).

accordingly, most populations would still contain A and a at equilibrium, the former being more frequent if the transgene confers an advantage.

It is of particular interest in the context of transgenic risk assessment to derive the probability u for a favourable escaped transgene A to become fixed under the conditions described above. The general solution for this problem is rather involved, but it is known that $u \approx 2\,s$ for large values of N and s ($4Ns > 3$). For the special case of intermediate fitness of the heterozygote ($w_{Aa} = 1 - 1/2\,s$), the probability u of fixation of as given by Kimura (1957) becomes

$$u(A) = \frac{1 - e^{-4Nsq}}{1 - e^{-4Ns}} \qquad [5]$$

which can be simplified to

$$u(A) = \frac{2\,s}{1 - e^{-4Ns}} \qquad [5a]$$

if $|s|$ is sufficiently small (Crow and Kimura, 1970). Figure 2 shows u as a function of population size and selective advantages of A, and reveals the importance of the initial frequency of A in this model; the probability of fixation is increased considerably if the rate at which transgenic hybrids are formed increases.

Quantitative models for the potential escape of designer genes – models derived from plant breeding theory

Before looking at the relevance of introgressive hybridization models for transgenic risk assessment, an issue must be adressed which was deliberately – albeit for good reason – avoided in the discussion on population genetic models above. As described above it is suggested that the most likely form of escape of a transgenic construct is through spontaneous interspecific hybridization with a wild relative species such as from *B. napus* into *B. campestris*. The genetic constitution of such hybrids will depend on the mode of hybridization. In the most closely related species, hybrids are formed more or less in the same way as intraspecific offspring, with normal meiotic pairing of and exchange between homologous chromosomes, and fully fertile backcrosses between (F_1) hybrids and either parental species. For a detailed discussion of cytogenetic aspects of crossability and gene exchange through hybridization see Martin and Jouve (1992). Hybridization between more distant relatives often relies on more complex mechanisms, such as chromosome doubling and the formation of amphidiploids (Khush and Brar, 1992), which not only reduces the rates at which such hybrids are formed spontaneously, but also the probability of backcrosses to either of the parents. Polyploidization also affects the dynamics of genetic changes even in fertile hybrids, for example by reducing the rate at which hybrids can shed deleterious genes (Hermsen, 1992).

For these reasons discussion of transgene escape through introgression will be restricted to the case of "simple" hybrids between closely related species. Here the fate of an escaped designer gene contained in an interspecific hybrid not only depends on the fitness (dis)advantage conferred by that gene, but also on the overall fitness of the hybrid. In discussion of population genetic models with this problem was dealt with rather crudely, because the models available often are unable to accomodate more than one locus at a time. Some of the models developed by plant breeders for studying introgressive hybridization, however, do address multilocus situations.

Crop species can generally be considered to have a lower overall fitness in the wild than the wild relative species itself. This is because under domestication crops have been selected for traits such as apical dominance, which reduces the ability to shade out competitors, or stable rachides in cereals and indehiscent pods in pulses, which restrict seed dispersal. Other examples for traits likely to reduce fitness in the wild are reduced seed dormancy, increased harvest index, or reproductive synchronization. Many of the apparently fitness-relevant differences between crops and their wild relatives are under the control of single major genes either coding for morphological traits (Gottlieb, 1984) or acting at the top of regulatory cascades, but others show quantitative inheritance.

If it is accepted that hybrids between crops and their wild relatives generally show reduced fitness in the wild, compared to the wild relative, then the "most dangerous genotype" ("worst-case genotype") in the context of this discussion would be the one consisting of wild-type (wild-relative) alleles at all loci plus the transgenic construct. The fitness of the worst-case genotype can be expressed as

$$\tilde{w}_T = \prod_i w_{H_i} \cdot w_T = \tilde{w}_H \cdot w_T \qquad [6]$$

where $\tilde{w}_H$ and w_T are the overall fitness of the hybrid, and the fitness associated with the presence of the transgenic construct, respectively. $\prod_i w_{H_i}$ represents the overall fitness of the hybrid, without the transgene, expressed as a product, with w_{H_i} being the fitness of the hybrid at its n individual loci. Furthermore, let $\tilde{w}_W$ be the overall fitness of the wild relatives and w_{W_i} the fitness values of its n loci of interest ($i = 1, \dots ,n$), so that $\tilde{w}_W = \prod_i w_{W_i}$.

Following the argument above,

$$\tilde{w}_W > \tilde{w}_H \qquad\qquad [7]$$

Hence the only case we need to consider here would be where

$$\tilde{w}_T = \tilde{w}_H \cdot w_T > \tilde{w}_W \qquad\qquad [8]$$

or in other words, where the fitness advantage w_T conferred by one allele (at a hemizygous locus) of the transgenic construct more than compensates for the general fitness disadvantage of the hybrid compared with the wild relative, e.g., by conferring increased resistance to insect herbivore pests or (viral) diseases. There is a possibility that an escaped transgene, though not compen–sating for the general reduction of fitness in the hybrid, is maintained in the wild-relative popula–tion through random genetic drift and/or a continuous re-supply through frequent hybridization (dealt with under the population genetic models above).

If the net fitness of the initial hybrid is greater than the average fitness of the wild relative, conditions are set for the "worst-case genotype" to develop, and as it does, the net fitness of hybrids and their progeny will increase continuously, as they can shed crop alleles through repea–ted cycles of selfing or backcrossing. Furthermore, as the fitness of its carriers increases, the escaped transgene is likely to spread throughout the wild-relative population. This is conditional, though, on the assumption that the environment is stable in terms of the advantage which the transgene confers. Genes coding for herbicide-tolerance are only advantageous if their carriers are exposed to that herbicide. Similarly, genes conferring disease resistance are fitness-neutral at best if the relevant pathogene is absent or low in abundance, and it is worth remembering that the incidence of many plant pathogenes can fluctuate strongly between seasons and years.

Let us now look at the dynamics of the introgressive hybridization that leads to the evolution of the "worst-case genotype". The scenario here is exactly the reverse of the situation faced by plant breeders who want to transfer a useful gene, e.g., for disease resistance, from a wild relative into a crop cultivar. Their goal is an individual or population containing crop alleles at all loci except the one coding for that particular resistance. The fastest and most efficient way of achieving this is through repeated cycles of backcrossing with strong selection for the crop phenotype and the disease resistance in question.

There are a number of factors, apart from fitness, which are relevant here. We have already mentioned the importance of fully functional meioses and normal recombination, and in this context we have to consider the locus at which the transgenic construct was integrated into the crop genome, and the number of copies present. Post-transformation selection usually aims to select for single-copy transgenics, and any transformed lines where the introduced gene is inserted *within* a functional gene are also likely to be discarded. Any transgenic construct, however, which happens to have been integrated *in close proximity* to a locus coding for a fitness-relevant trait in which crop and wild relative differ, will be slow to spread through the wild-relative population until that linkage is broken. Other factors are the heritability of the traits which make up the difference between crop and wild relative, and dominance relationships at the loci involved. Clearly high heritability will help the recovery of the wild-relative genotype from the hybrid, while dominance of wild-type alleles over crop alleles will slow down that process. As for the latter, Gottlieb (1984) reports that at a number of loci coding for (morphological) differences between crops and their wild relatives, wild-type alleles are indeed dominant, probably because under domestication "afunctional" alleles of genes coding for traits such as the presence of feeding deterrents (e.g., alkaloids), which are important in the wild, but undesirable in a crop have been selected. The other critical factor is the reproductive system; in strictly self-fertile species the initial hybrid would develop into a number of increasingly homozygous lines, whereas in obligatory out-crossing species the path to the "worst-case genotype" is through repeated cycles of backcrossing to the wild relative.

Selfing

Let us assume a hybrid between a crop and its wild relative, heterozygous at m loci and hemizygous for a transgene A. Each of the m loci codes for a single trait which contributes to the fitness difference between crop and wild relative in the wild. The relevant genetic make-up of the hybrid is $D_1d_1 \ldots D_md_m$ plus A-, with D_i and d_i designating the crop and wild-relative allele, respectively. There is no dominance except for the transgene, no linkage, and the fitness at each locus is $w_{DD} = 1 - s$, $w_{Dd} = 1 - s/2$ and $w_{dd} = 1$, with $s > 0$. Let the hybrid undergo self-fertilisation for n generations, then the probability u that a homozgous line thus developed contains two wild-relative alleles any one locus for large values of n approaches

$$u = \frac{1}{2}\left(1 + \frac{s}{2}\right) \qquad [9]$$

which yields approximately the same results as (5) above (Bailey and Comstock, 1976). The probability U that any line is homozygous for the wild-relative allele at *at least r* loci ($r \le m$) is given by the cumulative binomial distribution function

$$U = \sum_{k=r}^{m} \binom{m}{k} u^k (1 - u)^{n-k} \qquad [10]$$

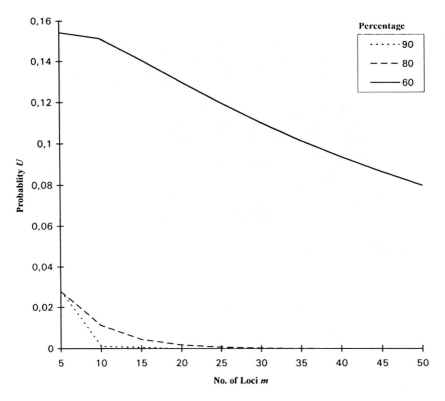

Figure 3. Combined probability U of fixing a transgene plus at least a set percentage of wild-relative alleles under selfing ($s = 0.1$).

(Bailey, 1977). Figure 3 shows the probability of both the transgene and a large proportion of the wild-relative genome becoming homozygous after repeated selfing. This probability is very low even for strong selection, so that the transgene A is likely to end up in homozygous lines which in their overall fitness at loci other than the transgene are below the average fitness of the wild relative. Hence if the transgene *per se* is selectively advantageous, it will "lose" some of that advantage it confers to make up for this difference. As for linkage, which has been excluded in the formal analyis above, it should increase the probability of recovering a large proportion of the wild-relative genome. In the present context, a suppression of recombination, resulting in substantial and persistent linkage, may well exist if the meiotic pairing of homologous chromosomes in the hybrid is imperfect.

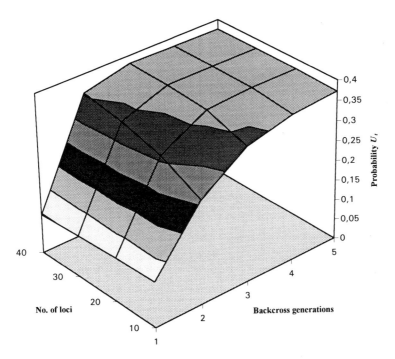

Figure 4. Combined probability U_t of fixing a transgene plus at least 80% of wild-relative alleles under back-crossing ($s = 0.1$).

Backcrossing

Here the notation used for the case of selfing above is retained, but its is assumed that the species in question is outcrossing repeatedly to members of the wild relative population. Again we assume that there is no dominance except for A, and no linkage. Fujimaki and Comstock (1977) extended this model to include linkage. Now the probability that the population derived from t generations of backcrossing is homozygous for the wild-relative allele at *at least r* loci ($r \leq m$) is given by the cumulative binomial distribution function

$$U_t = \sum_{k=r}^{m} \binom{m}{k} u_t^k (1 - u_t)^{n-k}$$

where   [11]

$$u_t = 1 + \left(\frac{1}{2}\right)^t \cdot \left(\left[\frac{1}{2} + \frac{s}{4}\right]^t - 1\right)$$

Figure 4 shows the probability of having both the transgene and two alleles of the wild relative at 80% of the loci involved after t generations of backcrossing. As t increases, this probability approaches the equilibrium frequency of A under backcrossing. It must be pointed out, however, that this model – while appropriate for enforced backcrossing in plant breeding – can describe the first few generations of random outcrossing after hybridization only. As the frequency of A rises, hybrid offspring A- will mate with other carriers of A, and the eventual maximum frequency of A becomes 1. Likewise, the probability of recovering 90% or more of the wild-relative genome *per se* approaches 1 for larger values of t.

The chance of a "worst-case genotype" evolving under the conditions described above is much higher for backcrossing than for selfing. Hence the chance for a successful transgene escape from an outcrossing crop species is higher not only because its capacity for pollen dispersal will increase the chances of spontaneous hybridization to occur, but also because a transgene contained in an initial hybrid has a higher chance to spread throughout the population of the wild relative, provided that the wild relative is outcrossing, too.

References

Adam, K.D., King, C.M. and Köhler, W.H. (1993) Potential ecological effects of escaped transgenic animals: Lessons from past biological invasions. *In*: K. Wöhrmann and J. Tomiuk (eds): *Transgenic Organisms: Risk Assessment of Deliberate Release*. Birkhäuser Verlag, Basel, pp 153–173.

Bailey, T.B. and Comstock, R.E. (1976) Linkage and the synthesis of better genotypes in self-fertilizing species. *Crop Science* 16: 363–370.

Bailey, T.B. (1977) Selection limits in self-fertilizing populations following the cross of homozygous lines. *In*: E. Pollack, O. Kempthorne and T.B. Bailey (eds): *Proceedings of the International Conference of Quantitative Genetics*. Iowa State University Press, Ames, pp 399–412.

Bodmer, W.F. and Cavalli-Sforza, L.L. (1976) *Genetics, Evolution, and Man*. Freeman, San Francisco.

Breyer, H. (1991) Interview on safety aspects of GMOs. *EC magazine* 11: 13.

Chèvre, A.M., Eber, F., Baranger, A., Kerlan, M.C., Barret, P., Festoc, G., Vallée, P. and Renard, M. (1995) Risk assessment of outcrossing of oilseed rape to related species. Workshop on *Brassica napus*, Helsingør, Denmark, *Book of abstracts* L-7.

Crow, J.F. and Kimura, M. (1970) *An Introduction to Population Genetics Theory*. Harper and Row, New York.

Dibner, M.D. (1991) Trecking trends in U.S. Biotechnology. *Bio/Technology* 9: 1335–1337.

Dietz, A. (1993) Risk assessment of genetically modified plants introduced into the environment. *In*: K. Wöhrmann and J. Tomiuk (eds): *Transgenic Organisms: Risk Assessment of Deliberate Release*. Birkhäuser Verlag, Basel, pp 209–227.

DLG (1995) Erste Freilandversuche mit gentechnisch veränderten Stärkekartoffeln – Erfahrungen mit Antragsverfahren und Öffentlichkeitsarbeit. *Niederschrift der 2. Sitzung des Ausschusses für Pflanzenzüchtung, Saatgut- und Versuchswesen am 12.1.1994 in Berlin*: 1-2, Deutsche Landwirtschafts-Gesellschaft, Frankfurt.

Downey, R.K., Klassen, A.J. and Stringam, G.R. (1980) Rapeseed and mustard. *In*: W.R. Fehr and H.H. Hadley (eds): *Hybridization of Crop Plants*. American Society of Agronomy and Crop Science, Madison, WI, pp 495–509.

Eber, F., Chèvre, A.M., Baranger, A., Vallée, P., Tanguy, X. and Renard, M. (1994) Spontaneous hybridization between a male sterile oilseed rape and two weeds. *Theor. Appl. Genet.* 88: 362–368.

Fraley, R.T., Frey, N.M. and Schell, J. (1988) *Genetic Improvement of Agricultural Important Crops*. Cold Spring Harbor Laboratory Press, Cold Spring Harbor, New York.

Fujimaki, H. and Comstock, R.E. (1977) A study of genetic linkage relative to success in backcross breeding program. *Japan. J. Breed.* 27: 105–115.

Gabriel, W. (1993) Technologically modified genes in natural populations: Some sceptical remarks on risk assessment from the view of population genetics. *In*: K. Wöhrmann and J. Tomiuk (eds): *Transgenic Organisms: Risk Assessment of Deliberate Release*. Birkhäuser Verlag, Basel, pp 109–116.

Gottlieb, L.D. (1984) Genetics and morphological evolution in plants. *Amer. Natur.* 123: 681–709.

Hartl, D. (1985) Engineered organisms in the environment: Inferences from population genetics. *In*: H.O. Halvorson, D. Pramer and M. Rogul (eds): *Engineered Organisms in the Environment. Scientific Issues*. American Society for Microbiology, Washington, pp 83–88.

Hartl, D.L. and Clark, A.G. (1989) *Principles of Population Genetics*, 2nd Edition. Sinauer, Sunderland.

Hauser, T.P., Jørgensen, R.B. and Ostergaard, H. (1995) Cross-compatibility between populations of weedy *Brassica campestris* and varieties of *B. napus*: Fitness of parental, hybrid and back-cross plants. Workshop on *Brassica napus*, Helsingør, Denmark, *Book of abstracts*: 3.4.

Hermsen, J.G.Th. (1992) Introductory considerations on distant hybridization. *In*: G. Kalloo and J.B. Chowdhury (eds): *Distant Hybridization of Crop Plants*. Springer-Verlag, Berlin, pp 1–14.

Hoyle, R. (1995) EPA okays first pesticidal transgenic plants. *Biotechnology* 13: 434–435.

Jørgensen, R.B. and Andersen, B. (1994) Spontaneous hybridization between oilseed rape (*Brassica napus*) and weedy *B. campestris* (Brassicaceae): A risk of growing genetically modified oilseed rape. *Amer. J. Bot.* 81: 1620–1626.

Jørgensen, R.B., Andersen, B., Mikkelsen, T., Landbo, L., Frello, S., Hansen, K. and Ostergaard, H. (1994) Risøs contribution to the programme focuses on gene dispersal in relation to risk assessment. *In: Risk Assessment in Relation to Release of Genetically Modified Plants*. Status Report, May 1994, Risø National Laboratory, Risø, Denmark, pp 2–5.

Jørgensen, R.B., Lambo, L. and Mikkelsen, T.R. Spontaneous hybridization between oilseed rape (*Brassica napus*) and weedy relatives. *Acta Horticulturae*; *in press*.

Kapteijns, A.J.A.M. (1993) Risk assessment of genetically modified crops. Potential of four arable crops to hybridize with the wild flora. *Euphytica* 66: 145–149.

Khush, G.S, and Brar, D.S. (1992) Overcoming the barriers of hybridization. *In*: G. Kalloo and J.B. Chowdhury (eds): *Distant Hybridization of Crop Plants*. Springer-Verlag, Berlin, pp 47–61.

Kimura, M. (1957) Some problems of stochastic processes in genetics. *Ann. Math. Statist.* 28: 882–901.

King, C. (1984) *Immigrant Killers*. Oxford University Press, Auckland.

Köhler, W. and Braun, P. (1995) Populationsgenetische und statistische Aspekte der Begleitforschung. *In*: S. Albrecht und V. Beusmann (eds): *Zur Ökologie transgener Nutzpflanzen*, Campus-Verlag, Frankfurt/M., pp 163–181.

Li, C.C. (1976) *First Course in Population Genetics*. The Boxwood Press, Pacific Grove, CA.

Linder, C.R. and Schmitt, J. (1994) Assessing the risks of transgene escape through time and crop-wild hybrid persistence. *Molec. Ecol.* 3: 23–30.

Martin, A. and Jouve, N. (1992) Cytogenetics of F_1 and their progenies. *In*: G. Kalloo and J.B. Chowdhury (eds): *Distant Hybridization of Crop Plants*. Springer-Verlag, Berlin, pp 82–105.

Miller, J.A. (1991) Biosciences and ecological integrity. *BioScience* 41: 206–210.

Naimon, J.S. (1991) Using expert panels to assess risks of environmental biotechnology applications. *In*: M. Levin and H. Strauss (eds): *Risk Assessment in Genetic Engineering*. McGraw Hill, New York, pp 319–353.

Nöh, I. (1994) Erfahrungen des Umweltbundesamtes bei der Risikobewertung von Freisetzungen. *In*: H. Gaugitsch and H. Torgersen (eds): *Umweltauswirkungen gentechnisch veränderter Organismen – Freisetzungskriterien international und in Österreich*. Bundesministerium für Umwelt, Jugend und Familie, Wien, pp 32–46.

Poulsen, G.S. and Fredshaven, J. (1995) Assessing competitive ability of transgenic oilseed rape. Workshop on *Brassica napus*, Helsingør, Denmark, *Book of abstracts* L-13.

Regal, P.J. (1994) Scientific principles for ecologically based risk assessment of transgenic organisms. *Molec. Ecol.* 3: 5–13.

Rufener Al Mazyad, P. and Ammann, K. (1995) *Brassica campestris*, a new weed for Switzerland? – A question to be studied in the risk assessment for the field releases of GMO's in Switzerland. Workshop on *Brassica napus*, Helsingør, Denmark, *Book of abstracts* 3.2.

Schuster, H.-J. (1991) Interview on safety aspects of GMOs. *EC magazine* 11: 10.

Seidler, R.J. and Levin, M. (1994) Potential ecological and nontarget effects of transgenic plant gene products on agriculture, silviculture, and natural ecosystems: General introduction. *Molec. Ecol.* 3: 1–3.

Sentker, A., Tomiuk, J. and Wöhrmann, K. (1994) Manipulierte Gene – Sicher unter menschlicher Kontrolle? *BIUZ* 24: 85–90.

Sperlich, D. (1988) *Populationsgenetik*. Fischer-Verlag, Stuttgart.

Wöhrmann, K., Tomiuk, J., Pollex, C. and Grimm, A. (1993) *Evolutionsbiologische Risiken bei Freisetzungen gentechnisch veränderter Organismen in die Umwelt*. Bundesminister für Umwelt, Naturschutz und Reaktorsicherheit, Umweltbundesamt, Berlin.

Wright S. (1969) *Evolution and the Genetics of Populations – Vol. 2: The Theory of Gene Frequencies*. The University of Chicago Press, Chicago.

Transmission of insect transposons into baculovirus genomes: An unusual host-pathogen interaction

J.A. Jehle

Department of Virology, Agricultural University Wageningen, Binnenhaven 11, NL-6709 PD Wageningen, The Netherlands

Summary. Baculoviruses are insect pathogenic viruses which have received considerable attention because of their potential as environmentally benign biological control agents of insects and as expression systems of eukaryotic genes. A peculiar characteristic of baculoviruses is that their genome can accommodate transposons derived from the insect hosts. So far, a number of different transposons including a *gypsy*-like retrotransposon, *Tc1/mariner*-like transposons and other class II transposons have been identified in the genomes of certain baculoviruses. Some of these elements were isolated from plaque morphology nucleopolyhedrovirus mutants in insect cell culture, some were obtained after *in vivo* infection of larvae by granuloviruses. This puzzling picture of horizontal escape of host transposons into baculovirus genomes provokes questions and speculations about the biological and evolutionary significance of this observation. Baculovirus mutagenesis by transposons may increase the genetic heterogeneity and could foster the flow of genetic information from a host genome into a virus genome. This interaction may also be used as a model to study horizontal transmission of insect transposons and may suggest that baculoviruses are one of the vectors involved in horizontal transposon transfer among insects. Possible consequences of transposon-mutagenesis on genetically engineered baculoviruses will be discussed.

Introduction

Baculoviruses are a group of occluded viruses which cause fatal natural epizootics in insect populations. To date, baculoviruses have been isolated from about 700 arthropod species, primarily from the orders Lepidoptera, Hymenoptera and Diptera (Martignoni and Iwai, 1986). Since in most cases the host range of baculoviruses is restricted to one or a few mostly closely related insect species these viruses obtain considerable attention as highly specific and effective biological control agents of pest insects in agriculture and forestry (Gröner, 1986; Huber, 1986).

The genome of baculoviruses is a circular, double stranded DNA molecule of 80–230 kb, capable of encoding about 120–150 genes. The proteinaceous occlusion bodies which occlude the enveloped rod-shaped (=baculum) virus particles consist mostly of a single 30 kDa polypeptide termed polyhedrin or granulin (Rohrmann, 1986). Baculoviruses are subdivided into two genera based on morphological and serological characteristics (Murphy et al., 1995): (i) Nucleopolyhedrovirus (NPV) forms polyhedra-like occlusion bodies with many virions each containing a single (SNPV) or multiple nucleocapsids (MNPV). (ii) Granulovirus (GV) has granule-like occlusion bodies containing only one virion with a single nucleocapsid (Fig. 1).

Pathogenesis and replication of baculoviruses has been extensively studied *in vivo* as well as *in vitro* (for review see, Granados and Williams, 1986; Blissard and Rohrmann, 1990; Volkman and Keddie, 1990). During virus replication in the nuclei of susceptible cells, two different viral phenotypes are generated (Fig. 2). Early in infection, the so-called budded viruses (BVs) are produced consisting of single virus particles which bud through the cell membrane into the extracellular fluid. This virus type circulates through the larval body and causes infection in susceptible host tissues. The BV phenotype is also infective to insect cell cultures. Late in infection, a second phenotype is produced and becomes enveloped by virus-coded proteins. This latter phenotype becomes embedded in the occlusion bodies and, hence, these viruses are called occlusion derived viruses (ODVs). They form the virus phenotype which is perorally infective for host larvae and is responsible for the dissemination of infection in the host population.

The polyhedrin gene (or the granulin gene in case of granuloviruses), which encode the matrix protein of the occlusion body, is under the control of a powerful promoter and is hyper-expressed very late in infection. It is not essential for virus replication. It can therefore be deleted or replaced with an exogenous gene without impairing viral replication. This nonessential nature of the polyhedrin gene or other very late genes such as *p10* has been exploited in the baculovirus expression

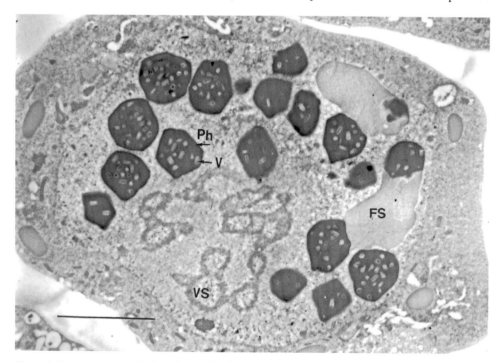

Figure 1. Electron microscopic dissection of a *Spodoptera frugiperda* cell infected with AcMNPV. (Ph = polyhedra, V = Virion, VS = virogenic stroma, FS = fibrillar structures, barr = 4 μm).

systems which allow ample synthesis of the foreign protein (Smith et al., 1983; Luckow and Summers, 1988; Vlak et al., 1990).

Transposable elements in baculoviruses

Transposons are mobile genetic elements ubiquitous in the genomes of prokaryota and eukaryota (Berg and Howe, 1989). Various genomic mutations, e.g., deletions, insertions or translocations,

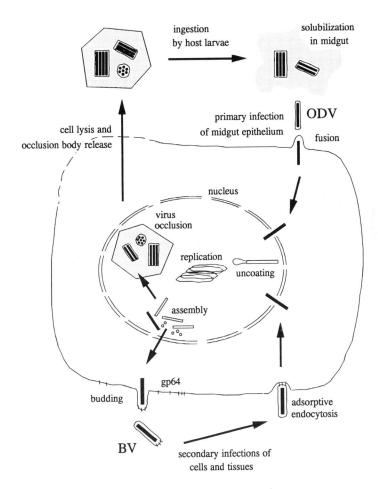

Figure 2. Schematic representation of the infection and replication cycle of AcMNPV in insects and insect cells. Note the two different virus phenotypes generated during infection. (ODV = occlusion derived virus, BV = budded virus).

have been attributed to the intragenomic mobility of transposons. In general, there are two main classes of transposons (Finnegan, 1992): *Class I elements* or retrotransposons which are mobilized using reverse transcription and usually have long terminal repeats. *Class II elements* or DNA transposons have inverted terminal repeats and transpose *via* a DNA intermediate. A characteristic of both classes is the transposition-mediated duplication of the target sequence (2–13 bp) flanking the transposon within the genome (Fig. 3).

The FP phenotype and transposable elements in nucleopolyhedroviruses

Upon serial passage of *Autographa californica* multiple nucleopolyhedrovirus (AcMNPV) in cell culture of the cabbage looper *Trichoplusia ni* (Noctuidae: Lepidoptera) or *Spodoptera frugiperda* (Noctuidae: Lepidoptera) virus mutants were found which caused the so-called FP (few polyhedra) phenomenon. These mutants were characterized by (i) a dramatically reduced number of occlusion bodies per infected cell and therefore less refractive plaques compared to wild-type virus; (ii) a propagation advantage in cell culture systems because of increased BV and decreased

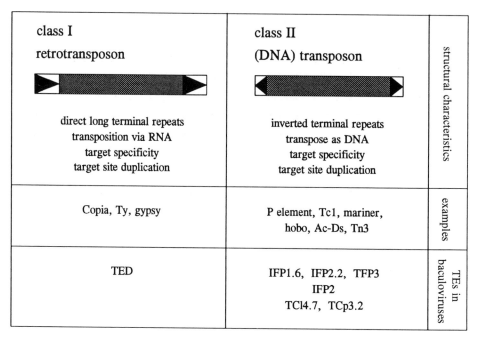

class I retrotransposon	class II (DNA) transposon	structural characteristics
direct long terminal repeats transposition via RNA target specificity target site duplication	inverted terminal repeats transpose as DNA target specificity target site duplication	
Copia, Ty, gypsy	P element, Tc1, mariner, hobo, Ac-Ds, Tn3	examples
TED	IFP1.6, IFP2.2, TFP3 IFP2 TCl4.7, TCp3.2	TEs in baculoviruses

Figure 3. Differences between the main transposon classes, retrotransposons and (DNA) transposons. (TEs = transposable elements).

Table 1. Comparison between different baculovirus transposons (modified after Friesen, 1993)

Transposon	Origin	Length (bp)	Target sequence	Inverted terminal repeat	Terminal triplet	Open reading frame	References
1. Retrotransposons							
TED (AcMNPV)	T. ni	7510	AATG	273[1]	-	+	1,2
2. Transposons							
IFP2 (GmMNPV)	T. ni	2746	TTAA	13 + 17	CCC/GGG	+	3
TFP3 (AcMNPV)	T. ni	782	TTAA	15	CCC/GGG	-	4
TFP3 (GmMNPV)	T. ni	830	TTAA	15	CCC/GGG	-	4
IFP2.2 (AcMNPV)	S. frugiperda	2164	GTTTTTC	-	-	?	5
IFP1.6 (AcMNPV)	S. frugiperda	1565	TTAA	14	CCT/AGG	?	5
E (AcMNPV)	S. frugiperda	630	TTAA	14	CCT/AGG	-	6
M5 (AcMNPV)	S. frugiperda	290	TTAA	13	CCG/CGG	-	7
TCl4.7 (CpGV)	C. leucotreta	4726	TA	29	CAG/CTG	+[2]	8
TCp3.2 (CpGV)	C. pomonella	3239	TA	756	CAG/CTG	+[2]	9

[1]Long terminal repeat (LTR), [2]ORFs probably defective. (1) Miller and Miller, 1982; (2) Friesen and Nissen, 1990; (3) Cary et al., 1989; (4) Wang et al., 1989; (5) Beames and Summers, 1990; (6) Schetter et al., 1990; (7) Carstens, 1987; (8) Jehle et al., 1995; (9) Jehle, 1994.

ODV production; and (iii) a reduced infectivity to larvae, a consequence of ineffective occlusion (for review see, Fraser, 1986).

One FP mutant isolate (named FP-D) harbored an insertion of a 7.5 kb retrotransposon (transposable element *D* or *TED*) (Miller and Miller, 1982). *TED*, which represents a middle repetitive sequence in *T. ni* and was apparently mobilized during the serial cell culture infection by AcMNPV, is closely related to the *gypsy*-like retrotransposons of *Drosophila* (Friesen and Nissen, 1990). The structural organization of *TED* is characterized by 273 bp identical direct terminal repeats and three long open reading frames resembling to *gag*, *pol* and *env* genes of retroviruses (Fig. 3). Expression of the *pol* gene of *TED* produced a reverse transcriptase and a protease which mediated the processing of the *gag* gene product and resulting in virus-like particles (Lerch and Friesen, 1992). Recently, it has been shown for the *gypsy* element of *Drosophila* that the virus-like particles are indeed infective and that *gypsy* is more of retrovirus-like than a retrotransposon (Kim et al., 1994; Pélisson et al., 1994). This might also hold true for *TED*. Due to the insertion of *TED*, an early viral gene (*p94*) at map unit 86.7 was disrupted and the transcription of neighboring genes was altered (Friesen and Miller, 1987; Friesen and Nissen, 1990).

Further studies on FP phenotypes resulted in the isolation of several AcMNPV mutants and *Galleria mellonella* multiple nucleopolyhedrovirus (GmMNPV) mutants most of which had insertions of class II transposons in their genomes (Fig. 3, Tab. 1). Despite their varying sizes of

290 – 2475 bp, all shared some common structural characteristics: (i) They had similar short inverted repeats of 13 – 15 bp; (ii) they were inserted at a 4 bp TTAA target sequence which was duplicated; (iii) all these transposable elements, except *IFP2*, lacked an open reading frame to encode a transposase. These common features suggest that these transposons belong to either the same or narrowly related transposon families and probably differ by the extent of internal deletions.

Surprisingly, most of these transposons were found to be inserted within a short region in the baculovirus genome (between map unit 36 – 37). By the insertion of these elements, the baculovirus gene *p25*, a late gene encoding a 25 kDa protein, is disrupted. The reason for this apparent accumulation of transposon insertions within a finite locus is due to the specific role of *p25* in baculovirus replication. This gene enhances the synthesis and nuclear localization of polyhedrin and appears to have a regulatory role in switching from BV to ODV production during the biphasic viral replication cycle. Its disruption by transposon insertion leads to an increased production of BVs and decreased formation of ODVs and occlusion bodies, typical of the FP mutants (Jarvis et al., 1992). BVs of AcMNPV are more than 1000-fold more infective for cultured *T. ni* cells than are ODVs (Volkman et al., 1976; for review see, Volkman and Keddie, 1990). Thus, the increased BV production of the FP mutants favors the replication of these mutants compared to wild-type virus under the specific conditions of *in vitro* replication in cell culture. The FP phenotype can be generated by any insertion (or deletion) disrupting *p25*, as shown by cloning the beta-galactosidase gene of *Escherichia coli* into this locus (Beames and Summers, 1989). Apparently, the FP phenotype provides a powerful and unique system for trapping host transposons in baculovirus genomes and selecting this phenotype against wild-type virus.

It can be expected that host transposon insertion may occur at various sites within the baculovirus genome and is not restricted to the FP locus. However, it will be difficult to identify many of these insertions because they do not have such a dramatic phenotypic effect as observed in FP mutants. For example, the isolate AcMNPV-E contains a transposon-like insertion of 630 bp within a non-translated genome region around map unit 81 and does not show FP plaque morphology. It differs from the parental strain AcMNPV-E2 by an additional *Eco*RI site and an additional size class of transcripts initiating within the insertion (Oellig et al., 1987; Schetter et al., 1990). In another mutant AcMNPV (M5) two short transposon-like sequences, each about 285 bp long, were identified. Both insertions, which most probably derived from middle-repetitive *S. frugiperda* sequences, were nearly identical and were also inserted at TTAA tetranucleotides as it is typical for many transposons found in NPVs (Carstens, 1987). Their insertion sites within the AcMNPV genome were mapped at 2.6 and 46 map units, respectively. When AcMNPV M5 is passaged in cell culture, defective genomes are produced which lack the region between both insertions, i.e., about 42% of the viral genome.

The only class II transposon in NPVs which seems to encode a complete tranposase gene is *IFP2* (insertion few polyhedra 2). This transposon is 2476 bp long, is flanked by a double set of inverted terminal repeats of 13 and 19 bp, respectively, and also possesses TTAA target specificity (Cary et al., 1989). It contains a single transcriptionally active ORF which can encode for a protein of 597 amino acids and a predicted molecular weight of 68.1 kDa (M.J. Fraser, personal communication). The viral copies of this element were obtained during the propagation of AcMNPV or GmMNPV in the TN-368 cell line and are also associated with FP phenotype (Fraser et al., 1983). However, in other cell lines or in some laboratory stock of *T. ni* larvae *IFP2* is not apparent, suggesting that it recently invaded the *T. ni* genome (Cary et al., 1989).

Tc1/mariner-like elements in granuloviruses

Escape of host transposons into baculovirus genomes is not solely due to the artificial conditions of insect cell culture but also occurs during infection of insect larvae by granuloviruses (Jehle et al., 1995). In a survey of possible recombination between two granuloviruses from *Cydia pomonella* (CpGV) and *Cryptophlebia leucotreta* (ClGV), larvae of *C. leucotreta* which are susceptible

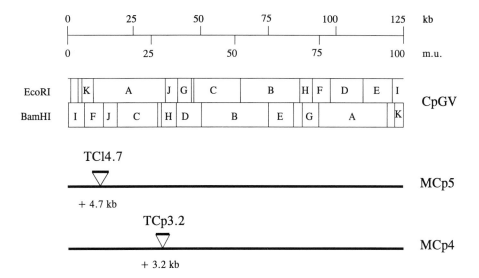

Figure 4. Restriction map of CpGV and the transposon carrying mutants MCp5 and MCp4. (kb=kilo bases, m.u=map unit)

for both viruses, were infected by adding the viruses to the artificial diet. Single genotypes of the virus offspring of these mixed infections were then isolated by an *in vivo* cloning procedure in larvae of *C. leucotreta* or *C. pomonella* and then characterized by restriction analysis. Out of 194 *in vivo* cloned CpGV and ClGV genotypes two CpGV mutants (MCp4 and MCp5) were identified to carry transposons of the class II type (Jehle, 1994; Jehle et al., 1995). Restriction analysis, sequencing and DNA hybridization revealed that the insertion in mutant MCp5 is 4726 bp long and has inverted terminal repeats of 29 bp (Fig. 4 and Fig. 5). This transposon, termed *TCl4.7*, inserted into a TA dinucleotide of a non-translated region near map unit 9.5 of the CpGV genome. It contained a defective open reading frame with amino acid sequence homology to *Tc1/mariner*-like elements (Fig. 6). *Tc1/mariner*-like transposons are characterized by amino acid sequence homology of the transposase gene, duplication of a TA target site and inverted terminal repeats of variable length (28–462 bp) (Radice et al., 1994; Robertson, 1995). They are widespread among nematodes, insects, fish and also show some sequence similarity to some bacterial and ciliate transposable elements (Doack et al., 1994).

By Southern hybridization it was shown that *TCl4.7* hybridized to middle repetitive sequences of genomic DNA of *C. leucotreta* but not to genomic DNA of *C. pomonella* indicating that *TCl4.7* inserted into the CpGV genome during the mixed infection experiment when CpGV was passaged in *C. leucotreta* (Jehle et al., 1995). Another transposon, isolated during these experiments, was apparently derived from low repetitive DNA of *C. pomonella*. It was termed *TCp3.2*,

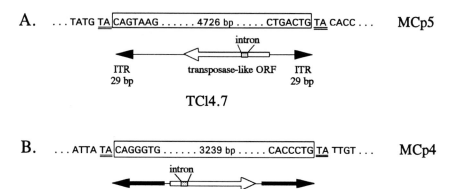

Figure 5. Schematic representation of (A) transposon *TCl4.7* and (B) transposon *TCp3.2*. The transposon sequences within MCp5 and MCp4, respectively, are boxed. The duplicated insertion sites are double underlined. (ITR = inverted terminal repeat).

had a size of 3239 bp and unusual long inverted repeats of 756 bp. It also inserted at a TA dinucleotide and contained a *Tc1/mariner*-like putative transposase gene (Fig. 5 and Fig. 6). These findings clearly demonstrated the presence of *Tc1/mariner*-like transposons in lepidopteran hosts and their mobilization during the infection by a virus. In contrast to the transposon-carrying FP phenotype of NPVs, the two GV mutants were identified only by a genotypic screening of 194 isolated genotypes by using restriction analysis of viral DNA. No morphological or biological differences to wt CpGV were found (Jehle et al., 1995).

Possible factors influencing frequency of transposition

The apparently high frequency of transposon-carrying mutants among progeny CpGV (2 out of 194) raised questions about the factors involved in transposon mobilization. The activity of most transposons is controlled by genetic or cellular host factors. For example, mobilization of transposon *Tc1* in *Caenorhabditis* occurs in a tissue- and strain-specific manner, whereas *P* elements of *Drosophila* transpose only in germ line cells and not in soma cells (Collins et al., 1987; Laski et al., 1986). Insect cell lines have been established exclusively for virus replication and, therefore, it cannot be expected that they provide normal transposition activity of all the transposons present

```
TCl4.7      EGQI....YL  VQDNSAVHRS  HVVQNWL...  ...SSQTDIT  VFDWPSKSPD
TCp3.2      DRRI....VF  QQDGCPAHWR  ITVREHLDNA  FPNSWIGRDG  PIPWPPRSPD
Ce-Tcl      RGF.....VF  QQDNDPKHTS  LHVRSWF...  ...QRRHVH   LLDWPSQSPD
Ce-Tc3      KDF.....RF  QQDNATIHVS  NSTRDYF...  ....KLKKIN  LLDWPARSPD
Dh-Uhu      RRTFRFYQDN  DQDNNQ#HKS  GLVPSWL...  ....IWNCPH  MII*PAQSPD
Dh-Minos    CGEF....TF  QQDGASSHTA  KRTKNWL...  ....QYNQME  VLDWPSNSPD
Dm-HBI      LKW.....TF  QQDNDQKRRC  KSAKNRF...  ....TQNRID  AMPWQAPPSH
Dm-Mariner  QHRV....IF  LHDNAPSHTA  RAVRDTL...  ....ETLNWE  VLPHAAYSPD
Hc-MLE      ..RS....LL  LHDNARPHTA  KQTTTKL...  ....NKLQLE  CLRHPPYSPD
Consensus   --R------F  -QDN---H--  -----WL---  ----------  -L-WP--SPD


TCl4.7      LNPIE.NLWG  QMVLNWDPTQ  VRSKKNLDEE  V.HRTWELLR  NTDTCSNMVT
TCp3.2      LAPLDFHIWG  RAKELVYATE  VESPEDLSQR  I.LAVFDVIK  GEIRMRTTTV
Ce-Tcl      LNPIE.HLWE  ELERRL.GGI  RASNADAKFN  Q.LENAWKAI  PMSVIHKLID
Ce-Tc3      LNPIE.NLWG  ILVRIVYAQN  KTYPTVASLK  QGILDAWKSI  PDNQLKSLVR
Dh-Uhu      VNVI*.NLWD  LLENNI.RNH  RSN....#KN  A.LLDEWSKI  SPETTRKLVS
Dh-Minos    LSPIE.NIWW  LMKNQL.RNE  PQRNISDLKI  K.LQEMWDSI  SQEHCKNLLS
Dm-HBI      LNPIE.NLYG  DIKQFV.SKK  SPTSKTQIWQ  V.VQDTWAKI  PPKPC*DLVD
DM-Mariner  LAPSDYHLFA  SMGHAL.AEQ  RF.DSYESVK  KWLDEWFAAK  DDEFYWRGIH
Hc-MLE      LAPIDYHFFR  NLDNFL.HGK  KF.NSYSVVQ  TAFKEFIDRR  PHAFFNKGIN
Consensus   L-PIE--LW-  -L--------  ----------  --L---W---  ------LV-
```

Figure 6. Alignment of the predicted amino acid sequences of the ORFs found in *TCl4.7* and *TCp3.2* with different transposases of the *Tc1/mariner* family. Ce-*Tc1* = *Caenorhabditis elegans Tc1* (Rosenzweig et al., 1983), Ce-*Tc3* = *C. elegans Tc3* (Collins et al., 1989), *Dh-Uhu* = *Drosophila heteroneura Uhu* (Brezinsky et al., 1990), *Dh-Minos* = *D. hydei Minos* (Franz and Savakis, 1991), Dm-*HB1* = *D. melanogaster HB1* (Brierley and Potter, 1985), Dm-*Mariner* = *D. mauritiana mariner* (Jacobson et al., 1986), Hc-*MLE* = *Hyalophora cecropia Mariner Like Element* (Lidholm et al., 1991).

in the insect genome. It is conceivable that when a virus replicates in different cell types within the larval host different host transposons could be caught by the viral DNA at a higher frequency than during replication in a specific cell line. The transposon-carrying mutants of CpGV, MCp4 and MCp5, were isolated after larvae of *C. leucotreta* were infected with a mixture of CpGV and a high dose of uv-irradiated ClGV. Endogenous or exogenous factors might have played a role to trigger transposon mobility during this experiment (Jehle, 1994; Jehle et al., 1995). It has been theorized that in *Drosophila* and plants, cellular stress brought about by heat shock, virus replication or uv-irradiation increases or enhances transposon activity (McClintock, 1984; Peterson, 1985; Walbot, 1992; Arnault and Dufournel, 1994).

So far, the extent of our knowledge about transposon activation during baculovirus infection is rather limited. But there are evidences that the escape of host transposons into baculovirus genomes is a much more frequent event than the few examples from the FP phenotype of NPVs replicated in cell culture. Transposons may insert at any site in the genome as long as there is a suitable target sequence. Hence, insertions may not exhibit phenotypical effect if they occur within intergenic regions. Such genotypes can be only identified by genotypic markers, e.g., by analyzing restriction profiles as done with MCp4 and MCp5. On the other hand, transposon insertion into an essential viral gene may result in non-viable progeny which cannot be isolated by conventional plaque purification. However, some of such non-viable genotypes could conceivably replicate and spread in presence of wildtype helper virus to provide a rescue of the defective function. The formation of occlusion bodies containing many virions, typical for MNPVs, may facilitate the propagation and survival of non-viable virus mutants under natural conditions. It is conceivable that a virus made non-viable by a transposon and a wildtype helper virus would become co-occluded, and hence, could co-infect and co-replicate and once again become co-occluded. Such mutants may exist at a low frequency within the naturally occurring wildtype virus populations, although, they cannot be isolated by conventional plaque purification or endpoint dilution methods.

Transposon-mutagenesis increases genetic variability of baculoviruses

The insertion of host transposable elements significantly affects the genetic integrity of baculovirus genomes. Transposon insertion may lead to the disruption of viral genes, as seen with the few *Polyhedra* mutants, where the genes encoding the 25 kDa or 94 kDa proteins were destroyed. However, when insertions occur at intergenic regions, e.g., the transposons *TCl4.7* and *TCp3.2* in CpGV were inserted at non-translated genome sites, less dramatic effects may occur. In many cases, the viral genomes are spontaneously endowed with a new genetic material, which may result in the incorporation of new functional genes or regulatory sequences. For example, because of the insertion of retrotransposon *TED*, the AcMNPV genome acquired several apparently functional genes, as *gag* (encoding the virus-like capsid protein), *pol* (protease and reverse transcriptase) and *env* (envelope-like glycoprotein) (Lerch and Friesen, 1992; Friesen, 1993).

On the other hand, transposon insertion can influence transcription of viral genes proximal to the insertion sites. Friesen et al. (1986) showed that a single LTR of *TED*, present in the mutant AcMNPV-FP-DS, is sufficient for the initiation of novel RNAs that extend into flanking viral sequences. Interestingly, the RNA transcription initiated in both directions at a CTTATAAG palindrome, which contains a bidirectionally oriented overlapping ATAAG motif. It is noteworthy to mention that the ATAAG sequence is the core promoter motif and RNA initiation site of late and very late baculovirus genes (Blissard and Rohrmann, 1990). The ATAAG motif within the LTR is obviously recognized by the virus-specific transcription machinery, since transcription initiation from this motif is not observed from *TED* elements in the genome of the host *T. ni* (Friesen and Miller, 1987).

Aberrant transcription patterns were also detected with several other transposon containing AcMNPV mutants. AcMNPV-*E* produces an additional size class of transcripts also initiating from an ATAAG motif, even though the insertion mutant did not show any phenotypic difference from wildtype virus (Oellig et al., 1987; Schetter et al., 1990). In the FP mutant AcFP875-2, the transposon *IFP2* was found to be inserted at a TTAA tetranucleotide between the coding region and the promoter of the 25 kDa gene, which is normally transcribed late from an ATAAG motif. Despite this disruption of coding and promoter sequences, transcription of the 25 kDa gene in AcFP875-2 was demonstrated (Beames and Summers, 1988, 1989). The transcripts in this mutant, however, did not initiate at its original RNA start site but at an ATAAG motif within the left terminus *IFP2* resulting in a 150 bp extension at the 5'-end of the transcripts. AcFP875-2 expressed a reduced level of the 25 kDa protein and produced more polyhedra than other FP mutants (Beames and Summers, 1989).

The influence of transposon insertion on transcription of genes adjacent to the insertion sites has been reported in several situations describing transposon-host-relationships and it was assumed that the change of temporal and spatial transcription patterns caused by transposon activity may play a major role in organismic evolution (for review see, McDonald, 1995). It appears that mutagenesis driven by transposons strongly affects the size and diversity of baculovirus genomes. Furthermore, the horizontal transposon escape into baculoviruses was suggested as a model for the flow of host genes or regulatory sequences which become functional in the viral genome (Blissard and Rohrmann, 1990).

Are baculoviruses involved in horizontal transmission of insect transposons?

Analysis of the phylogenetic distribution of many eukaryotic transposon families revealed the presence of closely related transposons in taxonomically widely separated host species. From the spatial distribution patterns of several insect transposons, like the *P* element in *Drosophila*, *mariner*- and *Tc1*-like transposons, *jockey* or *hobo*, a horizontal transmission across species barriers was postulated for many transposon families (Daniels et al., 1990; Maruyama and Hartl, 1991; Robertson, 1993, 1995; Mizrokhi and Mazo, 1990; Calvi et al., 1991). Despite the accumulating evidences supporting this hypothesis, the possible pathways and vectors involved in horizontal transposon transmission are mostly obscure. Shared parasites and pathogens have been suggested as a possible link. For example, *Proctolaelaps regalis*, a mite that infests *Drosophila*, was postulated to transmit *P* elements (Houck et al., 1991; Bachmann, this volume). However, an experimental substantiation is still lacking. The horizontal escape of host transposable elements into baculovirus genomes is the only clear example so far showing cross-species transmission (assuming that a virus is considered as a species). By entering a virus genome, transposons position themselves to invade other gene pools and thus, baculoviruses could be proposed as vectors involved in horizontal transposon transmission among insects. In addition to insertion into a baculovirus genome, several further prerequisites will have to be met for successful transposon transmission into other organisms. These are: (i) maintenance of transposon mobility after integration into the viral genome; (ii) a non-lethal infection of a host insect during which the transposon is mobilized again; (iii) transposon insertion into the germ line of the new host to establish a hereditable position within the newly invaded gene pool. Baculoviruses fulfill at least some of these conditions. The element *TED* found in the mutant AcMNPV-FP-D is transcriptionally active suggesting that it might be capable of retroid-like transposition after integration into the baculovirus genome (Lerch and Friesen, 1992). Transcription and transposition activity has been also shown for transposon *IFP2*, which is the only class II transposon found in NPVs and containing an open reading frame possibly encoding a transposase (Beames and Summers, 1988; M.J. Fraser, personal communication).

  The limited host range of viruses was occasionally raised as an argument against the hypothesis of a virus-based transposon shuttling among less related host species (Kidwell, 1992). On the other hand, host specificity may not be a real obstacle because there are several reports on abortive baculovirus infections of non-permissive hosts where a few baculovirus genes are transcribed and expressed without complete virus replication or lethal infection (for review see, Bilimoria, 1991). For example, AcMNPV is taken up by a wide range of non permissive insect cells but does not result in complete virus replication. Also, promoter-dependent gene expression was observed in non-permissive lepidopteran or dipteran cells (Carbonell et al., 1985; Guzo et al.,

1992; Morris and Miller, 1992). Friesen (1993) mentioned that the LTR of *TED* is also transcriptionally active in *D. melanogaster* cells which are not permissive for AcMNPV. Therefore, if transposon genes are sufficiently expressed upon an abortive infection, it is conceivable that a transposon can be mobilized to invade a new host genome. Although tempting, the idea that a transposon can invade an insect germ line during a non-lethal infection is still highly speculative and most difficult to prove. However, at least a physical interaction between baculovirus and reproductive organs during latent or carry-over infection of host insects has been reported. As a result of carry-over, about 15% mortality was observed in newly hatched larvae derived from parental insects that had survived a sublethal dose of a baculovirus (Smits and Vlak, 1988). Vertical transmission of latent NPVs present in host insects without apparent infection was demonstrated, although it is still not clear whether the latent infection is transovarially or intraovarially transmitted (Jurkovicova, 1979; Wood et al., 1986; Hughes et al., 1993). In summary, the occasional escape of host transposons into baculovirus genomes provides a model for the intergenomic mobility of genetic information and offers sound theoretical basis for the assumption that baculoviruses are one of the interspecies vectors of insect transposons.

Mutagenesis by transposon and biosafety of genetically engineered baculoviruses

As indicated earlier, baculoviruses have a considerable potential for the biological control of pest insects. In recent years the baculovirus expression system was also used to improve the insecticidal properties of baculoviruses with the aim of shortening the time span between infection and cessation of feeding by the host larvae pest. For this reason recombinant baculoviruses expressing insect-specific toxin or hormone genes were produced (for review see, Vlak, 1993). In 1993, the first deliberate release of a recombinant AcMNPV expressing an insect-specific scorpion toxin was performed (Cory et al., 1994). The idea to use genetically engineered baculoviruses for pest control had earlier provoked an intensive discussion on the biosafety of these viruses (Fuxa, 1990; Williamson, 1991, 1992; Hochberg and Waage, 1991; McClintock and Sjoblad, 1992). A major concern is the possibility of recombination with naturally occurring baculoviruses resulting in transfer of the insecticidal gene to other baculoviruses or in the transfer of factors extending the host range of the modified viruses and entailing unforeseen ecological consequences. In this forum, however, the possible consequences of transposon-mediated mutagenesis of genetically engineered baculoviruses were rarely considered.

Nearly all genes (e.g., toxin genes) introduced into baculoviruses for improving their insecticidal properties are under the control of the polyhedrin or the *p10* promoter (Vlak, 1993). The choice of these very late and strong promoters on the one hand directs high gene expression, and on the other hand restricts the effectiveness to only permissive hosts. As seen, naturally occurring mutagenesis by transposons apparently contributes to the genetic heterogeneity of baculoviruses by spontaneously introducing new genetic information into the viral genome. Analogous to the introduction of late promoter motifs, which directed transcription of new size classes of RNAs in AcMNPV-FP-DS, AcFP875-2 and other transposon-carrying baculovirus mutants, it is conceivable that sequences similar to early promoters or insect promoters are introduced. In the case of recombinant baculoviruses, this can result in an alteration of the originally desired regulation of the insecticidal gene. The results of Carbonell et al. (1985) and Morris and Miller (1992) indicated that genes under control of other than late or very late promoters are expressed in a wider range of insects than can be permissively infected. Hence, transposon-mediated introduction of promoter sequences upstream of an insecticidal gene may result in an extension of the host range where an insecticidal gene could be expressed and possibly harm others than target insects. Another possible implication of transposon-mutagenesis of baculovirus genomes is the acquisition of viral genes by a transposon, which together are then transferred to other baculovirus or host genomes. By this way new host range factors could enter a recombinant virus, or the reverse could happen where an insecticidal gene might be transferred to a naturally occurring baculovirus population. The probability of the latter of these scenarios will depend on the mobility of the transposons after they have been inserted into the baculovirus genome. In terms of the quality of the possible genomic alteration mediated by transposons, there is no difference in mutations which can be caused by non-homologous recombination, but if transposition events will increase the frequency of their occurrence has to be evaluated.

Acknowledgements
I thank Just M. Vlak and Basil M. Arif for valuable suggestions on the manuscript and Jan W.M. van Lent for providing Figure 1.

References

Arnault, C. and Dufournel, I. (1994) Genome and stresses: Reactions against aggressions, behavior of transposable elements. *Genetica* 93: 149–160.
Beames, B. and Summers, M.D. (1988) Comparison of host cell DNA insertions and altered transcription at the site of insertions in few polyhedra baculovirus mutants. *Virology* 162: 206–220.
Beames, B. and Summers, M.D. (1989) Location and nucleotide sequence of the 25K protein missing from baculovirus few polyhedra (FP) mutants. *Virology* 168: 344–353.
Beames, B. and Summers, M.D. (1990) Sequence comparison of cellular and viral copies of host cell DNA insertions found in *Autographa californica* nuclear polyhedrosis virus. *Virology* 174: 354–363.
Berg, D.E. and Howe, M.M. (1989) *Mobile DNA*. American Society of Microbiology, Washington, D.C.
Bilimoria, S.L. (1991) The biology of nuclear polyhedrosis viruses. *In*: E. Kurstak (ed.): *Viruses of Invertebrates*. Marcel Dekker Inc., New York, pp 1–72.
Blissard, G.W. and Rohrmann, G.F. (1990) Baculovirus diversity and molecular biology. *Annu. Rev. Entomol.* 35: 127–155.

Brezinsky, L., Wang, G.V.L., Humphreys, T. and Hunt, J. (1990) The transposable element *Uhu* from Hawaiian *Drosophila* – member of the widely dispersed class of *Tc1*-like transposons. *Nucleic Acids Res.* 18: 2053–2059.

Brierley, H.L. and Potter, S.S. (1985) Distinct characteristics of loop sequences of two *Drosophila* foldback elements. *Nucleic Acids. Res.* 13: 485–500.

Calvi, B., Hong, T.J., Findland, S.D. and Gelbart, W.M. (1991) Evidence for a common evolutionary origin in inverted repeat transposons in *Drosophila* and plants: *hobo*, *Activator*, and *Tam3*. *Cell* 66: 465–471.

Carbonell, L.F., Klowden, M.J. and Miller, L.K. (1985) Baculovirus-mediated expression of bacterial genes in dipteran and mammalian cells. *J. Virol.* 56: 153–160.

Carstens, E.B. (1987) Identification and nucleotide sequence of the regions of *Autographa californica* nuclear polyhedrosis virus genome carrying insertion elements derived from *Spodoptera frugiperda*. *Virology* 161: 8–17.

Cary, L.C., Goebel, M., Corsaro, B.G., Wang, H.-G., Rosen, E. and Fraser, M.J. (1989) Transposon mutagenesis of baculoviruses: Analysis of *Trichoplusia ni* transposon *IFP2* insertion within the *FP*-locus of nuclear polyhedrosis virus. *Virology* 172: 156–169.

Collins, J., Saari, B. and Anderson, P. (1987) Activation of a transposable element in the germ line but not in soma of *Caenorhabditis elegans*. *Nature* 328: 726–728.

Collins, J., Forbes, E. and Anderson, P. (1989) The *Tc3* family of transposable genetic elements in *Caenorhabditis elegans*. *Genetics* 120: 621–627.

Cory, J.S., Hirst, M.L., Williams, T., Hails, R.S., Goulson, D., Green, B.M., Carty, T.M., Possee, R.D., Cayley, P.J. and Bishop, D.H.L. (1994) Field trial of a genetically improved baculovirus insecticide. *Nature* 370: 138–140.

Daniels, S.B., Peterson, K.R., Strausbaugh, L.D., Kidwell, M.G. and Chovnick, A. (1990) Evidence for horizontal transmission of the *P* transposable element between *Drosophila* species. *Genetics* 124: 339–355.

Doack, T.G., Doerder, F.P., Jahn, C.L. and Herrick, G. (1994) A proposed superfamily of transposase genes: Transposon-like elements in ciliated protozoa and a common "D35E" motif. *Proc. Natl. Acad. Sci. USA* 91: 942–946.

Finnegan, D. (1992) Transposable elements. *Curr. Opinion Genet. Develop.* 2: 861–867.

Franz, G. and Savakis, C. (1991) *Minos*, a new transposable element from *Drosophila hydei*, is a member of the *Tc1*-like family of transposons. *Nucleic Acids Res.* 19: 6646.

Fraser, M.J., Smith, G.E. and Summers, M.D. (1983) Aquisition of host cell DNA sequences by baculoviruses: Relationship between host DNA insertions and FP mutants of *Autographa californica* and *Galleria mellonella* nuclear polyhedrosis viruses. *J. Virol.* 47: 287–300.

Fraser, M.J. (1986) Transposon-mediated mutagenesis of baculoviruses: Transposon shuttling and implications for speciation. *Ann. Entomol. Soc. Amer.* 79: 773–783.

Friesen, P.D., Rice, W.C., Miller, D.W. and Miller, L.K. (1986) Bidirectional transcription from a solo long terminal repeat of the retrotransposon *TED*: Symmetrical RNA start sites. *Mol. Cell. Biol.* 6: 1599–1607.

Friesen, P.D. and Miller, L.K. (1987) Divergent transcription of early 35- and 94-kilodalton protein genes encoded by the *Hind*III genome fragment of the baculovirus *Autographa californica* nuclear polyhedrosis virus. *J. Virol.* 61: 2264–2272.

Friesen, P.D. and Nissen, M.S. (1990) Gene organization and transcription of *TED*, a lepidopteran retrotransposon integrated within the baculovirus genome. *Mol. Cell. Biol.* 10: 3067–3077.

Friesen, P.D. (1993) Invertebrate transposable elements in the baculovirus genome: Characterization and significance. *In*: N.E. Beckage, S.N. Thompson and B.A. Federici (eds): *Parasites and Pathogens of Insects, Vol. 2: Parasites.* Academic Press, London, pp 147–178.

Fuxa, J.R. (1990) Fate of released entomopathogens with reference to risk assessment of genetically engineered microorganisms. *Bull. Entomol. Soc. Amer.* 35: 12–24.

Granados, R.R. and Williams, K.A. (1986) *In vivo* infection and replication of baculoviruses. *In*: R.R. Granados and B.A. Federici (eds): *The Biology of Baculoviruses, Vol. 1, Biological Properties and Molecular Biology.* CRC Press, Boca Raton, pp 89–108.

Gröner, A. (1986) Specificity and safety of baculoviruses. *In*: R.R. Granados and B.A. Federici (eds): *The Biology of Baculoviruses, Vol. 1, Biological Properties and Molecular Biology.* CRC Press, Boca Raton, pp 177–202.

Guzo, D., Rathburn, H., Guthrie, K. and Dougherty, E.M. (1992) Viral and host cellular transcription in *Autographa californica* nuclear polyhedrosis virus-infected *gypsy* moth cell lines. *J. Virol.* 66: 2966–2972.

Hochberg, M.E. and Waage, J.K. (1991) Control engineering. *Nature* 352: 16–17.

Houck, M.A., Clark, J.B., Peterson, K.R. and Kidwell, M.G. (1991) Possible horizontal transfer of *Drosophila* genes by the mite *Proctolaelaps regalis*. *Science* 253: 1125–1129.

Huber, J. (1986) Use of baculoviruses in pest management programmes. *In*: R.R. Granados and B.A. Federici (eds): *The Biology of Baculoviruses, Vol. 2, Practical Application for Insect Control.* CRC Press, Boca Raton, pp 181–202.

Hughes, D.S., Possee, R.D. and King, L.A. (1993) Activation and detection of a latent baculovirus resembling *Mamestra brassicae* nuclear polyhedrosis virus in *M. brassicae* insects. *Virology* 194: 608–615.

Jacobson, J.W., Medhora, M.M. and Hartl, D.L. (1986) Molecular structure of a somatically unstable transposable element. *Proc. Natl. Acad. Sci. USA* 83: 8684–8688.

Jarvis, D.L., Bohlmeyer, D.A. and Garcia, A. Jr. (1992) Enhancement of polyhedrin nuclear localization during baculovirus infection. *J. Virol.* 66: 6903–6911.

Jehle, J.A. (1994) *Sicherheitsaspekte der Gentechnologie: Verwandtschaft und Variabilität der Genome des Cryptophlebia leucotreta Granulosevirus und des Cydia pomonella Granulosevirus.* PhD thesis, TU Braunschweig.

Jehle, J.A., Fritsch, E., Nickel, A., Huber, J. and Backhaus, H. (1995) *TCl4.7*: A novel lepidopteran transposon found in *Cydia pomonella* granulosis virus. *Virology* 207: 369–379.

Jurkovicova, M. (1979) Activation of latent virus infections in larvae of *Adoxophyes orana* (Lepidoptera: Tortricidae) and *Barathra brassicae* (Lepidoptera: Noctuidae) by foreign polyhedra. *J. Invert. Pathol.* 34: 213–223.

Kidwell, M.G. (1992) Horizontal transfer. *Curr. Opinion Genet. Develop.* 2: 868–873.

Kim, A., Terzian, C., Santamaria, P., Pélisson, A., Prud'homme, N. and Bucheton, A. (1994) Retroviruses in invertebrates: The *gypsy* retrotransposon is apparently an infectious retrovirus of *D. melanogaster. Proc. Natl. Acad. Sci. USA* 91: 1285–1289.

Laski, F.A., Rio, D.C. and Rubin, G.M. (1986) Tissue specificity of *Drosophila P* element transposition is regulated at the level of mRNA splicing. *Cell* 44: 7–19.

Lerch, R.A. and Friesen, P.D. (1992) The baculovirus-integrated retrotransposon *TED* encodes *gag* and *pol* proteins that assemble into viruslike particles with reverse transcriptase. *J. Virol.* 66: 1590–1601.

Lidholm, D.-A., Gudmundsson, G.H. and Bomann, H.G. (1991) A highly repetitive, *mariner*-like element in the genome of *Hyalophora cecropia. J. Biol. Chem.* 266: 11518–11521.

Luckow, V.A. and Summers, M.D. (1988) Trends in the development of baculovirus expression vectors. *Bio/Technology* 6: 47–55.

Martignoni, M.E. and Iwai, P.J. (1986) *A Catalog of Viral Diseases of Insects, Mites, and Ticks.* Fourth Edition, Gen. Tech. Rep. PNW-195, USDA, Pacific Northwest Research Station, Portland.

Maruyama, K. and Hartl, D.L. (1991) Evidence for interspecific transfer of the transposable element *mariner* between *Drosophila* and *Zaprionus. J. Mol. Evol.* 33: 514–524.

McClintock, B. (1984) The significance of responses of the genome to challenge. *Science* 226: 792–801.

McClintock, J.T. and Sjoblad, R.D. (1992) Recombinant biopesticides. *Trends Ecol. Evol.* 7: 352.

McDonald, J.F. (1995) Transposable elements: Possible catalysts of organismic evolution. *Trends Ecol. Evol.* 10: 123–126.

Miller, D.W. and Miller, L.K. (1982) A virus mutant with an insertion of a *copia*-like transposable element. *Nature* 299: 562–564.

Mizrokhi, L.J. and Mazo, A.M. (1990) Evidence for horizontal transmission of the mobile element *jockey* between distant *Drosophila* species. *Proc. Natl. Acad. Sci. USA* 87: 9216–9220.

Morris, T.D. and Miller, L.K. (1992) Promoter influence on baculovirus-mediated gene expression in permissive and non-permissive insect cell lines. *J. Virol.* 66: 7397–7405.

Murphy, F.A., Fauquet, C.M., Bishop, D.H.L., Ghabrial, S.A., Jarvis, A.W., Martelli, G.P., Mayo, M.A. and Summers, M.D. (1995) *Virus Taxonomy. Sixth Report of the International Committee on Taxonomy of Viruses.* Springer-Verlag, Wien.

Oellig, C., Happ, B., Müller, T. and Doerfler, W. (1987) Overlapping sets of viral RNAs reflect the array of polypeptides in the *Eco*RI J and N fragments (map positions 81.2 to 85.9) of the *Autographa californica* nuclear polyhedrosis virus genome. *J. Virol.* 61: 3048–3057.

Pélisson, A., Sun, U.S., Prud'homme, N., Smith, P.A., Bucheton, A. and Corces, V.G. (1994) *Gypsy* transposition correlates with the production of a retroviral envelope-like protein under the tissue-specific control of the *Drosophila flamenco* gene. *EMBO J.* 13: 4401–4411.

Peterson, P.A. (1985) Virus-induced mutations in maize: On the nature of stress-induction of unstable loci. *Genet. Res.* 46: 207–217.

Radice, A.D., Bugaj, B., Fitch, D.H.A. and Emmons, S.W. (1994) Widespread occurrence of the *Tc1* transposon family: *Tc1*-like transposons from teleost fish. *Mol. Gen. Genet.* 244: 606–612.

Robertson, H.M. (1993) The *mariner* transposable element is widespread in insects. *Nature* 362: 241–245.

Robertson, H.M. (1995) The *Tc1-mariner* superfamily of transposons in animals. *J. Insect Physiol.* 41: 99–105.

Rohrmann, G.F. (1986) Polyhedrin structure. *J. Gen. Virol.* 67: 1499–1513.

Rosenzweig, B., Liao, L.W. and Hirsh, D. (1983) Sequence of the *C. elegans* transposable element *Tc1. Nucleic Acids Res.* 11: 4201–4209.

Schetter, C., Oellig, C. and Doerfler, W. (1990) An insertion of insect cell DNA in the 81-map-unit segment of *Autographa californica* nuclear polyhedrosis virus. *J. Virol.* 64: 1844–1850.

Smith, G.E., Summers, M.D. and Fraser, M.J. (1983) Production of the human beta interferon in insect cells infected with a baculovirus expression vector. *Mol. Cell. Biol.* 3: 2156–2165.

Smits, P.H. and Vlak, J.M. (1988) Biological activity of *Spodoptera exigua* nuclear polyhedrosis virus against *S. exigua* larvae. *J. Invert. Pathol.* 51: 107–114.

Vlak, J.M., Schouten, A., Usmany, M., Belsham, G.J., Klinge-Roode, E.C., Maule, A.J., van Lent, J.W.M. and Zuidema, D. (1990) Expression of cauliflower mosaic virus gene I using a baculovirus vector based upon the baculovirus *p10* gene and a novel selection system. *Virology* 179: 312–320.

Vlak, J.M. (1993) Genetic engineering of baculoviruses for insect control. In: J. Oakeshott and M.J. Whitten (eds): *Molecular Approaches to Fundamental and Applied Entomology.* Springer-Verlag, New York, pp 90–127.

Volkman, L.E., Summers, M.D. and Ching-Hsiu, H. (1976) Occluded and nonoccluded nuclear polyhedrosis virus grown in *Trichoplusia ni*: Comparative neutralization, comparative infectivity, and *in vitro* growth studies. *J. Virol.* 19: 820–832.

Volkman, L.E. and Keddie, B.A. (1990) Nuclear polyhedrosis virus pathogenesis. *Seminars in Virology* 1: 249–256.

Walbot, V. (1992) Reactivation of the mutator transposable elements of maize by ultraviolet light. *Mol. Gen. Genet.* 234: 353–360.

Wang, H.H., Fraser, M.J. and Cary, L.C. (1989) Transposons mutagenesis of baculoviruses: Analysis of *TFP3* lepidopteran transposon insertions at the *FP* locus of nuclear polyhedrosis viruses. *Gene* 81: 97–108.

Williamson, M. (1991) Biocontrol risks. *Nature* 353: 394.

Williamson, M. (1992) Environmental risks from the release of genetically modified organisms (GMOs) – The need for molecular ecology. *Molec. Ecol.* 1: 3–8.

Wood, H.A., Burand, J.P., Hughes, P.R., Flore, P.H. and Getting, R.R. (1986) Transovarial transmission of *Lymantria dispar* nuclear polyhedrosis virus. *In*: R.A. Samson, J.M. Vlak and D. Peters (eds): *Fundamental and Applied Aspects of Invertebrate Pathology*. Foundation of the Fourth International Colloquium of Invertebrate Pathology, Wageningen, p 405.

Transgenic Organisms – Biological and Social Implications
J. Tomiuk, K. Wöhrmann & A. Sentker (eds)
© 1996 Birkhäuser Verlag Basel/Switzerland

Influence of transgenes on coevolutionary processes

P.W. Braun

Department of Biometry and Population Genetics, Justus-Liebig University, Ludwigstraße 27, D-35390 Giessen, Germany

Summary. Coevolutionary processes have a decisive influence on the structure of ecosystems. Transgenic plants, upon their release, will be subject to this evolutionary biological process. This chapter aims to provide an overview of the present state of research into coevolution and of the possible consequences of coevolutionary processes upon interactions within ecosystems. With reference to two examples of horizontal gene transfer into culture-plants (*Bacillus thuringiensis*-toxin, vesicular-arbuscular *mycorrhiza*), the chances of success of pertinent plant breeding programs will be discussed.

Coevolution and evolutionary biology

One of the greatest challenges of evolutionary biology is the explanation of how interspecific interactions influence the evolution of the species involved, and how in turn evolution changes these interactions. This task requires a synthesis of the theories of genetic evolution and of population ecological interactions. In no way is this confined to a single realm of little practical relevance, as shown in Figure 1.

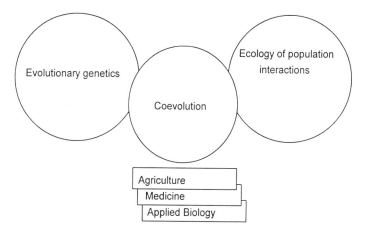

Figure 1. Research into coevolution demands knowledge not only from the area of evolutionary genetics, but also of the ecology of population interactions. Coevolutionary processes have great importance in agriculture and medicine, as well as in other fields of applied biology.

A major aim of plant breeding has always been the creation of pest- or disease-resistant culture plants. However, with each progression in plant breeding – whose success is seen in a new resistant type – begins a race between pest organisms and resistant plant varieties. The pest needs the plant for its sustenance and therefore seeks to overcome the resistance, while the plant meanwhile must ward off the pest, in order not to suffer life-threatening damage. In culture-plant systems this coevolutionary fight usually results in formerly resistant plants no longer being resistant. Their agricultural usefulness is therefore lost. The plant breeder is then faced with the demand for a new, further-resistant variety. It is therefore no wonder that the new possibilities made available by gene technology, for example the integration of resistance mechanisms from foreign species into culture plant varieties are praised as an enormous opportunity for the maintenance of plant health (Fischbeck, 1995). But just how do these "new" or "different" resistance mechanisms appear from the perspective of coevolution? Will they be more lasting or do they perhaps even present a danger for the ecosystems concerned? Prior experience with coevolutionary processes in biology helps to partially answer these questions.

The concept of coevolution was coined by Ehrlich and Raven (1964) as they described the apparently reciprocal influence of plants and herbivorous insects. In the following years the word coevolution increased in usage, however, the same meaning was not always applied to its use. A clear definition of the process became necessary (Thompson, 1989). Consequently it was suggested by Janzen (1980) that the concept of coevolution should be held as the interaction between

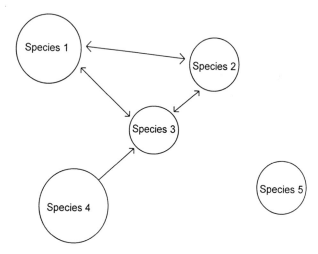

Figure 2. Ecological relationships between species. Species 1 and 2, 2 and 3, and 1 and 3 interact with one another. When these exchanges result in genetic alterations within the species, these species are liable to undergo coevolution. No exchange exists in the relationship between species 3 and 4, shown by the arrow pointing in only one direction. Species 5 has no relationship with the other species. Species 4 and 5, therefore, cannot contribute to any coevolutionary process in this network.

two or more species whose genetic composition changes in response to the genetic alteration of one-another (Fig. 2). This is the definition upon which this article is based.

A well documented example for coevolutionary processes is the chemical defence system of umbellate flowers. Particular species of the family Umbelliferae are protected by so-called linear furanocoumarins against almost all herbivorous insects. Some insects are nevertheless in a position to feed on plants containing this toxin (Berenbaum, 1983). These insects have however still not developed the ability to utilize the few Umbelliferae species which produce the equally toxic angled furanocoumarins. It can be assumed that the angled furanocoumarins are a later evolutionary acquisition than the linear furanocoumarins. Interestingly, these further-developed Umbelliferae have not replaced one protection system with the other but rather have added the new defence mechanism to the old (Futuyma, 1986). One might assume that these umbellate flowers had finally won the arms race in this herbivore-plant system. But plants with angled furanocoumarins are attacked by other specialized insects. These are not related to the species resistant to linear furanocoumarins (Berenbaum and Feeny, 1981). The angled furanocoumarins probably afforded the plants such good protection from their original herbivores that they remained at first as an unused resource in the corresponding ecosystem. This made them, however, attractive for other herbivore groups which bore the necessary genetic adaptation better than the former herbivores or which formerly had been supressed by the original pest. In this case the initial ecological release led to new herbivores which either occupied the vacant niche as secondary pests or which with time were able to colonize the unused resource due to a proper evolutionary adaptation. Thus the coevolutionary race continues.

Coevolution resulting from interactions

As may be inferred from previous chapters, coevolution can only be expected to arise from certain relationships in ecological systems. If we limit our investigation to two interacting species, only competition (– –), exploitation (+ –) and mutualism (+ +) fulfil the requirements for coevolutionary processes (Fig. 3). These interactions generally refer to the use of resources.

Interaction and resource usage

An important factor in evolution is resource usage. Selection favours with greater probability genotypes which use common rather than rare resources. When no resource type is common, generalist genotypes are favoured. According to the theory of optimal resource use, a consumer

Population A

Figure 3. Interactions between two populations A and B, indicated by a "+" or a "–" sign. If the interaction results in disadvantages for both species, the relationship is described as competition (– –). In the other two cases one speaks of mutualism (+ +) or exploitation (+ –).

can only afford to specialize on a single resource when that resource occurs in abundance, not, however, when it is rare (Emlen, 1966; MacArthur and Pianka, 1966). Similarly, generalists are favoured in situations where the quantity and quality of a resource varies in time and space. Cultured barley (*Hordeum vulgare*), for example, serves as a food resource for pathogenic mildew populations (*Erysiphe graminis* sp. *hordei*). European populations of the pathogen show considerable genetic homogeneity and specialization of their virulence characteristics closely correlated with the respective mildew-resistance characteristics of the predominant barley cultivars. The primary reason for this is the preference for planting of new mildew-resistant cultivars, as farmers take advantage of the disease-protection they offer. With the growing acreage of these new cultivars, however, increasingly only those mildew populations survive which possess the virulence factors necessary to overcome the mildew resistance. The result is generally a negation of the new resistance in the cultivars, often after only a few years (Wolfe, 1987). With large-area planting of single varieties, the consumer (*E. graminis*) can afford to be specialized to a low number of barley varieties, since its resource, a particular barley cultivar, is abundant. When wild populations of barley (*Hordeum spontaneum*) and mildew (*E. graminis*) are considered, we see that the corresponding pathogen populations are comparatively heterogeneous. This heterogeneity can be interpreted as resulting from selection pressure acting *via* heterogeneous distributions of mildew resistances within natural wild barley populations (Dinoor and Eshed, 1987). In addition, wild barley populations are generally small and widely spread. Gene flow between mildew populations, facilitated by large crop production areas or cultivation over seasons does not occur in regions where wild barley grows. Under these conditions it does not appear to be an evolutionarily sensible strategy for mildew populations on wild barley to specialize on a few combinations of mildew-resistant hosts. This host–parasite system illustrates the relationship between resource usage and coevolutionary processes.

Competition

When two species utilize the same resource, selection pressure emerges through the presence of the other species when the resource occurs in limited supply. This circumstance can be represented diagrammatically (Fig. 4).

A resource useful to both species can be represented by the function $K(x)$. This function describes the abundance of a resource at a site. Each species occupies a typical niche which can be represented by the usage functions $f_1(x)$ and $f_2(x)$ for the two species respectively. The competition between the species increases when d, the difference between their usage functions, decreases, or when w, the niche width, broadens. As a result of these considerations, MacArthur and Levins (1967) speculated that species only coexist when the degree of their niche-overlap

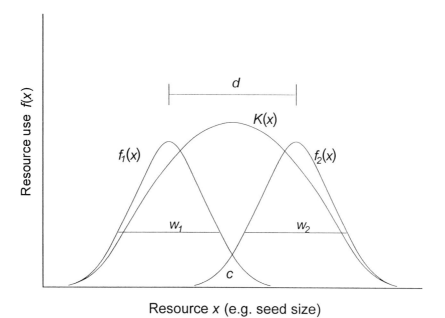

Figure 4. Usage functions $f_1(x)$ and $f_2(x)$ of two species. The curves show the use of a resource x by species 1 or 2. The occurrence of the resource is described by the resource function $K(x)$. The interspecific difference in resource use is indicated by the value d. The overlap of the usage functions (c) demonstrates the competition between the two species (Futuyma, 1986).

does not exceed a particular critical value. For the species involved it can therefore be evolutionarily advantageous when their usage functions diverge, a process leading to character displacement.
Without an accommodation of this kind and in the absence of migration to maintain the system, a
low value of d will lead to competitive exclusion of one species. The result of the ecological
release afforded by competitive exclusion can be a broadening of the remaining usage function or
also adaptive radiation of the remaining species. Adaptive radiation describes the evolutionary
divergence of members of a phylogenetic line into numerous different, adaptive forms. Normally
this divergence is related to diversification in the use of resources or habitats.

The ground finches (*Geospiza*-finches) of the Galapagos archipelago offer an impressive
example of the aforementioned coevolutionary processes (Fig. 5). The species *Geospiza fortis*

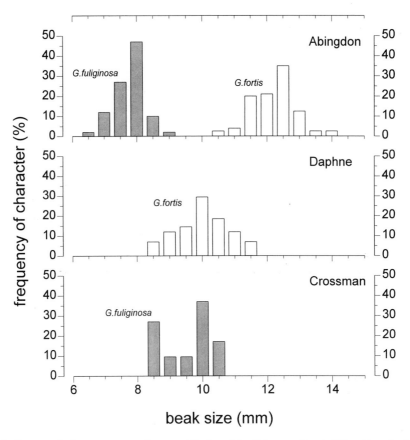

Figure 5. Character displacement among the ground finches of the Galapagos Islands. When *Geospiza fortis* and
Geospiza fuliginosa occur allopatrically (on the islands Daphne and Crossman), their beak size is similar. This
character is however distinguishable between the species on Abingdon where both species occur together, i.e.,
where they are sympatric (Futuyma, 1986).

and *Geospiza fuliginosa* have the same beak size on islands where one of these species occurs without the other (allopatry). On islands on which the species live sympatrically, however, the two species show different beak-sizes.

A relationship exists between beak-size in the finches and the size and hardness of the food source they utilize. It can be assumed that this adaptation arose as a consequence of competition avoidance (Futuyma, 1986).

Exploitation

The concept of exploitation is summarized here as reciprocal relationships in which one species profits from the presence of a second species which itself is disadvantaged through its interaction with the first species. It is up to the disadvantaged population in this interaction to endeavour

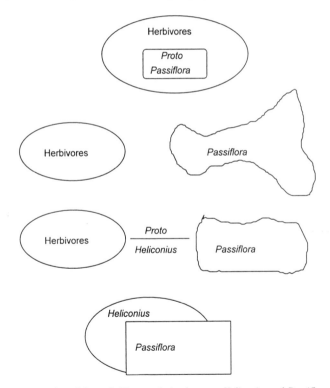

Figure 6. Graphical presentation of the probable coevolution between *Heliconius* and *Passiflora*. After the acquisition of resistance against its original herbivores the ancestors of today *Passiflora* escaped any damage through herbivore exploitation. Possibly after a phase of pest free evolution and presumably adaptive radiation, butterflies emerged that could graze on *Passiflora*. Now again *Passiflora* is an element of a herbivore–plant interaction and coevolutionary processes.

escaping its exploiter (predator, herbivore, parasite). When it is able to achieve this, it gains a selective advantage which the interaction partner in turn attempts to equilibrate by adaptation. In this way a coevolutionary contest between the two populations comes into being, which reaches a standstill only when the cost of an adaptation outweighs its usefulness. Predator–prey systems, host–parasite systems and herbivore–plant systems all fall under this classification.

The herbivore–plant system *Heliconius–Passiflora* is another impressive example of coevolution (Fig. 6). Members of the neotropical butterfly genus *Heliconius* feed as caterpillars on plant species from the family Passifloraceae (Benson et al., 1975). They number therefore among only very few insects which graze on *Passiflora*, since the plants of this group protect themselves from feeding damage by producing alkaloids and other secondary products. As a result of these substances, the ancestors of today's *Passiflora* were originally free from herbivores. This advantage over other plant groups was lost, however, as mechanisms were developed by butterflies of the genus *Heliconius* which enabled them to use *Passiflora* as a food source. For the butterflies these plants were a profitable resource on which they remained at first competitor-free. The reciprocal adaptation in this herbivore–plant system, however, continues. In the course of coevolution some *Passiflora* species acquired hook-equipped hairs (trichomes), which prevent the movement of freshly hatched caterpillars (Gilbert, 1971). *Passiflora* plants carrying these structures are generally protected from *Heliconius* larvae. Since this morphological adaptation is still not found in all or even most Passifloraceae, this represents what is probably a recent evolutionary innovation. The protection afforded to the plants by the trichomes, however, has already been overcome by the larvae of one *Heliconius* species, *Dione moneta* (Benson et al., 1975). This still rare adaptation of the herbivore to its host plant indicates that the apparently successful pest-defence offered by the trichomes presumably will be no long-term protection from *Heliconius* larvae. Other *Passiflora* species exhibit structures on their shoots and leaves which look like *Heliconius* eggs. *Heliconius* females, in order to reduce competition for their young, do not lay eggs at sites already containing eggs from the same species (Williams and Gilbert, 1981). By means of these egg-like structures the plants attain reduced grazing pressure from *Heliconius* larvae. The described selective exchange between the species appears to have contributed decidedly to species formation in both groups and can be further observed as coadaptation of different populations (Benson et al., 1975; Thompson, 1982).

Mutualism

The distinguishing feature of this interaction is that the populations involved are advantaged over populations exhibiting solitary growth. Because of this, populations which optimize the appro-

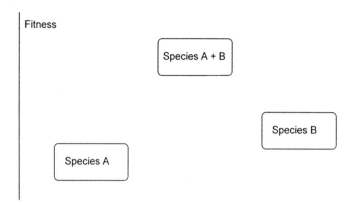

Figure 7. Characterization of a mutualistic relationship between two species. Coevolution takes place when mutualistic exchange leads to reciprocal genetic alteration of the species involved.

priate interactions achieve a selective advantage over populations which can not do this so well (Fig. 7).

The theoretical considerations related to this form of interaction are still only partially developed. The conditions for achieving mutualism, however, seem in the main to involve a close ecological relationship, for example exploitation or commensalism. Depending on ecological conditions, mutualistic interactions can be very stable. In the case of facultative mutualism the interaction can become unstable; however the extinction of one of the partners is rare (Boucher et al., 1982).

An example of a mutualism, which allows similarities to be drawn with an exploitation interaction, is the system yucca (Agavaceae) – yucca moth (Prodoxidae). Mutualistic moths of the genus *Tegeticula* lay their eggs in unfertilized ovules of yucca plants and thereby pollinate the young flowers. The emerging larvae eat a part of the unripe seed and afterwards pupate in the soil (Powell, 1992). In this system, both partners profit from the presence of the other interaction partner. In years with reduced flower number, however, the moths also lay eggs in more developed flowers where the interaction turns into an exploitation relationship (Feinsinger, 1983).

Transgenes and coevolutionary processes

During the discussion on the use of transgenic culture plants, the opportunity for accelerated resistance breeding in the traditional sense and also the introduction of genetic information from foreign species as protection against plant pests are commonly referred to (Fischbeck, 1995). An appraisal of these hopes, based on knowledge of coevolution, leads however to some scepticism, particularly when horizontal gene transfer is considered. Endeavours to intervene in the phytomedicinal field of exploitation interactions have progressed considerably in recent years. Presently, for example, much attention is being dedicated to the transfer of *Bacillus thuringiensis* toxins (*B.t.*-toxin) into crop plant genomes (Fischbeck, 1995; Leemans et al., 1990). This poison is largely pest-specific and has proved its worth as a biogenic weapon in the fight against pests during the past 30 years (Martinez-Ramirez et al., 1995). If this weapon could be integrated into a crop plant genome, expensive production and application costs of pesticide would no longer be necessary. Numerous work has shown expression under field conditions of the gene sequence for the *B.t.*-toxin and subsequent reduction in pest damage (Fischbeck, 1995; Leemans et al., 1990).

When this plan is viewed from the perspective of coevolution the gene technology-produced resistance character can be considered as being equivalent to a new mutation producing a substance for protection against herbivores. Since *B.t.*-toxin acts selectively, a close coevolution between the herbivore and plant populations involved can be expected. The herbivore population may either avoid this selection pressure or it could develop resistance. In the first case the system experiences so-called "ecological release". Perhaps the pest and also all other pests will no longer harm the transgenic plant. Examples of this are known for neophytes (Maynard Smith, 1989) which have, however, only recently been subjected to the evolutionary processes of their new ecosystem. With the transfer of *B.t.*-toxin, hopefully, no existing positive relationships will have been destroyed. If this were the case the plant would have a selective disadvantage, besides its advantage conferred by the new protection mechanism. If it were a lasting protection, without major fitness costs through the loss of beneficial interactions, this could indeed be viewed as the successful establishment of an important innovation. Many examples from ecology and especially from agricultural ecology, however, indicate that, as a result of an ecological release, other organisms may utilize the vacant food-niche. In the vacant niche, made available by the resulting gap in the food chain, a so-called "component community" can develop, superseding the previous occupiers of the niche, and within this component community adaptive radiation will occur (e.g., *Heliconius–Passiflora*-system). The possibility should also not be discounted that other pathogens may be promoted by a *B.t.*-toxin. Futuyma (1983) speculates that plants attacked by numerous pests do not develop specific defences against herbivores, since the presence of partial resistance would promote other herbivores. When the original pest ceases to be an exploiter, it can occur that secondary pests, which earlier had been suppressed by the original herbivores, may take its place as pest organisms. It follows from these considerations that, even in the case of lasting resistance of a crop against its original pest, the plant may still not be permanently protected.

In the second case, the development of resistance, the coevolutionary contest between the two interaction partners continues. The *B.t.*-toxin can be compared to a pesticide which exerts a con-

Table 1. Results of laboratory selection for resistance against *Bacillus thuringiensis* toxin (after Tabashnik, 1994)

Species	Origin[1]	Selection-duration[2]	Resistance ratio[3]
Lepidoptera			
Plutella xylostella	Germany (L)	30	1
Plutella xylostella	Hawaii (F)	9	66
Plodia interpunctella	Nebraska (L)	22	34
Plodia interpunctella	Kansas (F)	23	140
Coleoptera			
Leptinotarsa decemlineata	Michigan (F)	12	60
Diptera			
Culex quinquefasciatus	California (L)	3	70
Culex quinquefasciatus	California (F)	22	15

[1] Kept under laboratory conditions < 1 year (F) or > 1 year (L). [2] Duration of selection in generations. [3] Quotient of the LC_{50} value of the resistance-selected line *versus* the unselected line.

stant selection pressure throughout the entire vegetation period. According to all experience from agricultural ecosystems, the development of a resistance on the part of the parasite is therefore to be expected. Over the past decade, no reports have been published which show any decrease in effectiveness of *B.t.*-toxin. It is thus disquieting to hear of insect resistance to *B. thuringiensis* (Martinez-Ramirez et al., 1995; Tabashnik, 1994). Resistance against the *B.t.*-toxin can be found in selection experiments for the groups Lepidoptera, Coleoptera and Diptera (Tab. 1).

The risk of resistance against this compound is obviously not insignificant and also exists under simultaneous application of different *B.t.*-toxins (McGaughey and Johnson, 1992). A large-acreage plantation of crops armed with this toxin could not be expected to provide enduring herbivore relief. The principle of optimal resource use should be remembered here. Under this principle, a rapid evolution of resistance is expected when a resource is widespread. Maynard Smith (1989) mentions that the adaptation of herbivores to introduced sugarcane (*Saccharum* sp.) in some countries is closely coupled with the size of the respective plantation areas.

Further hopes for an economically advantageous gene-technology application are associated with an optimized phosphorus assimilation system in crop plants (Fischbeck, 1995). Vesicular–arbuscular mycorrhizal fungi (VAM) build a mutualistic interaction between plants and fungi which particularly contributes to improvement in phosphorus uptake by the plant. The plant–mycorrhiza interaction is, however, not always advantageous for both partners. Bethlenfalvay et al. (1982a,b) found that the interaction of a plant with VAM may negatively influence plant growth, when soil-phosphate is at medium levels. This symbiosis then takes on the character of an exploitation interaction. If then a plant is able to improve its phosphorus assimilation, as a result of molecular-biological alteration, it can be imagined that the mycorrhizae still present in the soil would take on a new role as root parasites. With respect to coevolution the possibility cannot be discounted that previously mutualistic interactions may transform, following gene-technological alterations, into exploitation interactions. It is therefore important to know, when dealing with mutualistic interactions, to what extent they originate from other interactions. On the other hand, it is known that the absence of VAM can reduce protection from root diseases (Werner, 1987). It appears as if this interaction is highly complex not only with respect to the advantages which the plant receives from that symbiosis but also in view to a shift of the system into another ecological relationship.

Coevolution and gene technology

As can be seen from the examples above, the use of applied gene technology in crop plants should be judged cautiously. Pathogen resistance generated by interspecific gene transfers into crop plant genomes will probably not be stable in the long term. The careless use of characters to combat pests in agricultural ecosystems could lead to a situation which no longer allows adequate ecological regulation of the system. In my opinion, the coevolutionary process in natural exploitation interactions should be analyzed and understood in much more depth than previously, in order to optimize plant production systems based on this knowledge. In chemical plant protection, it has long been known that pest organisms develop resistance against pesticides. "Pesticides at risk" are spoken of in this context, in order to make it clear that pesticide resistance endangers an important, rare and expensive tool in today's dominant agricultural production systems. Such weapons should thus be cautiously employed. At this point the question arises: What substitute do we have for "genes at risk"? With this comparison I want to point out that conventional procedures for management of agricultural production systems would be merely continued, unchanged, by the introduction of gene technology, rather than new ways being sought and applied. Under these conditions too little space remains for land management to accommodate resource-use strategies of ecological systems which have existed for millions of years. Although it is true that gene technology can also offer opportunities for ecologically appropriate land management (e.g., quantitative resistance with help from vertical gene transfer), for this to become a reality the coevolutionary processes in agricultural ecosystems must be better understood than now.

The probable long-term consequences for ecosystems through transgene-mediated evolutionary processes can today be only vaguely stated. This is due just in part to the imponderability of environmental development over long time-periods, but rather it is much more dependent on the lack of investigation into evolutionary biological processes in ecosystems. The provocative question, "So what?", with regard to unwanted gene flow from transgenic populations into living communities, is thus an indicator of ignorance in terms of possible side-effects of the new technology. It provides, however, no reason for an approach characterized by the words "everything will be okay". In fact, a long term monitoring of ecosystems should ensue, with the aim of quantifying the ecological and evolutionary genetic effects of transgenes.

References

Benson, W.W., Brown, K.S. and Gilbert, L.E. (1975) Coevolution of plants and herbivores: Passion flower butterflies. *Evolution* 29: 659–680.

Berenbaum, M. and Feeny, P. (1981) Toxicity of angular furanocoumarins to swallowtail butterflies: Escalation in a coevolutionary arms race? *Science* 212: 927–929.

Berenbaum, M. (1983) Coumarins and caterpillars: A case for coevolution. *Evolution* 37: 163–179.

Bethlenfalvay, G.F., Brown, M.S. and Pacovsky, R.S. (1982a) Parasitic and mutualistic associations between a mycorrhizal fungus and soybean: Development of the host plant. *Phytopathology* 72: 889–893.

Bethlenfalvay, G.F., Pacovsky, R.S. and Brown, M.S. (1982b) Parasitic and mutualistic associations between a mycorrhizal fungus and soybean: Development of the endophyte. *Phytopathology* 72: 894–897.

Boucher, D.H., James, S. and Keeler, K.H. (1982) The ecology of mutualism. *Annu. Rev. Ecol. Sys.* 13: 315–347.

Dinoor, A. and Eshed, N. (1987) The analysis of host and pathogen populations in natural ecosystems. *In*: M.S. Wolfe and C.E. Caten (eds): *Populations of Plant Pathogens*. Blackwell, Oxford, pp 75–88.

Ehrlich, P.R. and Raven, P.H. (1964) Butterflies and plants: A study in coevolution. *Evolution* 18: 586–608.

Emlen, J.M. (1966) The role of time and energy in food preference. *Amer. Natur.* 100: 611–617.

Feinsinger, P. (1983) Coevolution and pollination. *In*: D.J. Futuyma and M. Slatkin (eds): *Coevolution*. Sinauer, Sunderland, USA, pp 282–310.

Fischbeck, G. (1995) Biotechnologische Ansätze für die Züchtung gesunder Pflanzen und ihre Bedeutung für die Entwicklung umweltschonender Anbauverfahren. *In*: T. von Schell and H. Mohr (eds): *Biotechnologie – Gentechnik. Eine Chance für neue Industrien*. Springer-Verlag, Berlin, pp 181–200.

Futuyma, D.J. (1983) Evolutionary interactions among herbivorous insects and plants. *In*: D.J. Futuyma and M. Slatkin (eds): *Coevolution*. Sinauer, Sunderland, USA, pp 207–231.

Futuyma, D.J. (1986) *Evolutionary Biology*. Sinauer, Sunderland, USA.

Gilbert, L.E. (1971) Butterfly-plant coevolution: Has *Passiflora adenopoda* won the selectional race with heliconiine butterflies? *Science* 172: 585–586.

Janzen, D.H. (1980) When is it coevolution? *Evolution* 34: 611–612.

Leemans, J., Reynards, A., Höfte, H., Peferoen, M., van Maelert, H. and Joos, H. (1990) Insecticidal crystal proteins from *Bacillus thuringiensis* and their use in transgenic crops. *In*: R.R. Baker and P.E. Dunn (eds): *New Directions in Biological Control*. Alan R. Liss Inc., New York, pp 573–582.

MacArthur, R.H. and Pianka, E.R. (1966) On optimal use of a patchy environment. *Amer. Natur.* 100: 603–609.

MacArthur, R.H. and Levins, R. (1967) The limiting similarity, convergence, and divergence of coexisting species. *Amer. Natur.* 101: 377–387.

Martinez-Ramirez, A.C., Escriche, B., Real, D.M., Silva, F.J. and Ferre, J. (1995) Inheritance of resistance to a *Bacillus thuringiensis* toxin in a field population of diamondback moth (*Plutella xylostella*). *Pest. Sci.* 43: 115–120.

Maynard Smith, J. (1989) *Evolutionary Genetics*. Oxford University Press, New York.

McCaughey, W.H. and Johnson, D.E. (1992) Indianmeal moth (Lepidoptera: Pyralidae) resistance to different strains and mixtures of *Bacillus thuringiensis*. *J. Econ. Entomol.* 85: 1594–1600.

Powell, J.A. (1992) Interrelationships of yuccas and yucca moths. *Trends Ecol. Evol.* 7: 10–15.

Tabashnik, B.E. (1994) Evolution of resistance to *Bacillus thuringiensis*. *Annu. Rev. Entomol.* 39: 47–79.

Thompson, J.N. (1982) *Interactions and Coevolution*. Wiley, New York.

Thompson, J.N. (1989) Concepts of coevolution. *Trends Ecol. Evol.* 4: 179–183.

Werner, D. (1987) *Pflanzliche und mikrobielle Symbiosen*. Thieme Verlag, Stuttgart.

Williams, K.S. and Gilbert, L.E. (1981) Insects as selective agents in plant vegetative morphology: Egg mimicry reduces egg laying by butterflies. *Science* 212: 467–469.

Wolfe, M.S. (1987) Trying to understand and control powdery mildew. *In*: M.S. Wolfe and C.E. Caten (eds): *Populations of Plant Pathogens*. Blackwell, Oxford. pp 253–273.

Transgenic Organisms – Biological and Social Implications
J. Tomiuk, K. Wöhrmann & A. Sentker (eds)
© 1996 Birkhäuser Verlag Basel/Switzerland

The two strategies of biological containment of genetically engineered bacteria

T. Schweder

Institute of Microbiology and Molecular Biology, Ernst-Moritz-Arndt University, Jahnstraße 15, D-17487 Greifswald, Germany

Summary. The bacterium *Escherichia coli* played an important role in the development of genetic engineering techniques and applications. *E. coli* was also the pioneer organism used in the field of biological containment. Most of the model containment systems were developed exclusively for this bacterium. In this chapter the biological containment systems for *E. coli* are summarized and discussed. Furthermore, confinement systems for other bacteria will be presented. Finally, I describe what we can expect in the future in the field of biological containment.

Introduction

The use of genetically engineered microorganisms (GEMs) is increasingly being considered as an alternative solution to a variety of problems. This includes on the one hand the application of GEMs for the production of enzymes, peptides and other components in the fermentor and, on the other hand, the deliberate release of GEMs into the environment (see Teuber, this volume). The intentional release of GEMs is designed for the bioremediation of contaminated soil and groundwater, in biological control and as inoculants in agriculture, and as live vaccines in biomedicine.

Potential risk of using genetically engineered microorganisms

Despite a dramatic increase in the knowledge about the ecology of microorganisms in different environments, it is difficult to fully predict the behavior of GEMs or their DNA in the environment. Furthermore, despite many physical containment efforts, one cannot prevent an unintentional release of GEMs outside the laboratory with absolute security.

What are the properties of recombinant microorganisms, which we must consider if they are deliberately or unintentionally released into the environment? (i) Once bacteria are released into the environment it is not possible to remove them; (ii) in general, commonly used microorganisms are characterized by fast generation rates; (iii) most of these microorganisms are able to adapt to adverse environmental conditions; (iv) the exchange of genetic material between different species, called horizontal gene transfer is a very common traite of prokaryotes. For instance, bacteria are the only organisms capable of natural transformation (Lorenz and Wackernagel, this volume).

What are the potential risks of the release of GEMs into the environment? (i) Once in the environment, released GEMs could not only persist in soil or groundwater for a long time but also propagate; (ii) in the worst case, GEMs have a selective growth advantage or express toxic gene products which could cause a displacement of indigenous microorganisms; (iii) the foreign DNA could be transferred to indigenous microorganisms and could be established or even expressed in these microorganisms (the transfer of an antibiotic resistance gene would be an example of such an undesirable event); (iv) furthermore, undesired changes of the so-called ecological steady-state or even an influence on the evolution of microorganisms and viruses cannot be excluded in every case.

The concept of biological containment

In the past, there was a consensus between molecular biologists and ecologists that every disturbance of the ecological steady-state is an undesirable event. Furthermore, there was a fear that some GEMs could be harmful for humans. Because of the lack of predictable behavior of GEMs the decision was made to use only strains which cannot establish or persist in different environments. The concept of the biological containment was thus born. This concept demands that foreign genes must be introduced only into so-called safety strains. Such safety strains should not be able to transfer their foreign DNA to other organisms. Their ability of survival, propagation and spreading have to be restricted only to laboratory conditions.

To fulfill safety guidelines, any introduced genetic material of commercially used GEMs must be (i) limited in size to consist only of the gene(s) of interest; (ii) well characterized in the function of all the gene products; and (iii) free of certain sequences (e.g., gene products which are potentially toxic to other organisms).

The vectors used in gene technology should have features which fulfill the criteria described above. An important additional characteristic is to prevent the transfer of recombinant DNA by making the plasmids poorly mobilizable.

For the biological containment of GEMs there are two strategies. The initial strategy was to use chromosomal mutations which altered the bacteria so that they would poorly survive outside the laboratory. This mechanism can be regarded as a passive containment strategy. The second one is an active strategy which based on the construction of a suicide system. A simple suicide system consists of two parts. One part is the control sequence, which usally consists of a promoter and, if necessary, contains additional sequences involved in its regulation. The second part is a gene which codes for a product that is toxic for the cell. The choice of the promoter and the induction mechanisms strongly depends on the use of the appropriate microorganism.

Biological containment

Passive containment

In principle all of the *E. coli* strains used genetically in the laboratories are considered as enfeebled organisms. The best known so-called safety strains of *E. coli* are the strains K12 and Chi-1776. The strain Chi-1776, made by Curtiss (1976), is the prototype of a safety strain. This strain is absolutely not viable outside the laboratory, because in addition to other limitations this strain is not able to synthesize D-amino pimelic acid, an essential constituent of the bacterial cell wall not naturally occurring in the environment. Furthermore, the strain has additional mutations which make it especially sensitive to detergents, antibiotics and UV-light. The disadvantage using such an enfeebled strain is its very complicated handling in the laboratory. This strain does not survive in the fermentor satisfactorily. Therefore, the most frequently used host for recombinant DNA is not *E. coli* Chi-1776 but *E. coli* K12, a strain which is unable to colonize the intestine of humans and other animals under normal conditions but which does not possess so many disabling muta-

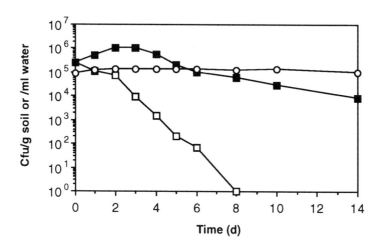

Figure 1. Survival of the *E. coli* K12 strain MG1655 in sterile nutrient rich soil (■), nutrient poor sand soil (□) and sterile river water (○).

tions. In general, this strain is considered to survive poorly in soil and water. However, survival experiments with *E. coli* K12 in soil or riverwater microcosms reveal that this strain remains viable under these conditions for a relativly long time (Fig. 1).

The *E. coli* K12 strain MG1655 showed a slow decline of viable cells (*cfu*) in a nutrient rich soil during the first 14 days. In fact, for the first two days the organisms even grew in the soil. However, in a very nutrient poor soil MG1655 cells survived for at least 8 days. In riverwater, the K12 strain also persisted over a long time with a negligible decline in viability. Similar results were reported for the survival of other typical *E. coli* laboratory strains in various natural environments by Devanas et al. (1986) and Chao and Feng (1990). Obviously, these survival experiments are not representative for every possible milieu. Each experiment reflects only one special environment. However, we have to consider all possibilities. In this connection it is worthwhile to bear in mind that the risk of transfer of foreign DNA from recombinant bacteria to indigenous bacteria will increase with increasing survival time of the recombinant microorganisms.

Which chromosomal mutations can minimize the survival of GEMs with minimal influence on the growth behavior of the bacteria in the fermentor? Furthermore, which chromosomal mutations can prevent the persistence of deliberately released GEMs in the environment after they have fulfilled their intended role?

For the bacterium *E. coli* a lot is known about its strategies to survive bad conditions, such as nutrient starvation, heat shock and osmotic stress. A whole cascade of genes are responsible for the fast adaptation of this bacterium to unfavorable conditions. The alarmon guanosinetetraphosphate (ppGpp) is a critical compound in the adaptation to nutrient starvation. It signals to the cell the changed environmental situation and enables it to respond quickly and effectively to the new conditions. This survival strategy is regulated by at least two genes (*relA* and *spoT*) and is called the stringent response (Cashel and Rudd, 1987). The loss of the stringent response is expected to reduce the cell's ability to survive prolonged periods of starvation in nature (Mach et al., 1989; Hofmann and Schweder, 1993). This hypothesis was confirmed by investigation of the survival of different *E. coli* mutant strains in soil microcosms (Fig. 2). The viabilities of the different *relA* mutant strains were significantly lower than those of the isoallelic *relA*[+] strains (Schweder, 1994).

Besides the stringent response there is another important survival strategy of *E. coli* which is driven by the alternative sigma factor σ^{38} and which seems to be closely connected to the stringent response (Gentry et al., 1993). σ^{38} is a major stress response switch in *E. coli* (McCann et al., 1991). This sigma factor has a central role in development of starvation-mediated general stress resistance in *E. coli*. McCann and coworkers showed that mutants of this sigma factor survive carbon and nitrogen starvation poorly in shake flasks. Survival experiments in a soil microcosms revealed a significantly lower viability of a deletion mutant of the σ^{38} gene (*rpoS*) than that of the isoallelic *rpoS*[+] strain (T. Schweder, unpublished result).

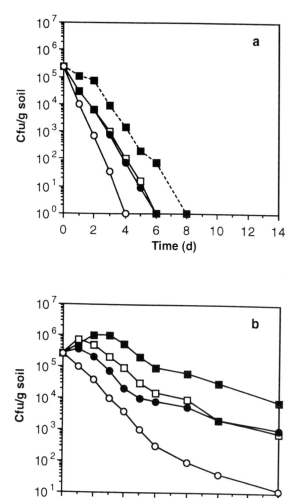

Figure 2. Comparison of the survival of an *E. coli relA/spoT* deletion mutant and an *E. coli relA* point mutant with their isoallelic partners (a) in a nutrient poor sand soil and (b) in a nutrient rich soil. (■) MG1655 = K12 wt strain (B. Bachmann); (□) CF1678 = MG1655(Δ*relA/spoT*) (Xiao et al., 1991); (●) CP78 (Fill and Friesen, 1968); (○) CP79 = CP78(*relA*) (Fill and Friesen, 1968).

The advantage of mutations in genes responsible for the survival under adverse conditions is that they have no or only a small influence on growth in a fermentor. Furthermore, this kind of mutations could prevent the prolonged persistence of intentionally released bacteria in soil or groundwater, and yet allow sufficient time for the recombinant microorganism to fulfill its engineered purpose. This strategy, hitherto only investigated in detail in *E. coli,* is easily transferable to other bacteria, because the stringent response and the σ^{38} regulon seems to be widespread among different bacteria.

For the biological containment of GEMs released in the environment, the use of *recA* mutants was also suggested (Molin et al., 1993). It is expected that such mutants living in soil or on plants, once exposed to irradiation, would not able to repair the DNA damages, preventing the establishment of GEMs under these conditions.

Beside the above approach, i.e., mutations in genes whose products are essential for long-term survival under natural conditions, Stocker (1990) described an approach, designed to disable recombinant bacteria used as live vaccines. The task of these GEMs is to serve as immunogens by presenting antigenic determinants to the host's immune system after oral application. To prevent their proliferation it was suggested that genes required for the synthesis of substrates which are of limited availability in the animal gut, such as genes for the synthesis of aromatic amino acids (e.g., *aroD*) (Stocker, 1990) or the *crp* gene responsible for the catabolite repression control (Curtiss et al., 1988) should be altered by mutation. This strategy was called attenuation of GEMs (Molin et al., 1993).

Active containment

Passive biological containment strategies, which guarantee an easy culturing of the GEMs in the laboratory, can also minimize the persistence of GEMs in soil or groundwater. However, this approach does not guarantee quick killing of unintentionally released GEMs. Furthermore, passive strategies that utilize debilitated hosts are obviously not always practical for the biological containment of planned environmental releases. Such GEMs must be able to compete success-fully for a time with indigenous microorganisms to perform their special tasks in soil or ground-water. Therefore, the idea was conceived to use so-called conditionally lethal biological contain-ment systems, which are induced under defined environmental conditions. Conditional suicide systems can be expected to produce a predictable killing of GEMs.

Some interesting model conditional suicide systems were developed during the last decade, which are summarized in Table 1. In principal, most of these systems are very simple. Nevertheless, there are some problems in constructing conditional suicide systems, which are discussed below.

The first and certainly most crucial step is to find a suitably controlled promoter. Such a promo-ter should have no or very low basal activity under permissive conditions but should be highly induced by a distinct signal, such as by temperature changes, metabolites, chemical inducers (e.g.,

Table 1. Model suicide systems of *E. coli* (P = promoter)

System	Killing by	Induction by	References
P_{trp}-*hok*	Collapse of the membrane potential	Lack of tryptophan	Molin et al., 1987
P_{lac}-*hok*	" – "	IPTG	Bej et al., 1989
P_{lac}-*relF*	" – "	IPTG	Knudson and Karlström, 1991
xylS/P_m-*lacI*/P_{tac}-*gef*	" – "	Lack of 3-methyl-benzoate	Contreras et al., 1991
P_R-*hok*/*sok*	" – "	Temperature shift from 40°C to 30°C	Thisted and Gerdes, 1992
P_{phoA}-*parB*	" – "	Phosphate limitation	Schweder et al., 1992
nptI/*sacR*/*B*	Cell lysis by levan accumulation	Sucrose	Recorbet et al., 1993
P_L-*nuc*	Decay of DNA and RNA	Temperature shift from 28° to 42°C	Ahrenholtz et al., 1994
P_{phoA}-T7Lys	Cell lysis by T7-lysozyme	Phosphate limitation	Schweder et al., 1995
fimB/*fimE*/P_{fimA}-*gef*	Collapse of the membrane potential	Stochastic switch on by inversion of the *fimA*-promoter	Klemm et al., 1995

IPTG) or nutrient limitation. The second part of a suicide system is a promoterless gene which codes for a host toxic protein.

One of the first suicide systems was constructed by Molin and coworkers in 1987 for *E. coli*. They used the lambda P_R promoter, the *c*1857 temperature-sensitive repressor gene and the *hok* (*host killing*) gene (Molin et al., 1987). The *hok* gene is part of the *parB* locus of the low-copy-number plasmid R1. The *hok* gene belongs to the *gef* gene family which codes for a class of highly homologous toxic proteins (Poulsen et al., 1989). The *parB* locus is an efficient natural plasmid stabilization system, which mediates a post-segregational killing of plasmid-free cells (Gerdes, 1988). Cells containing the above described suicide system grow normally under the permissive temperature of 30°C. But at 42°C the temperature-sensitive repressor is inactivated and the *hok* gene is expressed which causes a killing of the host cell. Another suicide system developed in this laboratory is based on the same killing gene (*hok*) but under the control of the *E. coli trp* promoter (Molin et al., 1987). In the fermentor, tryptophan in the medium represses the promoter. But once tryptophan is limited, the *trp* promoter is activated and causes *hok* synthesis, which consequently leads to host cell death. Another suicide system, also based on the *hok* gene, was constructed by Bej et al. (1989). They used the *lac* promoter as the control element. This promoter is either induced by lactose under glucose limitation or by the artificial substrate IPTG. A similar system was investigated by Knudson and Karlström (1991) using the *lac* promoter but *relF* as the killing gene, a gene belonging also to the *gef* gene family.

All four biological containment systems described above are considered model suicide systems. In terms of an applicable biological containment system, only the *trp*-controlled suicide system is quite promising, because the concentration of tryptophan in soil or groundwater is negligible. Even if it was theoretically possible to use the other three systems, which are heat sensitive, for the inactivation of GEMs in the fermentor, in every case, a pasteurization would be more reasonable. These systems are not adequate to prevent survival of GEMs outside the fermentor.

For the containment of unintentionally or deliberately released GEMs, there is need for environmentally regulated suicide systems. To find suitable environmentally regulated promoters, we investigated the feasibility of different promoters, which are regulated by: (i) Oxygen limitation *(fdhF* promoter; Birkmann et al., 1987); (ii) low temperature *(cspA* promoter; Goldstein et al., 1989); and (iii) phosphate limitation *(phoA* promoter; Wanner, 1987). All three promoters were fused to a promoterless *lacZ* gene on a multi-copy-number plasmid. Their activities were investigated by measuring β-galactosidase activity. The result of this preliminary experiment was that only the *phoA* promoter was suitable for a suicide system under these conditions (Schweder, 1994). The problem demonstrated by this experiment is not new. Many promoters lose their stringent regulation found on the chromosome once cloned into a multi-copy-number plasmid. The promoters frequently have higher basal activities, as in the cases of the *cspA* and the *fdhF* promoters, or can even lose their induction. The relativly low basal activity of the plasmid-located *phoA* promoter and its 20-fold induction after phosphate limitation indicates its potential usefulness for a suicide system. Furthermore, and this was the crucial point for choosing the *phoA* promoter, phosphate is frequently a growth-limiting nutrient in nature especially in soil because phosphates are bound in water-insoluble salts and are therefore not available for most microorganisms (Wanner, 1987).

For our first suicide system, we introduced the complete *parB* locus downstream of the *phoA* promoter of *E. coli* (Schweder et al., 1992). This system insures both plasmid stabilization during growth and killing of the cells upon phosphate limitation. The stability of the model plasmid used for the suicide system could be indeed dramatically increased, but a considerable number of cells survived the induction of suicide by this system.

We looked for other toxic genes which could be fused to the *phoA* promoter but which do not negatively influence the growth of *E. coli* cells in the fermentor. We chose the lysozyme gene

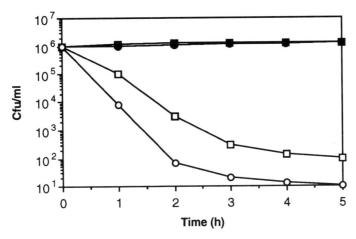

Figure 3. Killing of *E. coli* CP78 and CP79(*relA*) after induction of a suicide system based on the *phoA* promoter and the T7 lysozyme gene in phosphate free medium. (■) CP78 (without suicide system); (□) CP78 (with suicide system); (●) CP79 (without suicide system); (○) CP79 (with suicide system).

from phage T7 and replaced the *parB* locus with it (Schweder et al., 1995). Lysozyme plays an important role in lysis in T7-infected *E. coli* cells (Studier, 1991). This gene is compatible with the low basal level of the *phoA* promoter, because a small amount of lysozyme in the cell does not hurt it. But once this enzyme was overproduced, it caused very fast lysis of the host cell as shown in Figure 3. Surprisingly, this suicide system was much more effective in *E. coli relA* strains. The effective killing of *E. coli relA* mutants containing this suicide system was shown not only under phosphate limitation in shake flasks but also under environmental conditions in two different soil-microcosms (Schweder et al., 1995). These results suggest that the combination of the two different strategies of biological containment, chromosomal mutation and conditional suicide, can result in enhanced killing of recombinant *E. coli* cells.

Promoters which are regulated by repressors (e.g., *lac*, *tac* or lambda P_L promoter) are very stringently controlled. Thus, such promoters can be used for the construction of suicide systems with genes that encode very toxic products, as in the suicide system of Ahrenholtz et al. (1994). This suicide system consists of the *Serratia marcescens* endonuclease gene (*nuc*) under the control of the lambda P_L promoter localized on a plasmid and the thermosensitive lambda *c*1857 repressor gene localized on the chromosome. The nuclease of *S. marcescens* cleaves RNA and DNA to acid-soluble materials and introduces single- and double-stranded cuts into duplex DNA. The advantage of this kind of killing gene is that it causes not only the death of the recombinant host cell but also degrades the recombinant DNA, preventing any horizontal gene transfer. The induction mechanism of this system, a temperature shift from 28°C to 42°C, is of course only suitable for the inactivation of GEMs in the fermentor. However, the combination of the *nuc* gene with environmentally regulated promoters is promising.

As discussed above, the finding of an appropriate regulatory element to express the toxic gene is the crucial step, especially for the containment of bacteria designated for deliberate release. However, until now only few suitable promoters or regulatory systems were known.

In 1987 the idea of stochastic regulation of a suicide system was suggested by Molin et al. A practical stochastic killing system for biological containment of *E. coli* cells was presented by Klemm et al. (1995). This system was based on the invertible switch promoter of the *E. coli fimA* gene which is induced as a function of time. The *fim* promoter cassette was inserted in front of the promoterless *gef* gene. The *fim*-driven suicide system does not interfere much with cell growth and could be useful for containment system of deliberately released bacteria.

All the biological containment systems which discussed above were constructed in *E. coli* and worked best in this organism. However, some of these systems work partially in other bacteria as well. A substrate-dependent suicide system, described for *E. coli* by Contreras et al. (1991), was presented in 1993 by Jensen et al. in *Pseudomonas putida*. This system was based on two elements. One of which was a randomly integrated chromosomal fusion between a synthetic *lac* promoter and the *gef* gene. The second element was located on a plasmid. It consisted of the P_M promoter of the *P. putida* TOL plasmid-encoded *meta*-cleavage pathway operon for the catabolism of benzoate and derivatives, fused to the *lacI* gene (encoding the *lac* repressor). Furthermore the plasmid contained the *xylS* gene which codes for a positive regulator of the P_M promoter. This kind of containment system is designed to kill the released GEMs after they have fulfilled their purpose, which is the degradation of xenobiotics. The presence of the xenobiotic substrate prevents the induction of the suicide system. In this system, the effector for *xylS* is *m*-methylbenzoate, which activates the P_M-driven production of *lacI*, preventing the expression of the killing function. In the absence of the substrate, the expression of the *gef* gene is no longer repressed and the bacteria are killed.

In addition to concerns about the regulatory element, a further problem, perhaps the most troublesome, is the inactivation of the suicide function by mutations. Spontaneous mutations, which change the expression of the suicide system, can occur in the killing gene. It is impossible to completely prevent this kind of mutation; however, there are some strategies to mitigate this problem. One possibility is to increase the copy number of the containment system; e.g., by duplication of the suicide system on the chromosome (Knudson and Karlström, 1991; Jensen et al., 1993) or by its location on a multi-copy number plasmid which is stably propagated (Schweder et al., 1992). But mutations are also possible in the target of the toxic gene product which might render the host cell resistant against it, as it was shown for the GEF protein (Poulsen et al., 1992). To adress this problem, it was suggested that an additional different suicide system could be used. Approaches to prevent the accumulation of spontaneous mutations are to use very tight promoters which are marked by very low basal activity (e.g., promoters regulated by repressors), or alternately to use killing genes whose products are not strongly toxic in low concentrations; e.g., the lysozyme gene of phage T7 whose gene product is tolerated by the host cell in low concentrations (Schweder et al., 1995).

Outlook

The applications of GEMs in fermentation processes or during routine work in the laboratory have been practiced for several years (see Teuber, this volume). However, the deliberate release of GEMs in the environment, which has begun and will likely increase, is a new approach. There are some speculative risks concerning the ecological consequences of the release of GEMs. The risk assessment of the application of GEMs will promote a more scientifical view of these problems. An important aspect in this connection is that despite all physical containment efforts in the laboratories and biological containment of GEMs, no system can be hundred percent effective. Thus, what must be decided is whether the potential benefits outweigh the as yet not totally predictable risks.

However, the application of reasonable containment approaches would help to make the use of GEMs more safe. We can decrease the unpredictable risk by minimizing either the survivability of unintentionally released bacteria or preventing the prolonged persistence of deliberately released bacteria. Active biological containment is no longer only a theoretical possibility. There are strategies, investigated with the model bacterium *E. coli*, which can now be used with other bacteria designed for specific purposes in the natural environment. One active containment strategy constructed for a *P. putida* strain designed for bioremediation purposes is ready for field experiment trials (Ramos et al., 1995).

The chromosomal mutations in genes, which are responsible for survival strategies triggered by environmental conditions like nutrient limitation or high osmolarity, as suggested above, could be easily transfered to other bacteria and could be even suitable in some cases for the containment of bacteria dedicated for bioremediation or for other environmental tasks. In any case, the induction signal and the appropriate regulation of the suicide system must be suitable to the GEMs itself and to its purpose. The best choice of a toxic gene product in most cases is a nuclease, an enzyme which cannot only cause the killing of the host cell but also degradation of the foreign DNA. The construction of conditional suicide systems appropriate to the purpose of the GEMs remains very difficult and is not possible in every case. The continued increase in knowledge of the molecular biology and physiology of bacteria will yield new ideas for the regulation of such systems.

Acknowledgements
I am indebted to Abdul Matin and Cres Fraley for critical reading of the manuscript. This work was supported by the Bundesministerium für Forschung und Technology (0310308 A).

References

Ahrenholtz, I., Lorenz, M.G. and Wackernagel, W. (1994) A conditional suicide system in *Escherichia coli* based on the intracellular degradation of DNA. *Appl. Environ. Microbiol.* 60: 3746–3751.

Bej, A.K., Perlin, M.H. and Atlas, R.M. (1989) Model suicide vector for containment of genetically engineered microorganisms. *Appl. Environ. Microbiol.* 54: 2472–2477.

Birkmann, A., Zinoni, F. and Boeck, A. (1987) Factors affecting transcriptional regulation of the formate-hydrogen-lyase-pathway of *Escherichia coli. Arch. Microbiol.* 148: 44–51.

Cashel, M. and Rudd, K.E. (1987) The stringent response. *In*: F.C. Neidhardt, J.L. Ingraham, K.B. Low, B. Magasanik, M. Schaechter and H.E. Umbarger (eds): *Escherichia coli and Salmonella typhimurium. Cellular and Molecular Biology.* American Society of Microbiology, Washington, pp 1410–1438.

Chao, W.L. and Feng, R.L. (1990) Survival of genetically engineered *Escherichia coli* in natural soil and river water *J. Appl. Bact.* 68: 319–325.

Contreras, A., Molin, S. and Ramos, J.L. (1991) Conditional suicide system for bacteria which mineralize aromatics. *Appl. Environ. Microbiol.* 57: 1504–1508.

Curtiss, R. (1976) Genetic manipulation of microorganisms: Potential benefits and biohazards. *Annu. Rev. Microbiol.* 30: 507–533.

Curtiss, R., Goldschmidt, R.M., Fletchall, N.B. and Kelly, S.M. (1988) Construction and use of safer bacterial host strains for recombinant DNA research. *In*: W.A. Scott and R. Werner (eds): *Molecular Cloning of Recombinant DNA.* Academic Press, New York, pp 99–111.

Devanas, M.A., Rafaeli-Eshkol, D. and Stotzky, G. (1986) Survival of plasmid containing strains of *Escherichia coli* in soil: Effect of plasmid size and nutrients on survival of hosts and maintenance of plasmids. *Curr. Microbiol.* 13: 269–277.

Fill, N. and Friesen, J.D. (1968) Isolation of "relaxed" mutants of *Escherichia coli. J. Bacteriol.* 95: 729–731.

Gentry, D.R., Hernandez, V.J., Nguyen, L.H., Jensen, D.B. and Cashel, M. (1993) Synthesis of the stationary-phase sigma factor σ^s is positively regulated by ppGpp *J. Bacteriol.* 175: 7982–7989.

Gerdes, K. (1988) The *parB* (*hok/sok*) locus of plasmid R1: A general purpose plasmid stabilization system. *Bio/Technology* 6: 1402–1405.

Goldstein, J., Pollitt, N.S. and Inouye, M. (1989) Major cold shock protein of *Escherichia coli. Proc. Natl. Acad. Sci. USA* 87: 283–287.

Hofmann, K. and Schweder, T. (1993) *Escherichia coli* host/plasmid systems providing biological containment. *In*: K. Wöhrmann and J. Tomiuk (eds): *Transgenic Organisms: Risk Assessment of Deliberate Release.* Birkhäuser Verlag, Basel, pp 193–208.

Jensen, L.B., Ramos, J.L., Kaneva, Z. and Molin, S. (1993) A substrate-dependent biological containment system for *Pseudomonas putida* based on the *Escherichia coli gef* gene. *Appl. Environ. Microbiol.* 59: 3713–3717.

Klemm, P., Jensen, L.B. and Molin, S. (1995) A stochastic killing system for biological containment of *Escherichia coli. Appl. Environ. Microbiol.* 61: 481–486.

Knudson, S.M. and Karlström, O.H. (1991) Development of efficient suicide mechanism for biological containment of bacteria. *Appl. Environ. Microbiol.* 57: 85–92.

Mach, H., Hecker, M., Hill, I., Schroeter, A. und Mach, F. (1989) Physiologische Bedeutung der "Stringent Control" bei *Escherichia coli* unter extremen Hungerbedingungen. *Z. Naturforschung* 44: 838–844.

McCann, M.P., Kidwell, J.P. and Matin, A. (1991) The putative σ factor *KatF* has a central role in development of starvation-mediated general resistance in *Escherichia coli. J. Bacteriol.* 173: 4188–4194.

Molin, S., Klemm, P., Poulsen, L.K., Biehl, H., Gerdes, K. and Anderson, P. (1987) Conditional suicide system for containment of bacteria and plasmids. *Bio/Technology* 5: 1315–1317.

Molin, S., Boe, L., Jensen, L.B., Kristensen, C.S., Givskov, M., Ramos, J.L. and Bej, A.K. (1993) Suicidal genetic elements and their use in biological containment of bacteria. *Annu. Rev. Microbiol.* 47: 139–166.

Poulsen, L.K., Larsen, N.W., Molin, S. and Anderson, P. (1989) A family of genes encoding a cell-killing function may be conserved in all Gram-negative bacteria. *Mol. Microbiol.* 3: 1463–1472.

Poulsen, L.K., Larsen, N.W., Molin, S. and Anderson, P. (1992) Analysis of an *Escherichia coli* mutant strain resistent to the cell-killing function encoded by the *gef* gene family. *Mol. Microbiol.* 6: 895–905.

Ramos, J.L., Anderson, P., Jensen, L.B., Ramos, C., Ronchel, M.C., Diaz, E., Timmis, K.N. and Molin, S. (1995) Suicide microbes on the loose. *Biotechnology* 13: 35–37.

Recorbet, G., Robert, C., Givaudan, A., Kudla, B., Normand, P. and Faurie, G. (1993) Conditional suicide system of *Escherichia coli* released into soil that uses the *Bacillus subtilis sacB* gene. *Appl. Environ. Microbiol.* 59: 1361–1366.

Schweder, T., Schmidt, I., Herrmann, H., Neubauer, P., Hecker, M. and Hofmann, K. (1992) Construction of an expression vector system providing plasmid stabilization during fermentation processes and conditional suicide of plasmid containing cells of *Escherichia coli* in the environment. *Appl. Microbiol. Biotechnol.* 38: 91–93.

Schweder, T. (1994) *Konstruktion und Untersuchung von alternativen biologischen "containment" Systemen für Escherichia coli.* Ph.D. thesis, University of Greifswald, Greifswald, Germany.

Schweder, T., Hofmann, K. and Hecker, M. (1995) *Escherichia coli* K12 *relA* strains as safe hosts for expression of recombinant DNA. *Appl. Microbiol. Biotechnol.* 42: 718–723.

Stocker, B.A.D. (1990) Aromatic-dependent *Salmonella* as live vaccine presenters of foreign inserts in flagellin. *Res. Microbiol.* 141: 787–796.

Studier, E.W. (1991) Use of bacteriophage T7 lysozyme to improve an inducible T7 expression system. *J. Mol. Biol.* 219: 37–44.

Thisted, T. and Gerdes, K. (1992) Mechanism of post-segregational killing by *hok/sok* system of plasmid *R1*. *J. Mol. Biol.* 223: 41–54.

Wanner, B.L. (1987) Phosphate regulation of gene expression in *Escherichia coli*. *In*: F.C. Neidhardt, J.L. Ingraham, K.B. Low, B. Magasanik, M. Schaechter and H.E. Umbarger (eds): *Escherichia coli and Salmonella typhimurium. Cellular and Molecular Biology*. American Society of Microbiology, Washington, pp 1326–1333.

Xiao, H., Kalman, M., Ikehara, K., Zemel, S., Glaser, G. and Cashel, M. (1991) Residual guanosine 3',5'-bispyrophosphate synthetic activity of *relA* null mutants can be eliminated by *spoT* null mutations. *J. Biol. Chem.* 266: 5980–5990.

Transgenic Organisms – Biological and Social Implications
J. Tomiuk, K. Wöhrmann & A. Sentker (eds)
© 1996 Birkhäuser Verlag Basel/Switzerland

Monitoring genetically modified organisms and their recombinant DNA in soil environments

K. Smalla and J.D. van Elsas[1]

Federal Research Centre for Agriculture and Forestry, Messeweg 11-12, D-38104 Braunschweig, Germany,
[1]IPO-DLO, Soil Biotechnology, P.O. Box 9060, NL-6700 GW Wageningen, The Netherlands

Summary. Important aspects of monitoring microorganisms introduced in soil environments are reviewed. Adequate, representative soil sampling is pinpointed as crucial for monitoring the persistence of GMOs and their potential effects. Different approaches to detect GMOs and the recombinant DNA as well as their applications for monitoring field tests are discussed with respect to their sensitivity, specificity and feasibility. Recently developed molecular tools to follow microbial community shifts and their application for studying effects of GMOs on the microbiota are critically evaluated.

Introduction

The large-scale application of genetically modified microorganisms in the environment, for bioremediation and improvement of plant growth and protection has raised concerns about potential environmental impacts. Assessment of potential risks associated with the environmental release of genetically modified organisms (GMO) requires adequate methods of monitoring the fate of the genetically modified organisms in the environment. The major challenge for the development of suitable monitoring techniques is the fact that only a minor fraction of the total bacterial community in the environment is accessible to cultivation techniques. Recently, highly sensitive and specific detection methods for GMOs, in particular molecular techniques have become available.

Monitoring of field sites after the deliberate release of GMOs will generally focus on two levels: (i) Monitoring the GMO, its persistence and establishment in the respective soil, its ability to colonize crops and the dissemination of the recombinant DNA within indigenous soil-bacteria; and (ii) monitoring of potential environmental impacts of the release which can be positive or negative. Impacts might be changes of the microbial community with respect to taxonomic composition or metabolic pattern in response to the released GMO.

The intention of this chapter is to briefly review important aspects of monitoring genetically modified organisms following their release: sampling, sample transport, processing and analysis (Fig. 1). Problems of soil sampling and most approaches for detection of microorganisms or

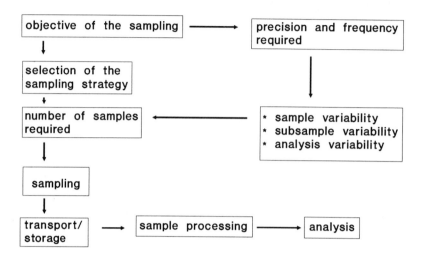

Figure 1. Aspects of monitoring GMOs.

specific DNA in soil environments are not specific for tracking GMOs but will be discussed. Hence most information given is of general importance for studies on the structure, function and activities of soil microbial communities and individual populations within those communities.

Sampling and transport

Adequate representative sampling of soils is crucial for tracking the survival and distribution of genetically modified microorganisms after a field release. Soil represents a highly complex and heterogeneous environment with respect to the soil matter and organism distribution. The extent of soil heterogeneity varies for each soil. The soil microbiota are localized in close association with soil particles, mainly clay–organic matter complexes and often embedded in a polysaccharide matrix (Foster, 1988). Heterogeneity of soil poses particular problems for any soil sampling. The samples obtained from a field site should be representative of the whole. The objective of the study and the required level of precision of the data will determine the sampling requirements (sampling strategy, sample size, number of samples needed). A decision has to be made on the type of samples to take, depending on the objectives of the sampling and the characteristics of the field site. Several sample types are distinguished (EPA 1992; Wollum, 1994): Judgement samples, simple random samples, stratified random samples, systematic samples and random samples taken in blocks (Fig. 2). Judgement samples are inherently biased and cannot be recommended for monitoring field released GMOs.

Pretests are required to estimate the variability and the expected distribution of the population under study. The number of samples needed can be determined by statistical methods, as outlined by McIntosh (1990), Rasch et al. (1992), and Wollum (1994). Both the type I error – the probability that an effect is detected which does not exist, and the type II error – the probability that an effect which exists is not detected, should be taken into consideration for monitoring GMOs. Obviously, effects due to GMO releases which are not detected might be more critical. The more samples available, the better the estimate of the mean. The number of samples taken is usually a compromise between statistical requirements and purely practical consideration. To increase the number of samples which still can be handled in the laboratory in a reasonable time and with justifiable cost, samples obtained randomly are often bulked and mixed. Subsamples of the resulting composite samples are subsequently analyzed (Hoffmann, 1991).

Drying of soils or roots or their exposure to sunlight might result in decreased numbers of culturable bacteria. Samples should therefore be transported in darkened containers and at temperatures and at a water content corresponding to those under field conditions. Extreme conditions

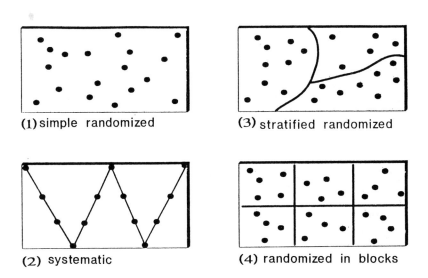

(1) simple randomized

(3) stratified randomized

(2) systematic

(4) randomized in blocks

Figure 2. Monitoring strategies.

should be avoided. The time between sampling, processing, and analysis should be as short as possible to avoid changes of the microbial composition and activity of the soil sample. Wollum (1994) reported that short-term storage (7–21 days) of soils did not result in changes of several soil properties such as total microbial biomass, enzymatic activities and bacterial counts.

Before further processing, soil samples are often sieved through a 2–4 mm mesh sieve to remove plant material, stones, small soil animals, and to homogenize the soil sample. Effects of sieving, storage, and incubation temperature on the phospholipid fatty acid (PLFA) profiles of a soil community have been studied by Petersen and Klug (1994). Significant shifts of the PLFA composition were observed only for storage at 25°C.

Several biotic and abiotic factors (see below) affect the survival and establishment of introduced GMOs. Furthermore, some of these factors might influence the degree of interaction of microorganisms or free DNA with the soil matrix. Recording of or information on the following biotic and abiotic factors might be of importance for monitoring the survival, persistence and gene transfer of microorganisms in soils or sediments (EPA, 1992):

Biotic factors

- Host microorganism (ability to compete with indigenous microorganisms, survival under field conditions, distribution in the field)
- Predators/parasites
- Vectors of microbial transport (for example earthworms)
- Type and variability of vegetation.

Abiotic factors

Physical factors

- Temperature
- Humidity
- Oxygen
- Proportion of organic substances
- Soil type (proportion of sand, clay, silt)
- Proportion of humic substances
- Pore size distribution.

Chemical factors

- pH
- Nutrient availability
- Conductivity
- Chemical contamination
- Cation exchange capacity
- Fungicide application.

Processing of soil samples

Before analyzing microorganisms in soil, rhizosphere or sediment samples using cultivation techniques, immunofluorescence microscopy or *in situ* hybridizations, microbial cells have to be dislodged from soil or sediment particles. The problems of the recovery of bacterial cells have been recently reviewed by Bakken and Lindahl (1995) and Smalla (1995a,b). Efficient recovery of cells may be problematic due to (i) heterogeneous distribution of microorganisms in different soil fractions, and (ii) different strength of interaction of microorganisms with the soil matrix, depending on the soil type. Introduced microorganisms are generally more easily extracted from soil than are indigenous bacteria (Hopkins et al., 1991). This observation might be an advantage when monitoring the fate of introduced GMOs. The microbial fraction can be extracted from the soil matrix using the following steps: (i) Dispersion of soil aggregates and dislodgement of microorganisms; (ii) separation of soil particles from the microbial fraction; and (iii) purification of the microbial fraction.

Dispersion of soil aggregates and dislodgement of cells can be achieved by resuspending soil or sediment samples in liquids such as sodium pyrophosphate (Strickland et al., 1988), saline or phosphate-buffered saline, and homogenization by shaking in shake flasks with beads or gravel, Waring or Stomacher blending. Furthermore, mild ultrasonication can be used to dislodge bacteria from the soil matrix (Strickland et al., 1988). Excessive ultrasonication was reported to destroy cells and reduce the culturability of bacterial cells (Ramsay, 1984). An efficient dispersion of soil particles and dislodging of cells can be achieved by cation exchange resins (DOWEX A1, CHELEX 100) combined with detergents and blending (MacDonald, 1986; Herron and Wellington, 1990, 1992; Hopkins et al., 1991; Hopkins and O'Donnell, 1992). Electrostatic bonds between soil particles as well as between soil particles and microorganisms are interrupted by exchanging polyvalent ions. In addition, detergents like deoxycholate are used to break up polymer bonds.

For cultivation techniques the dispersion of soil aggregates and dislodgement of cells from soil particles is usually achieved by stomaching or shaking with sterile glass beads or gravel and subsequent plating of serial dilutions of the cell–soil particle mixed suspension. However, the application of microscopic methods or total DNA or RNA extraction from the bacterial fraction requires the separation of the bacterial fraction from the soil particles. Separation can be achieved by blending and shaking followed by low speed centrifugation steps (Fægri et al., 1977; Bakken, 1985; Holben et al., 1988; Steffan et al., 1988). The recovery of microorganisms depends on the efficiency of disaggregation of soil particles as well as the loosening of bacteria from the soil surface. Cells are easier dislodged and separated from a sandy soil than from a clay soil. Even repeated homogenization/centrifugation steps were found not to recover 100% of the soil bacteria. Fægri et al. (1977) estimated that 50–80% of the soil bacteria could be recovered. Obviously, the recovery rate will be influenced not only by the soil type but also by the characteristics of the microorganisms. These problems have to be taken into consideration when comparing limits of detection for introduced GMOs.

Detection of the GMO and its DNA

The detection and enumeration of previously released GMOs presents many challenges. Molecular approaches for environmental monitoring of microorganisms have been recently reviewed by Atlas et al. (1992). Traits such as antibiotic resistances, bioluminescence or other enzymes encoded by the genetic construct can be used for the detection of the GMO in the presence of the indigenous microbial population. The availability of information on the genetically modified organism and the genetic modification (marker genes, promoter sequences) is a prerequisite for the development of specific detection techniques.

Two general approaches are used for detecting GMOs; (i) cultivation-based methods and (ii) direct methods obviating cultivation, such as total DNA extraction followed by analysis or immunofluorescence microscopy. It should be emphasized that often only a combination of different techniques can provide the information required for an adequate assessment of environmental releases of GMOs.

Cultivation-based detection of GMOs

Detection methodologies for GMOs often rely on cultivation techniques as the primary tool. The selection and enumeration of the GMO by plating dilution series of resuspended soil samples on media containing, for example, antibiotics is easy and simple. Reporter or selective marker genes are often used to facilitate counting the colony forming units (cfu) of the GMO. However, cultivation-based methods are known to be biased since only culturable bacteria are accessible.

Selective cultivation

The detection of the GMO by selective cultivation often takes advantage of antibiotic, heavy metal or herbicide resistances encoded by the GMO. The application of selective cultivation techniques improves the limit of detection since the natural background is reduced. Selectivity is of particular importance for the detection of GMOs present at low numbers. The most frequently used antibiotic resistance marker gene is the *npt*II gene conferring a kanamycin and neomycin resistance to its host. Although selective plating onto kanamycin-containing media only reduces the natural background from approx. 10^7 to $10^4 - 10^5$ cfu/g soil the GMOs can be unequivocally detected by colony hybridization using the *npt*II gene as a probe. Obviously, background kanamycin resistance in soil bacteria is encoded by other genes or cell wall impermeability since the *npt*II gene has not been detected in culturable soil bacteria so far (Smalla et al., 1993, 1995; Smalla, 1995 a,b). Plating of soil suspensions onto selective media might also be biased due to the fact that the resistance gene is expressed only by a proportion of the population containing the resistance gene. This may lead to an underestimation of the cfu number of normally resistant bacteria (Genthner et al., 1990; Tebbe et al., 1992). In a study on the stability of antibiotic resistance markers in *Bacillus cereus* Halverson et al. (1993) suggested that the stability of antibiotic resistance markers should be assessed in the ecosystem in which they will be studied.

The deliberate release of GMOs marked with antibiotic resistance genes is not desirable when the respective antibiotics are of medical importance. Spontaneous antibiotic-resistant mutants which are based on changes in cellular constituents rather than active resistance mechanisms, e.g., streptomycin, rifampicin, nalidixic acid resistant strains (Kluepfel et al., 1991) or genes encoding resistance to heavy metals (Bale et al., 1987) and herbicides (Ramos et al., 1991) may be preferred.

Reporter genes

Reporter genes are defined as genes conferring distinctive phenotypic properties which allow the marked organism to be tracked in the presence of the indigenous microbiota (Prosser, 1994). Efficient marker systems are extremely useful tools in microbial ecology since they make it possible to track specifically introduced bacteria in the presence of a natural background to study population dynamics, ability to colonize plant roots or gene dissemination (Kloepper and Beau-champ, 1992; Beauchamp et al., 1993).

Detection of GMOs containing reporter genes such as *luc*, *lux*, *lacZY*, *xylE* or *gusA* by plating onto selective media is highly sensitive and specific. GMOs containing reporter genes were used in several release experiments. Recently, molecular marker systems for the detection of geneti-cally engineered microorganisms in the environment were reviewed by Prosser (1994).

Stable maintenance and expression of the marker gene within the host are required for the reliable detection of the marked GMO. However, both maintenance and expression of a specific marker gene may vary depending on the host, its localization on the chromosome or a plasmid, and the environmental conditions (Cook et al., 1993; Prosser, 1994).

The *lacZY* genes from *E. coli* coding for a β-galactosidase and a lactose permease have been successfully exploited as marker genes for fluorescent pseudomonads (Drahos et al., 1988; Kluepfel et al., 1991, 1995; Kluepfel, 1993; Ryder et al., 1994; Wendt-Potthoff et al., 1994). The expression of the marker gene can be detected by plating on an X-Gal (5-chloro-4-bromo-3-indolyl β galactopyranosidase) resulting in blue colonies. Advantage was taken of the inability of pseudomonads to utilize lactose as sole carbon source. Long-term field releases have been performed with genetically modified *Pseudomonas aureofaciens* strains containing chromoso-mally inserted *lacZY* flanked with the left and the right *Tn7* termini (Kluepfel et al., 1991, 1995; Kluepfel, 1993). The *Tn7:lacZY* chromosomal insert was shown to be stable under nonselective conditions. In addition to the natural ability to fluoresce under UV the root colonizing *P. aureo-faciens Tn7:lacZY* was resistant to rifampicin and nalidixic acid. The detection of the three diffe-rent phenotypes could be used for sensitive and specific detection of the GMO by selective plating. The same construct was used to monitor the survival of *P. corrugata* in wheat rhizo-sphere and soil in field tests in Australia (Millis, 1992). The ability of the *lacZY* modified *P. corrugata* to grow on lactose as sole source of carbon, its distinctive DNA restriction pattern, colony morphology and *in vitro* inhibition of take-all enabled an unequivocal detection of the strain. A sugarbeet phylloplane isolate of *P. aureofaciens* was chromosomally modified for monitoring purposes by the insertion of two gene cassettes, the *lacZY* and the Km^r-*xylE* coding for kanamycin resistance and a catechol 2,3-dioxygenase (Bailey, 1995). The marker genes were chosen to facilitate detection of the released strain in the sugar beet phyllosphere and rhizosphere by simple plating or enrichment methods. The detection system applied was sensitive, allowing one GMO/g leaf material to be identified using MPN (most probable number) assays and selec-tive cultivation. The GMO was isolated from a background of 10^{12} cfu/g. The *xylE* reporter gene has been used to mark different Gram-negative and Gram-positive bacteria for tracking the fate of the marked strains under environmental conditions (Morgan et al., 1989; Winstanley et al., 1991; Wipat et al., 1991). Thermoregulated expression of the *xylE* gene was applied to reduce the metabolic burden due to the expression of the marker gene under environmental conditions. To study *in situ* gene transfer from indigenous soil bacteria, a genetically modified *Rhizobium legumi-nosarum* biovar *viciae* marked with the *gus*A gene has been used in a field release in England (Hirsch et al., 1995).

Other authors used the *lux*AB and *lac*ZY reporter genes which allow viable cells to be detected by light emission or blue colony colour in the presence of n-decanal and X-Gal, respectively, to monitor the fate of introduced GMOs in soil habitats (Flemming et al., 1994). Bioluminescence marker systems are particularly valuable since negligible background was found in terrestrial and fresh water ecosystems. Two different bioluminescence systems have been successfully applied to monitor the fate of introduced microorganisms, the bacterial luciferase encoded by the *lux*AB genes from *Vibrio* species (Boivin et al., 1988; Rattray et al., 1990; de Weger et al., 1991) and the eukaryotic firefly luciferase (*luc*) (Palomares et al., 1989; Selbitschka et al., 1992; Cebolla et al., 1993; Möller et al., 1994). Flemming et al. (1994) found *lux*AB reporter genes to be suitable for tracking a GMO in soil, because they are stable under soil conditions, easily and rapidly assayed, unique to the GMO and pose little metabolic burden to its host. Molecular biology of bacterial bioluminescence was recently reviewed by Meighen (1991). Bioluminescent GMO can be detected by several means, including visual identification, luminometry, X-ray-film and CCD imaging (Prosser, 1994). Bioluminescence offers considerable advantages over other marker systems in that it can provide means of *in situ* detection in environmental samples. However, luminescence only provides a reliable *in situ* measure of viable cell counts when cells are active, either utilizing substrates available or following cultivation (Meikle et al., 1992; Duncan et al., 1994). Luminescence of starved (modified) *E. coli* and *Vibrio harveyi* strains was reported to fall below background levels as assessed by luminometry (Duncan et al., 1994). Cebolla et al. (1993) and Keller et al. (1995) used the firefly *luc* gene for stable tagging of *Rhizobium meliloti* strains studied in soil habitats. Light is emitted through ATP-dependent conversion of luciferin in the presence of oxygen.

Gene probes and PCR

The survival, establishment and dissemination of GMOs introduced into soil environments can be specifically followed at the DNA level. Most GMOs carry unique DNA stretches which make their specific detection possible by means of gene probes or the polymerase chain reaction. Therefore, DNA–DNA hybridization can be applied to confirm the presence of the genetic construct in colonies expressing a certain antibiotic resistance or respective reporter genes. Colonies or overnight broth of picked colonies or isolated DNA are transferred to a membrane (e.g., Nylon

Hybond). The cells are lysed by enzymatic attack (lysozyme, mutanolysin or lysostaphin) and denatured using SDS and alkaline conditions. The single-stranded DNA is subsequently fixed to the membrane using UV or 2 hours treatment at 80°C. DNA–DNA hybridizations are performed under high stringency conditions using single-stranded ^{32}P or nonradioactively (digoxigenin) labeled gene probes (Sambrook et al., 1989). Gene probes consist mostly of parts of or the whole genetic construct. Dot blot cell hybridizations enable the simultaneous analysis of 94 picked colonies. GMOs containing DNA segments which do not naturally occur in microorganisms such as the *luc* gene of the American firefly (Selbitschka et al., 1992) or the patatin gene of potato (van Elsas et al., 1991) or the bovine aprotinin gene (Tebbe and Vahjen, 1993) can be unequivocally detected by DNA–DNA hybridization. Specific detection might be also possible when the respective DNA segment is absent in the microbial population of the habitat under investigation, e.g., the *lux* gene was detected only in marine vibrios, which should allow the fate of a *lux*-containing GMO to be followed in soil environments (Shaw et al., 1992; Prosser, 1994). However, when the natural presence of DNA homologous to the gene probe in the microbial soil population cannot be excluded, hybridization-positive colonies should be confirmed by PCR (polymerase chain reaction). The application of PCR in environmental microbiology has been reviewed by Steffan and Atlas (1991). PCR permits *in vitro* DNA amplification by melting the double-stranded DNA (94–95°C), primer annealing (50–60°C, depending on the optimal annealing temperature) and primer extension at 72°C by the action of a Taq polymerase. Defined gene segments can be amplified to detectable amounts by repeats of the process (usually 20–35 times) in a thermal cycler.

The specific and unequivocal detection of recombinant DNA is made possible by an appropriate selection of adequate primer systems, even in the presence of the naturally occurring genes. Information on the sequence of the construct and/or the flanking region as well as the uniqueness of these as compared to other sequences are prerequisites for a successful primer design.

Strains containing the construct are compared with the original host using hybridization or PCR-based fingerprinting techniques (for review see, Tichy and Simon, 1995). In case the resulting pattern differs from the original host, identification using BIOLOG or 16S rDNA sequencing would identify the recipient of the construct.

Immunological detection

Alternative methods for detecting GMOs are immunology based (Schmidt, 1974; van Vuurde and van der Wolf, 1995). Prerequisite for immunological approaches is the availability of specific antibodies (specific for the bacterial host or a gene product encoded by the recombinant DNA).

Furthermore, the antigen should be expressed stably and at levels sufficient for detection. The absence of the antigen in the indigenous microbial community is needed to ensure a high specificity of the antigen–antibody-reaction for the GMO. GMOs have been detected by immunological methods using colony or Western blots or ELISA (enzyme-linked immunosorbent assays). Antibodies coupled to magnetic beads have been applied for selective recovery of *Pseudomonas putida* cells (Morgan et al., 1991). Immunological methods for the detection of GMOs recovered by cultivation have been less frequently applied, compared to selective plating combined with reporter gene or gene probing. However, immunological methods might be valuable tools for tracking the fate of GMOs producing nonbacterial proteins, e.g., interferon, insulin or aprotin (Tebbe, 1995).

Direct detection methods obviating cultivation

A well-known problem in microbial ecology is that only a minor fraction of bacteria is commonly accessible after a cultivation step (Roszak and Colwell, 1987; Torsvik et al., 1990). The development of detection techniques obviating cultivation steps, such as (i) microscopic methods using specifically labeled antibodies or oligonucleotide probes or (ii) direct extraction of total community DNA or RNA from environmental samples, and subsequent analysis has given new insights into microbial diversity.

Microscopic methods

Several authors used antibodies labeled by fluorescence for direct counting of GMOs by immunofluorescence microscopy (Brettar and Höfle, 1992; van Elsas et al., 1991). Immunofluorescence-based detection of GMOs recovered from soil is impaired by laborious recovery of cells, sometimes high background fluorescence and a detection limit around $10^4 - 10^5$ cells per gram of soil (van Elsas et al., 1991).

In situ hybridization of whole cells using fluorescently labeled oligonucleotides targeted to the 16S rRNA or 23S rRNA (for review see, Amann et al., 1995) allows for the detection of microorganisms in their natural microhabitat. The microscopic identification of individual cells provides information on the cell morphology, the spatial distribution, and the growth rate, independently of their culturability. However, the detection of cells recovered from soil by *in situ* hybridization might be difficult due to a low rRNA content of dormant cells, and background problems. *In situ* hybridization can only be used for tracking GMOs when the introduced microorganism is absent in the indigenous microflora and the cells remain metabolically active. The sensitivity of oligo-nucleotides targeting the gene modification still has to be increased to allow the specific *in situ* detection of the genetic construct.

Direct DNA extraction from soil samples

Direct DNA extractions became a valuable tool for tracking the fate of GMOs in soil environ-ments because they allow detection of the construct (i) in GMOs which became nonculturable due to environmental stress; (ii) in bacteria which are not accessible to cultivation techniques; (iii) persisting as free DNA adsorbed to soil particles.

The methodology of well-established protocols for nucleic acid extraction from bacterial iso-lates has been adapted to application on environmental samples. Critical points for nucleic acid extraction from environmental samples, e.g., difficulties in lysing Gram-positive bacteria or spores, are well-known from the work with bacterial isolates. Methods of nucleic acid extraction from environmental samples have two approaches: (i) The cells are lysed directly within the environmental sample; and (ii) the cells are lysed after recovery of the bacterial fraction from soil or sediment particles.

Direct lysis

Direct extraction of total microbial community DNA is often based on the original protocol of Ogram et al. (1987). Several protocols have since then been published for direct DNA extraction from soil (Steffan et al., 1988; Pillai et al., 1991; Porteous and Armstrong, 1991; Tsai and Olson, 1991, 1992; Picard et al., 1992; Selenska and Klingmüller, 1992; Dijkmans et al., 1993; Smalla et al., 1993). A general scheme illustrating the different steps needed for direct DNA extraction is seen in Figure 3. Soil is directly subjected to cell lysis conditions using, for instance, freezing/thawing, ultrasonication, microwave, bead beater and/or lysozyme treatment steps followed by alkaline SDS treatment. The DNA is usually recovered after phenol, phenol–chloro-form extraction or potassium acetate precipitation steps. The degree of shearing of DNA and the degree of contamination with coextracted substances (humic and fulvic acids) is not only deter-mined by the protocol applied but also by the soil type. Several methods based on gel electropho-resis or gel filtration, salt precipitation, ion exchange or cesium chloride/ethidium bromide gradi-ents have been published for purification of crude DNA (Porteous and Armstrong, 1991; Tsai and Olson, 1991, 1992; Smalla et al., 1993; Tebbe and Vahjen, 1993; Young et al., 1993). The purpose of purification procedures is to remove humic acids and other impurities which impair

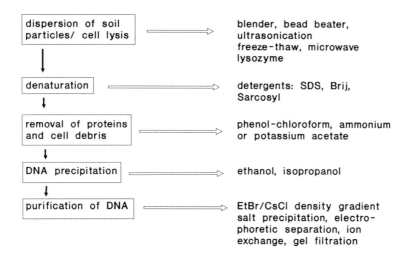

Figure 3. Scheme of the direct DNA extraction from soil.

the efficient application of molecular techniques, and to attain quantitative DNA recovery. Presently, there is no general method for purification of soil DNA which can successfully be applied to each environmental sample.

Incomplete lysis of cells, especially of so-called dwarf cells or spores (Moré et al., 1994), and losses of DNA as a consequence of several purification steps might be pinpointed as the main limitation of the approach. The DNA recovered by direct lysis contains fungal and nonmicrobial DNA in addition to bacterial DNA. Bacterial DNA obtained by direct lysis of cells in environmental samples can originate from metabolically active cells but also from dormant or dead cells or free DNA adsorbed to soil particles. The analysis of directly extracted DNA by gene probing or PCR is the most sensitive way to follow the presence of free recombinant DNA which might persist adsorbed to clay minerals over long periods of time (for review see, Lorenz and Wackernagel, 1994).

Recovery of the bacterial fraction followed by nucleic acid extraction

The bacterial fraction is usually recovered by application of mechanical agitation using blending, shaking or ultrasonication and low speed centrifugation steps as described above. Isolation of bacterial DNA from soil after prior recovery of the bacterial fraction was pioneered by Torsvik (1980). Several groups have used the DNA extraction from the bacterial fraction (Holben et al., 1988; Steffan et al., 1988; Cresswell et al., 1991; Jacobsen and Rasmussen, 1992). Compared to direct lysis protocols, DNA extraction from bacterial fraction is more time-consuming and laborious while less DNA is recovered. However, the DNA obtained is less contaminated by co-extracted substances and should mainly consist of bacterial DNA. Incomplete dislodgement of cells from soil or sediment particles and lysis and loss of DNA during different extraction and/or purification steps are known to be critical.

Total community DNA obtained by both approaches can be analyzed by dot or slot blot hybridizations (limit of detection $10^4 - 10^6$ cfu/g soil), using specific probes or after specific amplification by PCR (limit of detection 10 to 5×10^3 cfu/g). Detection of GMOs in total community DNA extracts by gene probing and by PCR has been applied by several authors (e.g., Steffan and Atlas, 1988; Cresswell et al., 1991; Wipat et al., 1991; Genthner et al., 1992; Picard et al., 1992; Smalla et al., 1993; Tebbe et al., 1992; Flemming et al., 1994; Wendt-Potthoff et al., 1994). Reported limits of detection (for review see, Tebbe, 1995) are difficult to compare since they are based on cfu/g or cells or target sequences. Furthermore they might be, independent of the extraction and purification protocol applied, influenced by several factors such as soil type, species of introduced microorganism, inoculation method, time of inoculation, number of target sequences per cell and PCR conditions (efficiency of the PCR).

Monitoring environmental impacts of GMOs

The development of appropriate methodologies to measure potential environmental impacts of GMOs is still in its infancy. Ecological effects of GMOs have been studied for a few years only (for review see, Seidler, 1992; Leung et al., 1994). Recently, Seidler (1992) evaluated research from several laboratories on the detection of ecological effects from GMOs. A list of endpoints for studying perturbations caused by GMOs was evaluated. Perturbations induced by GMOs were usually transient (Seidler, 1992). Parameters that can be monitored to study soil ecosystems have also been summarized by Smit et al. (1992) (Tab. 1).

Effects caused by the release of GMOs could range from (i) displacement of certain species; (ii) major changes in the community structure and function; (iii) perturbation of ecological balance; (iv) accumulation of toxic metabolites or (v) increased microbial activity due to nutrient input (Smit et al., 1992). In contrast to methods for detecting GMOs and their DNA where it is relatively easy to determine the specificity and sensitivity of the detection method, the situation is more complicated for methods to measure ecological impacts. The intellectual challenge is to

Table 1. Parameters that can be monitored in study of the soil ecosystem (after Smit et al., 1992)

---

* Microbial populations (based on cultivation):
  via plating on specific media to enumerate certain groups of microorganisms, enumeration of microorganisms by most probable number (MPN) techniques.
  These methods can be combined with the use of specific DNA probes.

* Microscopical counts of total bacteria or fungi with general dyes, specific species or taxonomic groups using antibodies or oligonucleotides coupled to fluorescent dyes.

* Microbial processes:
  Substrate-induced respiration
  Respiration
  Nitrification
  Denitrification
  Sulphur oxidation

* Enzyme activities

* Biomass determinations

* ATP measurements

* DNA (RNA) extraction:
  Diversity determinations
  Specific sequence detection with probes or PCR

* Lipopolysaccharide (LPS) extraction and analysis

* Muramic acid analysis

* Fatty acid and lipid analysis

* Food-web interactions

---

identify correct and appropriate experimental endpoints. The practical challenge is to develop methods sensitive enough to detect taxonomic or functional shifts of the microbial community. On the other hand, such methods should not require too much cost and time, allowing their application for monitoring large-scale field releases of GMOs. Furthermore, detected shifts in terms of taxonomic or functional changes of the microbial community have to be evaluated in an ecological context, e.g., of community shifts caused by normal agricultural practice such as ploughing (which has a major impact on inundation).

Important parameters to be measured should be determined according to the released organism and its expected potential impacts. Measurements of microbial biomass, total community enzyme activities or specific microbial activities like denitrification or nitrification have been used to evaluate effects of pesticides on the soil microbial community (Alef, 1991). Crawford et al. (1993) demonstrated that inoculation of soils with a recombinant *Streptomyces lividans* increased the rate of organic carbon mineralization, affected the pH value, nitrogen cycling, the relative population of some microbial groups, and enzyme activities. Although statistically significant, Crawford et al.

(1993) questioned the ecological impact due to the transiency of the effects. Plate counts, measurement of CO_2 and N_2O emissions in soils inoculated with a genetically modified and a nonmodified *Rhizobium leguminosarum* strain as well as effects on the mycorrhizae of pea roots were chosen as experimental endpoints to monitor environmental impacts of a field release in Italy (Nuti). Significant difference could not be detected between soils inoculated with the GMO or the nonmodified organism. Recently, new methods have been developed for studying shifts of microbial communities which might have the potential to become useful tools for monitoring environmental impacts of GMOs. Community level approaches have been developed based on the analysis of fatty acid methyl esters (Federle et al., 1986; Bååth et al., 1992; Haack et al., 1994; Peterson and Klug, 1994) or BIOLOG metabolic pattern (Garland and Mills, 1991, 1994; Winding, 1994; Zak et al., 1994; Ellis et al., 1995) from the whole community. Furthermore, total community DNA or RNA analysis using hybridization studies (Ritz and Griffiths, 1994), low-molecular-weight-RNA analysis (Höfle, 1992) or profiling of microbial populations by denaturing gradient gel electrophoresis of PCR amplified segments of the 16S rDNA (Muyzer et al., 1993) might be applicable for monitoring taxonomic shifts within microbial communities. Other methods for measuring the soil microbial diversity have been recently reviewed by Leung et al. (1994). The methods have to be assessed with respect to their accuracy, sensitivity and interpretation. Metabolic fingerprinting using BIOLOG MicroPlates has been assessed as a method for measuring perturbation of microbial communities associated with sugarbeets from a GMO inoculation (Ellis et al., 1995).

Conclusion

The development of new molecular techniques for the detection of microorganisms and their DNA in environmental samples has greatly improved our abilities to specifically track the fate of GMOs, the persistence of their DNA, and gene transfer processes at high levels of sensitivity. The analysis of directly isolated microbial community DNA enables us to follow the persistence of the recombinant DNA independent from cultivation. Since most methods described in this chapter suffer from biases, multiphasic approaches should be favoured. However, the potential effects of GMOs on the indigenous microbiota are difficult to assess. The development of new tools to enable sensitive and reliable measurement of structural and functional changes in microbial communities is presently in progress. The application of these new approaches will not only prove extremely useful for evaluating risks associated with the deliberate release of GMOs but will also allow new insights and a better understanding of microbial ecology.

Acknowledgements
This work was supported by the EU BIOTECH grant BIO2CT-920491.

References

Alef, K. (1991) *Methodenhandbuch Bodenmikrobiologie*. ecomed, Landsberg/Lech.

Amann, R.I., Ludwig, W. and Schleifer, K.-H. (1995) Phylogenetic identification and *in situ* detection of individual microbial cells without cultivation. *Microbiol. Rev.* 59: 1, 143–169.

Atlas, R.M., Sayler, G., Burlage, R.S. and Bej, A.K. (1992) Molecular approaches for environmental monitoring of microorganisms. *BioTechniques* 12: 706–717.

Bååth, E., Frostegård, Å. and Fritze, H. (1992) Soil bacterial biomass, activity, phospholipid fatty acid pattern, and pH tolerance in an area polluted with alkaline dust deposition. *Appl. Environ. Microbiol.* 58: 4026–4031.

Bailey, M.J. (1995) Extraction of DNA from the phyllosphere. *In*: J.T. Trevors and J.D. van Elsas (eds): *Nucleic Acids in the Environment*. Springer-Verlag, Heidelberg, pp 89–109.

Bakken, L.R. (1985) Separation and purification of bacteria from soil. *Appl. Environ. Microbiol.* 49: 1482–1487.

Bakken, L.R. and Lindahl, V. (1995) Recovery of bacterial cells from soil. *In*: J.T. Trevors and J.D. van Elsas (eds): *Nucleic Acids in the Environment*. Springer-Verlag, Heidelberg, pp 9–27.

Bale, M.J., Fry, J.C. and Day, M.J. (1987) Plasmid transfer between strains of *Pseudomonas aeruginosa* on membrane filters attached to river stones. *J. Gen. Microbiol.* 133: 3099–3107.

Beauchamp, C.J., Kloepper, J.W. and Lemke, P.A. (1993) Luminometric analyses of plant root colonization by bioluminescent pseudomonads. *Can. J. Microbiol.* 39: 434–441.

Boivin, R., Chalifour, F.-P. and Dion, P. (1988) Construction of a Tn5 derivative encoding bioluminescence and its introduction in *Pseudomonas*, *Agrobacterium* and *Rhizobium*. *Mol. Gen. Genet.* 213: 50–55.

Brettar, I. and Höfle, M.G. (1992) Influence of ecosystematic factors on survival of *Escherichia coli* after large-scale release into lake water mesocosms. *Appl. Environ. Microbiol.* 58: 2201–2210.

Cebolla, A., Ruiz-Berraquero, F. and Palomares, A.J. (1993) Stable tagging of *Rhizobium meliloti* with the firefly luciferase gene for environmental monitoring. *Appl. Environ. Microbiol.* 59: 2511–2519.

Cook, N., Silcock, D.J., Waterhouse, R.N., Prosser, J.I., Glover, L.A. and Killham, K. (1993) Construction and detection of bioluminescent strains of *Bacillus subtilis*. *J. Appl. Bacteriol.* 75: 350–359.

Crawford, D.L., Doyle, J.D., Wang, Z., Hendricks, C.W., Bentjen, S.A., Bolton, H. Jr., Fredrickson, J.K. and Bleakley, B.H. (1993) Effects of a lignin peroxidase-expressing recombinant, *Streptomyces lividans* TK23.1, on biogeochemical cycling and the numbers and activities of microorganisms in soil. *Appl. Environ. Microbiol.* 59: 508–518.

Cresswell, N., Saunders, V.A. and Wellington, E.M.H. (1991) Detection and quantification of *Streptomyces violaceolatus* plasmid DNA in soil. *Letters Appl. Microbiol.* 13: 193–197.

de Weger, L.A., Dunbar, P., Mahafee, W.F., Lugtenberg, B.J.J. and Sayler, G.S. (1991) Use of bioluminescence markers to detect *Pseudomonas* spp. in the rhizosphere. *Appl. Environ. Microbiol.* 57: 3641–3644.

Dijkmans, R., Jagers, A., Kreps, S., Collard, J.-M. and Mergeay, M. (1993) Rapid method for purification of soil DNA for hybridization and PCR analysis. *Microbial Releases* 2: 29–34.

Drahos, S.J., Barry, G.F., Hemming, B.C., Brandt, E.J., Skipper, H.D., Kline, E.L., Kluepfel, D.A., Hughes, T.A. and Gooden, D.T. (1988) Pre-release testing procedures: US field test of a *lacZY*-engineered soil bacterium. *In*: M. Sussman, C.H. Collins, F.A. Skinner and D.E. Stewart-Tull (eds): *Release of Genetically Engineered Microorganisms*. Academic Press, London, pp 181–191.

Duncan, S., Glover, L.A., Killham, K. and Prosser, J.I. (1994) Luminescence-based detection of activity of starved and viable but nonculturable bacteria. *Appl. Environ. Microbiol.* 60: 1308–1316.

Ellis, R.J., Thompson, I.P. and Bailey, M.J. (1995) Metabolic profiling as a means of characterizing plant-associated microbial communities. *FEMS Microbiol. Ecol.* 16: 9–18.

EPA (1992) Monitoring small-scale field tests of microorganisms. USEPA.

Fægri, A., Torsvik, V.L. and Goksøyr, J. (1977) Bacterial and fungal activities in soil: Separation of bacteria and fungi by a rapid fractionated centrifugation technique. *Soil Biol. Biochem.* 9: 105–112.

Federle, T.W., Livinston, R.J., Wolfe, L.E. and White, D.C. (1986) A quantitative comparison of microbial community structure of estuarine sediments from microcosms and the field. *Can. J. Microbiol.* 32: 319–325.

Flemming, C.A., Lee, H. and Trevors, J.T. (1994) Bioluminescent most-probable-number method to enumerate lux-marked *Pseudomonas aeruginosa* UG2Lr in soil. *Appl. Environ. Microbiol.* 60: 3458–3461.

Foster, R.C. (1988) Microenvironments of soil microorganisms. *Biol. Fertil. Soils* 6: 189–203.

Garland, J.L. and Mills, A.L. (1991) Classification of heterotrophic microbial communities on the basis of patterns of community-level sole-carbon-source utilization. *Appl. Environ. Microbiol.* 57: 2351–2359.

Garland, J.L. and Mills, A.L. (1994) A community-level physiological approach for studying microbial communities. *In*: K. Ritz, J. Dighton and K.E. Giller (eds): *Beyond the Biomass*. British Society of Soil Science, Wiley-Sayce, pp 77–83.

Genthner, F.J., Upadhyay, J., Campbell, R.P. and Sharak Genthner, B.R. (1990) Anomalies in the enumeration of starved bacteria on culture media containing nalidixic acid and tetracycline. *Microbial Ecol.* 20: 283–288.

Genthner, F.J., Campbell, R.P. and Pritchard, P.H. (1992) Use of a novel plasmid to monitor the fate of a genetically engineered *Pseudomonas putida* strain. *Mol. Ecol.* 1: 137–143.

Haack, S.K., Garchow, H., Odelson, D.A., Forney, L.J. and Klug, M.J. (1994) Accuracy, reproducibility, and interpretation of fatty acid methyl ester profiles of model bacterial communities. *Appl. Environ. Microbiol.* 60: 2483–2493.

Halverson, L.J., Clayton, M.K. and Handelsman, J. (1993) Variable stability of antibiotic-resistance markers in *Bacillus cereus* UW85 in the soybean rhizosphere in the field. *Mol. Ecol.* 2: 65–78.

Herron, P.R. and Wellington, E.M.H. (1990) New method for extraction of *Streptomyces* spores from soil and application to the study of lysogeny in sterile amended and nonsterile soil. *Appl. Environm. Microbiol.* 56: 1406–1412.

Herron, P.R. and Wellington, E.M.H. (1992) Extraction of Streptomyces spores from soil and detection of rare gene transfer events. *In*: E.M.H. Wellington and J.D. van Elsas (eds): *Genetic Interactions Among Microorganisms in the Natural Environment.* Pergamon Press Ltd., pp 91–103.

Hirsch, P., Mendum, T., Jording, D. and Selbitschka, W. (1995) Field release of genetically modified *Rhizobia* in the U.K. *In*: D.D. Jones (ed.): *Proceedings of the 3rd International Symposium on the Biosafety Results of Field Tests of Genetically Modified Plants and Microorganisms,* 13–16 November 1994, Monterey, California, pp 407–417.

Höfle, M.G. (1992) Bacterioplankton community structure and dynamics after large-scale release of nonindigenous bacteria as revealed by low-molecular-weight-RNA analysis. *Appl. Environ. Microbiol.* 58: 3387–3394.

Hoffmann, G. (1991) Verband Deutscher Landwirtschaftlicher Untersuchungs- und Forschungsanstalten: *Methodenbuch Band I – Die Untersuchung von Böden.* VDLUFA-Verlag.

Holben, W.E., Jansson, J.K., Chelm, B.K. and Tiedje, J.M. (1988) DNA probe method for the detection of specific microorganisms in the soil bacterial community. *Appl. Environ. Microbiol.* 54: 703–711.

Hopkins, D.W., Macnaughton, S.J. and O'Donnell, A.G. (1991) A dispersion and differential centrifugation technique for representatively sampling microorganisms from soil. *Soil Biol. Biochem.* 23: 217–225.

Hopkins, D.W. and O'Donnell, A.G. (1992) Methods for extracting bacterial cells from soil. *In*: E.M.H. Wellington and J.D. van Elsas (eds): *Genetic Interactions Among Microorganisms in the Natural Environment.* Pergamon Press Ltd., pp 104–112.

Jacobsen, C.S. and Rasmussen, O.F. (1992) Development and application of a new method to extract bacterial DNA from soil based on separation of bacteria from soil with cation-exchange resin. *Appl. Environm. Microbiol.* 58: 2458–2462.

Keller, M., Selbitschka, W., Dammann-Kalinowski, T., Niemann, S., Hagen, M., Dresing, U., Quester, I., Pühler, A., Tichy, H.V., Simon, R., Schwieger, F. and Tebbe, C. (1995) Field release of two bioluminescent *Rhizobium meliloti* strains: Survival, spread and interaction with the indigenous soil microflora. *In*: D.D. Jones (ed.): *Proceedings of the 3rd International Symposium on the Biosafety Results of Field Tests of Genetically Modified Plants and Microorganisms,* 13–16 November 1994, Monterey, California, pp 419–433.

Kloepper, J.W. and Beauchamp, C.J. (1992) A review of issues related to measuring colonization of plant roots by bacteria. *Can. J. Microbiol.* 38: 1219–1232.

Kluepfel, D.A., Kline, E.L., Skipper, H.D., Hughes, T.A., Gooden, D.T., Drahos, D.J., Barry, G.F., Hemming, B.C. and Brandt, E.J. (1991) The release and tracking of genetically engineered bacteria in the environment. *Phytopathology* 81: 348–352.

Kluepfel, D.A. (1993) The behavior and tracking of bacteria in the rhizosphere. *Annu. Rev. Phytopathol.* 31: 441–472.

Kluepfel, D.A., Lamb, T.G., Snyder, W.E. and Tonkyn, D.W. (1995) Six years of field testing a *lac*ZY modified fluorescent pseudomonad. *In*: D.D. Jones (ed.): *Proceedings of the 3rd International Symposium on the Biosafety Results of Field Tests of Genetically Modified Plants and Microorganisms,* 13–16 November 1994, Monterey, California, pp 169–176.

Leung, K., England, L.S., Cassidy, M.B., Trevors, J.T. and Weir, S. (1994) Microbial diversity in soil: Effect of releasing genetically engineered microorganisms. *Mol. Ecol.* 3: 413–422.

Lorenz, M.G. and Wackernagel, W. (1994) Bacterial gene transfer by natural genetic transformation in the environment. *Microbiol. Rev.* 58: 563–602.

MacDonald, R.M. (1986) Sampling soil microfloras: Dispersion of soil by ion exchange and extraction of specific microorganisms from suspension by elutriation. *Soil Biol. Biochem.* 18: 399–406.

McIntosh, M.S. (1990) Statistical techniques for field testing of genetically engineered microorganisms. *In*: M. Levin and H. Strauss (eds): *Risk Assessment in Genetic Engineering – Environmental Release of Organisms.* McGraw-Hill Inc., New York, pp 219–239.

Meighen, E.A. (1991) Molecular biology of bacterial bioluminescence. *Microbiol. Rev.* 55: 123–142.

Meikle, A., Killham, K., Prosser, J.I. and Glover, L.A. (1992) Luminometric measurement of population activity of genetically modified *Pseudomonas fluorescens* in the soil. *FEMS Microbiol. Letters* 99: 217–220.

Millis, N.F. (1992) Australian experience in the release of live modified organisms. *In*: R. Casper and J. Landsmann (eds): *Proceedings of the 2nd International Symposium on the Biosafety Results of Field Tests of*

Genetically Modified Plants and Microorganisms. Biologische Bundesanstalt für Land- und Forstwirtschaft, Braunschweig, Germany, pp 81–89.

Möller, A., Gustafsson, K. and Jansson, K. (1994) Specific monitoring by PCR amplification and bioluminescence of firefly luciferase gene-tagged bacteria added to environmental samples. *FEMS Microbiol. Ecol.* 15: 193–206.

Moré, M.I., Herrick, J.B., Silva, M.C., Ghiorse, W.C. and Madsen, E.L. (1994) Quantitative cell lysis of indigenous microorganisms and rapid extraction of microbial DNA from sediment. *Appl. Environm. Microbiol.* 60: 1572–1580.

Morgan, J.A.W., Winstanley, C., Pickup, R.W., Jones, J.G. and Saunders, J.R. (1989) Direct phenotypic and genotypic detection of a recombinant pseudomonad population released into lake water. *Appl. Environ. Microbiol.* 55: 2537–2544.

Morgan, J.A.W., Winstanley, C., Pickup, R.W. and Saunders, J.R. (1991) Rapid immunocapture of *Pseudomonas putida* cells from lake water by using bacterial flagella. *Appl. Environ. Microbiol.* 57: 503–509.

Muyzer, G., de Waal, E.C. and Uitterlinden, A.G. (1993) Profiling of complex microbial populations by denaturing gradient gel electrophoresis analysis of polymerase chain reaction-amplified genes coding for 16S rRNA. *Appl. Environ. Microbiol.* 59: 695–700.

Ogram, A., Sayler, G.S. and Barkay, T.J. (1987) DNA extraction and purification from sediments. *J. Microbiol. Meth.* 7: 57–66.

Palomares, A.J., DeLuca, M.A. and Helinski, D.R. (1989) Firefly luciferase as a reporter enzyme for measuring gene expression in vegetative and symbiotic *Rhizobium meliloti* and other Gram-negative bacteria. *Gene* 81: 55–64.

Petersen, S.O. and Klug, M.J. (1994) Effects of sieving, storage, and incubation temperature on the phospholipid fatty acid profile of a soil microbial community. *Appl. Environ. Microbiol.* 60: 2421–2430.

Picard, C., Ponsonnet, C., Paget, E., Nesme, X. and Simonet, P. (1992) Detection and enumeration of bacteria in soil by direct DNA extraction and polymerase chain reaction. *Appl. Environ. Microbiol.* 58: 2717–2722.

Pillai, S.D., Josephson, K.L., Bailey, R.L., Gerba, C.P. and Pepper, I.L. (1991) Rapid method for processing soil samples for polymerase chain reaction amplification of specific gene sequences. *Appl. Environ. Microbiol.* 57: 2283–2286.

Porteous, L.A. and Armstrong, J.L. (1991) Recovery of bulk DNA from soil by a rapid, small-scale extraction method. *Curr. Microbiol.* 22: 345–348.

Prosser, J.I. (1994) Molecular marker systems for detection of genetically engineered microorganisms in the environment. *Microbiol.* 140: 5–17.

Ramos, J.L., Duque, E. and Ramos-Gonzalez, M.I. (1991) Survival in soils of an herbicide-resistant *Pseudomonas putida* strain bearing a recombinant TOL plasmid. *Appl. Environ. Microbiol.* 57: 260–266.

Ramsay, A.J. (1984) Extraction of bacteria from soil: Efficiency of shaking or ultrasonication as indicated by direct counts and autoradiography. *Soil Biol. Biochem.* 16: 475–481.

Rasch D., Guiard, V. and Nürnberg, G. (1992) *Statistische Versuchsplanung: Einführung in die Methoden und Anwendung des Dialogsystems*. G. Fischer, Stuttgart, Jena, New York.

Rattray, E.A.S., Prosser, J.I., Killham, K. and Glover, L.A. (1990) Luminescence-based nonextractive technique for *in situ* detection of *Escherichia coli* in soil. *Appl. Environ. Microbiol.* 56: 3368–3374.

Ritz, K. and Griffiths, B.S. (1994) Potential application of a community hybridization technique for assessing changes in the population structure of soil microbial communities. *Soil Biol. Biochem.* 26: 963–971.

Roszak, D.B. and Colwell, R.R. (1987) Survival strategies of bacteria in the natural environment. *Microbiol. Rev.* 51: 365–379.

Ryder, M.H., Pankhurst, C.E., Rovira, A.D., Correll, R.L. and Ophel Keller, K.M. (1994) Detection of introduced bacteria in the rhizosphere using marker genes and DNA probes. *In*: F. O'Gara, D. Dowling and B. Boesten (eds): *The Molecular Ecology of the Rhizosphere*. VCH Publishers, Weinheim, Germany, pp 83–98.

Sambrook, J., Fritsch, E.F. and Maniatis, T. (1989) *Molecular Cloning: A Laboratory Manual*, Second Edition. Cold Spring Harbor, New York.

Schmidt, E.L. (1974) Quantitative autecological study of microorganisms in soil be immunofluorescence. *Soil Science* 118: 141–149.

Seidler, R.J. (1992) Evaluation of methods for detecting ecological effects from genetically engineered microorganisms and microbial pest control agents in terrestrial systems. *Biotech. Adv.* 10: 149–178.

Selbitschka, W., Pühler, A. and Simon, R. (1992) The construction of *rec*A-deficient *Rhizobium meliloti* and *R. leguminosarum* strains marked with *gus*A or *luc* cassettes for use in risk-assessment studies. *Mol. Ecol.* 1: 9–19.

Selenska, S. and Klingmüller, W. (1992) Direct recovery and molecular analysis of DNA and RNA from soil. *Microbial Releases* 1: 41–46.

Shaw, J.J., Dane, F., Geiger, D. and Kloepper, J.W. (1992) Use of bioluminescence for detection of genetically engineered microorganisms released into the environment. *Appl. Environ. Microbiol.* 58: 267–273.

Smalla, K. (1995a) Extraction of microbial DNA from sewage and manure slurries. *In*: A.D.L. Akkermans, J.D. van Elsas and F.J. de Bruijn (eds): *Molecular Microbial Ecology Manual*. Kluwer Academic Publishers, Dordrecht, pp 1–11.

Smalla, K. (1995b) Probensammeln und Probenvorbereiten. *In*: Dechema e.V., Frankfurt (ed.): *Materialien und Basisdaten für gentechnisches Arbeiten und für die Errichtung und den Betrieb gentechnischer Anlagen, Band 5: Monitoring*, pp 139–184.

Smalla, K., van Overbeek, L.S., Pukall, R. and van Elsas, J.D. (1993) Prevalence of *npt*II and *Tn5* in kanamycin-resistant bacteria from different environments. *FEMS Microbiol. Ecol.* 13: 47–58.

Smalla, K., Gebhard, F., van Elsas, J.D., Matzke, A. and Schiemann, J. (1995) Bacterial communities influenced by transgenic plants. *In*: D.D. Jones (ed.): *Proceedings of the 3rd International Symposium on the Biosafety Results of Field Tests of Genetically Modified Plants and Microorganisms*, 13–16 November 1994, Monterey, California, pp 157–167.

Smit, E., van Elsas, J.D. and van Veen, J.A. (1992) Risks associated with the application of genetically modified microorganisms in terrestrial ecosystems. *FEMS Microbiol. Rev.* 88: 263–278.

Steffan, R.J. and Atlas, R.M. (1988) DNA amplification to enhance detection of genetically engineered bacteria in environmental samples. *Appl. Environ. Microbiol.* 54:2185–2191.

Steffan, R.J., Goksøyr, J., Bej, A.S. and Atlas, R.M. (1988) Recovery of DNA from soils and sediments. *Appl. Environ. Microbiol.* 54: 2908–2915.

Steffan, R.J. and Atlas, R.M. (1991) Polymerase chain reaction: Application in environmental microbiology. *Annu. Rev. Microbiol.* 45: 137–161.

Strickland, T.C., Sollins, P., Schimel, D.S. and Kerle, E.A. (1988) Aggregation and aggregate stability in forest and range soils. *Soil Seience Soc. Amer. J.* 52: 829–833.

Tebbe, C.C., Ogunseitan, O.A., Rochelle, P.A., Tsai, Y.-L. and Olson, B.H. (1992) Varied responses in gene expression of culturable heterotrophic bacteria isolated from the environment. *Appl. Microbiol. Biotechnol.* 37: 818–824.

Tebbe, C.C. and Vahjen, W. (1993) Interference of humic acids and DNA extracted directly from soil in detection and transformation of recombinant DNA from bacteria and a yeast. *Appl. Environ. Microbiol.* 59: 2657–2665.

Tebbe, C.C. (1995) Durchführung und Ergebnisse von Monitoringprogrammen. *In*: Dechema e.V., Frankfurt (ed.): *Materialien und Basisdaten für gentechnisches Arbeiten und für die Errichtung und den Betrieb gentechnischer Anlagen, Band 5: Monitoring*, pp 575–624.

Tichy, H.-V. and Simon, R. (1995) Nukleinsäuretechniken. *In*: Dechema e.V., Frankfurt (ed.): *Materialien und Basisdaten für gentechnisches Arbeiten und für die Errichtung und den Betrieb gentechnischer Anlagen, Band 5: Monitoring*, pp 237–301.

Torsvik, V.L. (1980) Isolation of bacterial DNA from soil. *Soil Biol. Biochem.* 12: 15–21.

Torsvik, V., Goksøyr, J. and Daae, F.L. (1990) High diversity in DNA of soil bacteria. *Appl. Environ. Microbiol.* 56: 782–787.

Tsai, Y.L. and Olson, B.H. (1991) Rapid method for direct extraction of DNA from soil and sediments. *Appl. Environ. Microbiol.* 57: 1070–1074.

Tsai, Y.L. and Olson, B.H. (1992) Rapid method for separation of bacterial DNA from humic substances in sediments for polymerase chain reaction. *Appl. Environ. Microbiol.* 58: 2292–2295.

van Elsas, J.D., van Overbeek, L.S. and Fouchier, R. (1991) A specific marker, *pat*, for studying the fate of introduced bacteria and their DNA in soil using a combination of detection techniques. *Plant Soil* 138: 49–60.

van Vuurde, J.W.L. and van der Wolf, J.M. (1995) Immunofluorescence colony staining. *In*: A.D.L. Akkermans, J.D. van Elsas and F.J. de Bruijn (eds): *Molecular Microbial Ecology Manual*. Kluwer Academic Publishers, Dordrecht, pp 1–19.

Wendt-Potthoff, K., Backhaus, H. and Smalla, K. (1994) Monitoring the fate of genetically engineered bacteria sprayed on the phylloplane of bush beans and grass. *FEMS Microbiol. Ecol.* 15: 279–290.

Winding, A. (1994) Fingerprinting bacterial soil communities using Biolog microtitre plates. *In*: K. Ritz, J. Dighton and K.E. Giller (eds): *Beyond the Biomass*. British Society of Soil Science, Wiley-Sayce, pp 85–94.

Winstanley, C., Morgan, J.A.W., Pickup, R.W. and Saunders, J.R. (1991) Use of *xyl*E marker gene to monitor survival of recombinant *Pseudomonas putida* populations in lake water by culture on non-selective media. *Appl. Environ. Microbiol.* 57: 1905–1913.

Wipat, A., Wellington, E.M.H. and Saunders, V.A. (1991) *Streptomyces* marker plasmids for monitoring survival and spread of streptomycetes in soil. *Appl. Environ. Microbiol.* 57: 3322–3330.

Wollum, A.G. II (1994) Soil sampling for microbiological analysis. *In*: R.W. Weaver, S. Angle, P. Bottomley, D. Bezdicek, S. Smith, Tabatabai, A. and A. Wollum (eds): *Methods of Soil Analysis Part 2 – Microbiological and Biochemical Properties*. Soil Science Society of America, Inc., pp 1–14.

Young, C.C., Burghoff, R.L., Keim, L.G., Minak-Bernero, V., Lute, J.R. and Hinton, S.M. (1993) Polyvinylpyrrolidone-agarose gel electrophoresis purification of polymerase chain reaction-amplifiable DNA from soils. *Appl. Environ. Microbiol.* 59: 1972–1974.

Zak, J.C., Willig, M.R., Moorhead, D.L. and Wildman, H.G. (1994) Functional diversity of microbial communities: A quantitative approach. *Soil Biol. Biochem.* 26: 1101–1108.

Transgenic Organisms – Biological and Social Implications
J. Tomiuk, K. Wöhrmann & A. Sentker (eds)
© 1996 Birkhäuser Verlag Basel/Switzerland

Recent advances in ecological biosafety research on the risks of transgenic plants: A trans-continental perspective

I.M. Parker and D. Bartsch[1]

Department of Botany, University of Washington, Seattle, USA
[1]*Department of Biology V (Ecology, Ecotoxicology and Ecochemistry), Technical University of Aachen, Worringer Weg 1, D-52056 Aachen, Germany*

Summary. Until recently, ecological tests of invasiveness for transgenic plants have investigated traits with limited ecological relevance, and so the fact that most have shown no additional risk of invasion is not surprising. Similarly, results to date provide us with little confidence in the opinion held by many in the biotechnology industry that genetic engineering is unconditionally safe. The experimental biosafety research has resulted in a situation where companies in some countries have gained permission to grow potentially "risky" transgenic plants on a commercial scale, without substantial ecological data quantifying the risk they pose. Perhaps experiments on transgenic organisms so far have been more useful in generating new questions and new concepts of ecological risk assessment than in providing generalisations about organisms themselves. In cases where predictions of risk have a high degree of uncertainty, or if the sheer volume of field releases overwhelms this stage of biosafety assessment, monitoring potential ecological long-term effects is the final method of choice. This situation underscores the need for solid monitoring programs in the future.

Introduction

After nearly a decade of theoretical discussion, ecologists are beginning to come to a consensus about how the ecological risks of genetically engineered organisms (GEOs) should be assessed (Mooney and Bernardi, 1990; Rissler and Mellon, 1993; Tiedje et al., 1989; Williamson, 1992, 1993). However, empirical work in this area has lagged behind, giving us little opportunity so far to ask how well our ideas hold up under the challenge of the real process of risk assessment. There are a variety of reasons for this lag in empirical studies. The first is that the development of new GEOs has followed an exponential trajectory (Kareiva, 1993), and given the time-intensive nature of ecological experiments, ecological assessment of these new products is falling far behind. Secondly, there is a "catch-22" involved in risk assessment of GEOs: The truly risky products are the ones for which we most need ecological information, yet they are also the products for which it is most difficult to get permission to do the kinds of studies necessary to assess risk (Parker and Kareiva, 1995). A third problem with current biosafety research, at least in the United States, is that most studies are conducted by the biotechnology companies themselves, whose scientists are often trained as agronomists rather than ecologists. The studies presented by these companies to argue against risk are usually descriptive and often anecdotal, and even very

Table 1. Petitions for deregulation of transgenic crop species in the United States

	FLAVR SAVR TOMATO Calgene 5/92	VIRUS-RESISTANT SQUASH Upjohn 7/92	BXN COTTON Calgene 7/93	ROUNDUP-READY SOYBEAN Monsanto 9/93	LAURATE-OILSEED RAPE Calgene 3/94	B.t. POTATO Monsanto 9/94	B.t.k. COTTON Monsanto 1/95
Total pages	36	41	96	64	140	384	269
Pages on weediness	7.5	2	12	0.5	35	3.5	4
Pages on hybridization	1.5	2	2.5	1.5	22	14	19

INVASIVENESS/WEEDINESS

Explicitly defined "weed"	+	−	+	−	+	−	−
Used Baker's list (Baker, 1965) to argue against a risk of weediness	+	+	+	+	+	−	+
Used "not reported as a weed" to argue against risk	+	+	+	−	+	+	−
Used varation for similar traits in this or other species to argue against risk	−	+	+	+	+	−	−
Quantitative data on volunteerism in field trails	−	−	+	−	+	−	−
Some form of ecological (in contrast to agronomic) data collected from field trails	−	−	+	+	+	−	+
Direct experiments to assess weediness	0	0	2	0	5	3	1

HYBRIDISATION

Presence of weedy or naturalized relatives of the crop	+	+	+	+	+	+	+
"No known reports of hybrids" as an argument against risk	+	−	+	+	+	−	−
Quantitative data on frequency of interspecific crossing	−	−	−	−	+	+	+
Direct experiments to assess weedines of modified hybrids	0	1	0	0	2	0	0
Used ability to control with herbicides or tillage to argue against risk	−	+	+	+	+	+	+

Characteristics of and information contained in each of the seven petitions for deregulation of genetically engineered crops submitted to the United States Department of Agriculture. Information alluded to in a qualitative way without documentation was scored as "−". Sources: Asgrow (1992), Calgene (1992, 1993, 1994), Monsanto (1993, 1994, 1995). Taken from Parker and Kareiva (in press).

good studies tend to be presented without appropriate statistics or sample sizes, etc., which makes them difficult to interpret (Tab. 1; see also Parker and Kareiva, 1995). Finally, logistical constraints plague biosafety research both in the US and in Europe, although these constraints are quite different. In the US, most biosafety research is conducted or partially funded by biotechnology companies, a situation which represents an inherent conflict of interests; in Europe, ecological biosafety experiments are ironically the target of frequent sabotage by radical anti-biotechnology activists. Given the current climate for biosafety research, it is not surprising that ecological work has been slow to answer the important questions we have about the environmental safety of genetically novel organisms.

The aim of this chapter is to summarise current approaches and results of ecological biosafety research. First, we discuss the relevant ecological attributes of transgenic plants; that is, what should be the focus of biosafety research? Secondly, we review recent studies on the ecology of transgenic plants, pointing out their strengths and weakness as well as their general findings. Finally, we consider the future of biosafety research – in particular, the role of monitoring – in the post-commercialisation era.

Relevant attributes of transgenic plants

In assessing the risk of genetically modified plants, one consideration that must be addressed is the effect of the modified trait on non-target organisms. This is primarily a concern for plants that are engineered to produce some novel substances, for example *Bacillus thuringiensis* (*B.t.*) endo-toxin. Regarding the effects of these novel substances on non-target organisms, scale is an important issue. Using *B.t.* endotoxin (see Braun, this volume) as an example, it is often stated that because *B.t.* is a "natural" product, and because it is currently used in its natural form on organic farms to control pests, plants producing *B.t.* must not represent an additional risk. However, levels of *B.t.* incorporated in the soil from plant remains expressing the *B.t.* gene may be an order of magnitude higher than one would obtain from traditional use (P. Kareiva, unpublished data). Companies marketing GEOs generally do attempt to assess the effects on non-target organisms (Jones, 1995); however, their experimental methods tend to focus on short-term effects and to neglect impacts on micro- and macro-fauna for the long-term or at the appropriate scale (e.g., Monsanto, 1994).

Aside from non-target effects, the main ecological concern over genetic engineering in agriculture is whether it will create new invasion problems, and the rest of this article will focus on the question of invasiveness. Transgenic plants are considered potentially risky if they contain a trait that confers a large fitness advantage in natural situations. The trait may transform the target plant itself into an ecological threat, or it may move into closely related weeds and exacerbate current problems.

Tiedje et al. (1989) present a comprehensive review of attributes of GEOs and of potential host environments that should be considered in risk assessment procedures. For transgenic plants, phenotypic changes that represent a cause for concern include: (i) A shift in environmental limits to growth or reproduction and (ii) an increase in resistance to disease, parasitism, or herbivory.

These phenotypic changes should result in increased fitness and may also boost the plant's tendency toward invasiveness or weediness. Such traits would result in a rise in the finite rate of population increase, a composite measure that can reflect a variety of ecological factors. One example would be the enhancement of competitive ability (Crawley, 1990, 1992), another would be "ecological release" from natural enemies (predators, diseases, etc.), which has long been postulated as a major cause for the success of invasive exotic species (Dobson and May, 1986; Pimentel, 1986; Schierenbeck et al., 1994).

There are different ways to measure invasiveness. One way, first suggested by R. Manasse (Rissler and Melon, 1993), uses "replacement studies", in which known proportions of transgenic and non-transgenic seeds are sown and then the proportion of transgenic plants is monitored in later generations. If the transgene confers invasiveness, it should spread rapidly through the experimental population. A second way to quantify invasiveness is to measure it directly using the finite rate of increase, which for an annual plant is expressed as the number of seeds obtained next year as a function of the number sown (and should incorporate seed bank, overwintering, etc.; Crawley et al., 1993). Again because of the "catch-22" of tight regulations on overwintering of transgenic plants, few studies have quantified invasiveness using either replacement studies or direct measurement. In the limited number of cases where one environmental factor clearly prohibits invasion, one may be able to test simply for an effect of the transgene on that factor. For example, in some regions of the United States, potato is unable to overwinter because of the harsh freezing conditions. For these regions (but not in regions where potato does successfully overwinter), one may conclude that a trait with no effect on cold tolerance should not increase the invasiveness of potato (Monsanto, 1994).

In the design of experimental studies of invasiveness, one should consider two points related to the performance of a modified organism. The first is that care should be taken to investigate conditions where the modified trait offers a real ecological advantage. In the case of herbicide resistance, plants should be tested in habitats influenced by herbicide use (examples are railway rights of way and street embankments); pathogen resistance should be tested in the presence of the pathogen, and so on. The second point is that experiments should move toward testing explicitly whether differences between plant breeding lines are due to the recombinant DNA, or to some aspect of the genetic background. For example, in a 1993 field trial, Bartsch et al. (1995)

Table 2. Characteristics, strengths and weaknesses of some recent examples of ecological biosafety research

Study	Target plant	Trait	Strength of study	Weakness of study
Crawley et al., 1993	Oilseed rape	Herbicide resistance	Quantified lambda	Trait with limited ecological significance
			Long time scale (3 years)	No selective advantage of the experimental conditions
			Spatial scale and replication	
Parker and Kareiva, in press	Oilseed rape	Seed oil modification	Quantified lambda	Trait with limited ecological significance
			Compared transgenic to parental line	Unclear selective advantage of the experimental conditions
			Comparative greenhouse and field studies	
BRIDGE 1991, 1995	Oilseed rape Sugar beet	Herbicide resistance	Measured competitiveness	Trait with limited ecological significance
			Explicitly considered genetic background of oilseed rape	No selective advantage of the experimental conditions
				Did not quantify lambda
Anonymous 1991, 1993a,d (Maribo); Madsen, 1994	Sugar beet	Herbicide resistance	Measured competitiveness in the field and in a wild beet habitat	Trait with limited ecological significance
			Explicitly considered genetic background	No selective advantage of the experimental conditions
				Did not quantify lambda
Bartsch et al., 1995	Sugar beet	Virus resistance	Ecologically relevant trait	Did not quantify lambda
			Measured competitiveness	
			Explicitly considered genetic background	

found that transgenic lines of beet (*Beta vulgaris*) consistently performed less well than parental lines. They hypothesised that this result might be due to a two-fold increase in inbreeding caused by the process of generating the transgenic line. When new, fully hybrid lines were created with perfectly equivalent genetic backgrounds for transgenic and non-transgenic lines, the transgenic line showed an ecological advantage in the environment where such an advantage would be predicted.

Another factor to be considered is the presence of wild, weedy or feral relatives, and the possibility of outcrossing with the transgenic crop (Anonymous, 1991, 1993a; Bartsch et al., 1993; de Vries et al., 1992; Raybould and Gray, 1993). Much of the early biosafety research has concerned outcrossing distances of transgenic pollen, the use of "trap crops", etc. (Anonymous, 1991, 1993a; Klinger et al., 1992; Manasse, 1992; Morris et al., 1994). Although deregulation petitions often include this kind of information (Monsanto, 1994), it is more appropriate for asking about the safety of field trials – meaning our ability to isolate them – than the long-term safety of a modified organism. Under the conditions and scale of commercial release, gene flow into wild relatives will occur if it possibly can occur. Therefore in this case, biosafety research must also deal with the ecological behaviour of outcrossed transgenic relatives.

Has past biosafety research on invasiveness considered the relevant attributes of transgenic plants?

As we discussed above, there are different ways to assess invasiveness in a genetically modified crop, and there are several factors that should play a role in the experimental design of invasiveness studies. How well have studies to date handled these critical factors? Only two crop species have been the target of extensive research on invasiveness potential so far, oilseed rape (*Brassica napus*) and sugar beet (*Beta vulgaris*). Therefore we concentrate our review on these two crops, highlighting both the strengths and weaknesses of current biosafety research (Tab. 2).

Oilseed rape

In the first, pioneering study on the potential invasiveness of a transgenic crop, Crawley et al. (1993) compared lines of non-transgenic oilseed rape with lines transformed for tolerance to the herbicide glufosinate-ammonium (and also for kanamycin resistance). Several aspects of this study, sponsored by the Prosamo group in the United Kingdom, made it a model for invasiveness studies. The finite rate of increase was used as a response variable, and was measured over three years in twelve different sites, under a barrage of experimental treatments. Then lambdas for transgenic and non-transgenic lines were compared using ANOVA statistics to ask whether the introduction of the transgene itself had a significant effect on oilseed rape demography. No significant advantage was found for the transgenic lines, and oilseed was able to replace itself (lambda > 1) only in cultivated treatments. The major weakness of the Prosamo study was that the transgenes had no explicit adaptive significance in the field. Because herbicide and kanamycin were not applied as treatments, the engineered traits would not be expected to have an effect on the ecological behaviour of oilseed rape. Although it addressed concerns that unexpected ecological changes might accompany the transformation process, this study must remain primarily a model experiment. However, the impressive size and time frame of the Prosamo study make it a useful resource for testing statistical approaches to quantifying uncertainty in risk assessment for invasion (Kareiva et al., in press). Monte Carlo simulations indicate that the probability of oilseed invading in the UK is very low without a substantial increase in fitness.

Further studies on transgenic oilseed rape have been carried out by the BRIDGE project sponsored by the European Community (BRIDGE, 1991, 1995). Several experiments on glufosinate-resistant plants were performed to assess aspects of invasiveness risk. The various scientific groups involved in BRIDGE used multiple lines, multiple sites or seasons and plants with identical genetic background. Because no herbicide treatment was used, no change of the competitive ability in the transgenic plants was found (e.g., Fredshavn et al., 1995). The release conditions were focused on agricultural fields.

In the US, much of the work on oilseed rape has been sponsored and co-ordinated by the Calgene Company, and has focused on Calgene's lines with modified seed oil content. Seeds were modified to produce either laurate, or high levels of stearate (see also Friedt and Ordon, this volume). Because oil content may determine many important aspects of seed ecology, such as dormancy, early establishment, and seed predation rates (Linder, 1994), Calgene chose to address concerns about invasiveness directly with experiments.

The first round of studies, by Linder and colleagues (Adler et al., 1993; Linder and Schmitt, 1994), concerned dormancy, germination, and seed viability in the field and under different temperature conditions in the greenhouse. Linder's work suggested that laurate-modified seeds may germinate earlier and that high stearate seeds may have a higher dormancy rate, with possible implications for population dynamics of oilseed rape. It was suggested to Calgene that they follow up with more field studies to determine whether these changes would be reflected in lifetime fitness. Kareiva and Parker then designed a set of experiments to assess the relative performance, over the remainder of the life cycle, of transgenic and parental lines of oilseed rape (Calgene, 1994; Parker and Kareiva, in press). As in Crawley et al. (1993), lambda was used to compare lines; however, modified lines were compared directly to the parental lines from which they were derived. Greenhouse experiments showed a slight increase in performance of the laurate line, but in field experiments in the southeastern US, including cultivated and uncultivated habitats and two planting seasons, all lines were equally unable to replace themselves or even survive to maturity. Frick (in Calgene, 1994) continued with similar experiments in Saskatchewan,

Canada, where oilseed rape commonly volunteers along roadsides and railways. In Saskatchewan experimental populations were able to replace themselves in several cases, although almost exclusively under the cultivated treatment (Parker and Kareiva, 1995). In 7 out of 8 cases, the parental line performed better than the transgenic (Calgene, 1994).

One weakness of Parker and Kareiva's (1995) experiments with oilseed rape was that because of a lack of multiple seed lots or multiple insertion events (Tab. 2), they were not able to eliminate the possibility that differences between the lines were caused by other genetically – or environmentally – determined factors (e.g., Damgaard and Loeschcke, 1993).

In summary, research with genetically engineered oilseed rape has generated initial guidelines for setting up invasiveness studies, and has provided us with some limited answers about the nature of invasiveness (Tab. 2). However, because neither herbicide tolerance nor seed oil modification represent the kind of genetic changes that concern us most, we must be careful not to extrapolate from these early findings to questions of risk for all transgenic crops. As we stated above, the types of traits of most concern are those which clearly modify the response to limiting environmental factors or increase resistance to natural enemies.

Sugar beet

Beet (*Beta vulgaris* L.) has provided another valuable model system for invasiveness studies (Anonymous, 1993c, 1994; Bartsch et al., 1994; BRIDGE, 1991, 1995); several characteristics contribute to its prevalence and importance. First, beet is an important cash crop in Europe and a common target of recombinant DNA technology. Secondly, gene flow has been demonstrated between cultivated sugar beets (*Beta vulgaris* subsp. *vulgaris* provar. *altissima* DÖLL) and wild beets (*Beta vulgaris* subsp. *maritima* ARCANG.), as evidenced by the introgression of the annual habit into cultivated beets (Boudry et al., 1993) and the corresponding introgression of genes from seed beet into wild beet populations (Santoni and Berville, 1992). Thirdly, crosses between wild beets and sugar beet breeding plants result in a hybrid form ("weed beet") that can bolt in a single season, while growing among biennial sugar beet varieties (Fig. 1). These annual weed beets are a serious problem in parts of Europe, including Belgium, Germany, and northern France, and recent reports suggest that weed beet may also be feral in California (Carsner, 1928; N. Ellstrand, personal communication). If the bolting plants are not removed immediately, stable weed beet complexes quickly form and are difficult to eradicate.

The BRIDGE group has assessed the invasiveness of herbicide-resistant beet in a way similar to their work on oilseed rape, described above (BRIDGE, 1991, 1995). In addition, Maribo Seed in Denmark has been involved in risk assessment of herbicide-resistant beet (Anonymous, 1991,

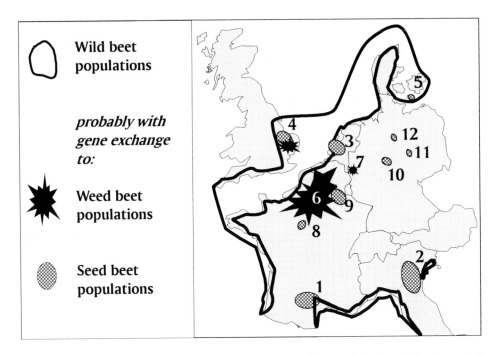

Figure 1. Overview: The distribution of breeding districts (with seed beet populations), agricultural areas (with weed beet populations) and natural habitats (with wild beet populations) is shown for some examples in Central Europe. An overlap of seed beet/wild beet populations is given in numbers 1–5, some weed beet populations are illustrated in 4, 6, and 7 and a few probably isolated locations of breeding companies are found in 8–12. Data are extracted from Boudry et al. (1993), Doney et al. (1990), Frese (1991), Hornsey and Arnold (1979), Letschert (1993), Longden (1974, 1989).

1993a, 1993d). Strengths of this work include fitness studies of comparable transgenic and nontransgenic hybrids on competitiveness in the field and in a coastal wild beet habitat.

Although the Maribo study included data on frost tolerance, these data were collected in the greenhouse and are not transferable to field conditions (Anonymous, 1993b, 1994). Young conventional and transgenic beets in the greenhouse experienced 100% mortality after growth at constant –5°C, whereas 1993–95 field studies with conventional beets in Germany showed that fluctuating daily temperatures reaching much lower than that (–1°C to –25°C produced only 10–90% mortality (D. Bartsch, unpublished data).

Bartsch et al. (1995) used beets as a model system to carry out the first rigorous investigation of a transgenic plant carrying a gene conferring protection from natural enemies. They compared the performance of transgenic beets resistant to beet necrotic yellow vein virus (BNYVV) to that of unmodified material from the same breeding line. They also grew these lines side by side with a conventionally bred variety carrying a similar phenotypic trait, the first ecological experiment to make this explicit comparison. Field experiments in 1993 and 1994 quantified seedling emergence and competitiveness during the early life stages, when BNYVV is likely to have some of its largest effects (Abe and Tamada, 1986). The potential ecological advantage of virus resistance was assessed under conditions of virus infection and no virus infection. At the site where virus was present, results showed a small but detecable ecological advantage of the genetically engineered trait. At the site where virus was absent, no significant difference between lines was found; that is, no "cost of resistance" was detected. Although genetically engineered virus resistance did confer a detectable ecological advantage, the conventional tolerant variety in fact performed best in all competition treatments in both years.

Finally, Bartsch et al. (1995) attempt to assess what the likely fitness consequence might be of introgression of the transgene into wild beets growing near Italian seed beet production areas (Fig. 1). It was hypothesised that wild beets in Italy, where BNYVV was first isolated (Faccioli and Giunchedi, 1974), would be most likely to suffer form BNYVV infection and therefore might benefit from the transfer of the transgene. However, in 1994 no virus infection was observed in a representative number of these coastal populations of wild beet, suggesting that an increase in wild beet fitness is unlikely at these sites – but not necessarily at all sites (Whitney, 1989).

The work of Bartsch et al. on beets represents the next big step in biosafety research, a step which takes us into an area where our experience with transgenic organisms may teach us as much about ecology as ecology has taught us about risk assessment of transgenic organisms.

Monitoring the commercialisation of transgenic crops

We are now at the stage where the first commercialisation of transgenic crops is taking place world-wide (Stone, 1994). In the US, several petitions for deregulation of transgenic crop lines have already been approved, and these products are already in production (Tab. 1). In Europe, virus-resistant tobacco, herbicide-resistant oilseed rape, and soon herbivore-resistant maize (producing *B.t.* endotoxin) will also be commercialised (Landsmann and Casper, 1995). In China, large-scale field trials of transgenic tobacco, tomatoes and rice on thousands of hectares are already common (Stone, 1994). Given the speed with which the development of transgenic crops is occurring, the future of ecological biosafety research will rest not just on risk assessment, but on the monitoring of previously released organisms. Because monitoring is a resource-intensive process, we will have to develop guidelines for prioritising releases and target the ecologically "riskier" organisms. If risk assessment procedures suggest a significant risk of invasion, but positive considerations (such as social or economic factors) outweigh these ecological concerns, commercial release should only occur if the benefits are great enough to justify and fund a sound monitoring program.

A complete monitoring program will require the collection of baseline data ("indirect monitoring"; Bartsch and Hücking, 1995), as well as "direct monitoring", which includes the measurement of factors influencing plant invasiveness. Indirect monitoring requires the identification of biological indicators and global system parameters that could be impacted by unexpected "escape" of transgenic crops or their transgenes (Fig. 2). Direct monitoring will include observation of the spread (or lack of spread) of transgenic organisms, and should provide direct quantitative data on plant multiplication in the field, similar to that used as the basis for risk analysis. Direct monitoring should also consider the spread of the genes into close relatives, and can include experimental manipulations like the introduction of pollen-sterile plants in an area to study pollen flow and hybridization with transgenic crops. A very useful survey of monitoring methods is given by Kjellsson and Simonsen (1994).

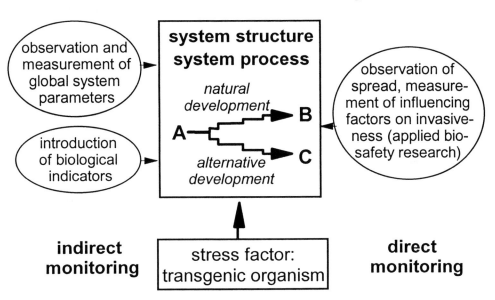

Figure 2. Possible influence of transgenic organisms on biological system dynamic. Ecosystem dynamics and evolution will probably change the system structure and processes (from level A to level B). It is necessary to differentiate the clear impact of transgenic plant ("stress factor"), e.g., the alternative system developement from level A to level C. Will the ecological long-term effect be the displacement of natural individuals with a loss of genetic variance in a wild polulation? And, more importantly, is the difference between the resulting levels B and C an ecological damage? Do we interpret a simple introgression of the genetically engineered trait in wild populations as no damage, if the genetic structure/variance has not been affected?

Conclusions and unanswered questions

Perhaps experiments on transgenic organisms so far have been more useful in generating new questions and new concepts of ecological risk assessment than in providing generalisations about organisms themselves. The studies on oilseed rape by Parker and Kareiva (1995) found support for two expectations regarding assessments of invasiveness. The first is that, for traits that should not show an explicit genotype-by-environment effect (i.e., that do not alter a response to biotic stress), greenhouse comparisons will be most sensitive and therefore conservative in assessing relative performance. Secondly, for traits that do alter a biotic stress, such as drought or frost response, herbivory, or disease, greenhouse studies will not do a good job of assessing the risk of invasion in the field. The Maribo beet experiments to quantify overwintering survival are an example of this principle (Anonymous, 1993b, 1994).

New results may alter the important debate over whether there is a fundamental difference between traits incorporated using traditional breeding methods and similar traits incorporated using the new technology (Mellon and Rissler, 1995; Regal, 1994; Williamson, 1992, 1993). One suggestion is that by passing the normal the evolutionary process, genetic engineering may be able to circumvent phylogenic, developmental, or physiological constraints. For example, it is generally believed that resistance to pathogens or protection from predators comes at some evolutionary "cost"; that is, organisms cannot have something for nothing (Bergelson, 1994). However, the recent studies done on transgenic beets appear to show no cost of virus resistance at all.

Until recently, ecological tests of invasiveness for transgenic plants have investigated traits with limited ecological relevance (Tab. 2), and so the fact that most have shown no additional risk of invasion is not surprising. Similarly, results to date provide us with little confidence in the opinion held by many in the biotechnology industry that genetic engineering is unconditionally safe. The "catch-22" of experimental biosafety research has resulted in a situation where companies in some countries have gained permission to grow potentially "risky" transgenic plants on a commercial scale, without substantial ecological data quantifying the risk they pose (Tab. 1; Parker and Kareiva, 1995). This situation underscores the need for solid monitoring programs in the future.

References

Abe, H. and Tamada, T. (1986) Association of beet necrotic yellow vein virus with isolates of *Polymyxa betae* Keskin. *Ann. Phytopath. Soc. Japan* 52: 235–247.

Adler, L.S., Wickler, K., Wyndham, P.S., Linder, C.R. and Schmitt, J. (1993) Potential for persistence of genes escaped form canola: Germination cues in crop, wild, and crop-wild hybrid *Brassica rapa. Funct. Ecol.* 7: 736–745.

Anonymous (1991) Reports on 1990 trials with transgenic glyphosate tolerant sugar beet. *Reports to the National Agency for Environmental Protection, Denmark.*

Anonymous (1993a) Reports on 1992 trials with transgenic glyphosate tolerant sugar beet. *Reports to the National Agency for Environmental Protection, Denmark.*

Anonymous (1993b) Report results from frost resistance trials with sugar beet (*Beta vulgaris* L.) transformed with glyphosate resistance genes. *Reports to the National Agency for Environmental Protection, Denmark.*

Anonymous (1993c) Competitive ability of transgenic sugar beet 1992–1993. *Report to the National Agency for Environmental Protection, Denmark.*

Anonymous (1993d) Report of trials with establishment of transgene sugarbeet. *Report to the National Agency for Environmental Protection, Denmark.*

Anonymous (1994) Report on results from indoor freezing trials 1993/94 with sugar beet (*Beta vulgaris* L.) transformed with glyphosate-tolerance genes. *Report to the National Agency for Environmental Protection, Denmark.*

Asgrow Seed Company (1992) *Petition for determination of regulatory status of Cucurbita pepo L. cultivar YC77E ZW-20.* United States Department of Agriculture, APHIS.

Baker, H. (1965) Characterictics and modes of origin of weeds. In: H.G. Baker and G.L. Stebbins (eds): *The Genetics of Colonizing Species.* Academic Press, New York. pp 147–168.

Bartsch, D., Sukopp, H. and Sukopp, U. (1993) Introduction of plants with special regard to cultigens running wild. In: K. Wöhrmann and J. Tomiuk (eds): *Transgenic Organisms: Risk Assessment of Deliberate Release.* Birkhäuser Verlag, Basel, pp 135–151.

Bartsch, D., Haag, C., Morak, C., Pohl, M. and Witte, B. (1994) Autecological studies of the competitiveness of transgenic sugar beets. *Verhandlungen der Gesellschaft für Ökologie (München-Weihenstephan)* 23: 435–444.

Bartsch, D., Pohl-Orf, M., Schmidt, M. and Schuphan, I. (1995) Naturalization of transgenic (BNYV-Virus resistant) sugar beet in agricultural and non-agricultural areas. *Proceedings of the 3rd International Symposium on the Biosafety Results of Field Test of Genetically Modified Plants and Microorganisms.* Monterey, California, Conference Proceedings, pp 353–361.

Bartsch, D. and Hücking, C. (1995) Future aspects of ecological biosafety research. In: J. Landsmann and R. Casper (eds): *Key Biosafety Aspects of Genetically Modified Organisms. Mitteilungen aus der Biologischen Bundesanstalt für Land- und Forstwirtschaft* 309: 71–79.

Bergelson, J. (1994) Changes in fecundity do not predict invasiveness: A model study of transgenic plants. *Ecology* 75: 249–252.

Boudry, P., Mörchen, M., Saumitou-Laprade, P., Vernet, P. and van Dijk, H. (1993) The origin and evolution of weed beets: Consequences for the breeding and release of herbicide resistant transgenic sugar-beets. *Theor. Appl. Genet.* 87: 471–478.

BRIDGE (1991) *Biotechnology R and D in the EC.* In: I. Economidis (ed.): *Biotechnology Action Programme (BAP). Part II. Detailed Final Report of BAP Contractors in Risk Assessment (1985–1990)* Commission of the European Communities, Brussels.

BRIDGE (1995) Safety Assessment of Genetically Modified Plants – BRIDGE 1992–1994. Practical Information and Programmes. Commission of the European Communities, Brussels.

Calgene (1992) *Petition for Determination of Nonregulated Status: Flavr Savr Tomato.* United States Department of Agriculture, APHIS.

Calgene (1993) *Petition for Determination of Nonregulated Status: BXN Cotton.* United States Department of Agriculture, APHIS.

Calgene (1994) *Petition for Determination of Nonregulated Status: Laurate Oilseed Rape.* United States Department of Agriculture, APHIS.

Carsner, E. (1928) The wild beet in California. *Facts Sugar* 23: 1120–1121.

Crawley, M.J. (1990) The ecology of genetically modified organisms. In: H.A. Mooney and G.A. Bernardi (eds): *Introduction of Genetically Modified Organisms into the Environment.* John Wiley and Sons, New York, pp 133–150.

Crawley, M.J. (1992) The comparative ecology of transgenic and conventional crops. In: R. Casper and J. Landsmann (eds): *Proceedings of the 2rd International Symposium on the Biosafety Results of Field Test of Genetically Modified Plants and Microorganisms.* May 11–14, 1992. Goslar, Germany (Biologische Anstalt für Land- und Forstwirtschaft, Braunschweig), pp 43–52.

Crawley, M.J., Hails, R.S., Rees, M., Kohn, D. and Buxton, J. (1993) Ecology of transgenic oilseed rape in natural habitats. *Nature* 363: 620–623.

Damgaard, C. and Loeschcke, V. (1993) Inbreeding depression and dominance-suppression competition after inbreeding in rapeseed (*Brassica napus*). *Theor. Appl. Genet.* 88: 321–323.

de Vries, E.T., van der Meijden, R. and Brandenburg, W.A. (1992) Botanical files – A study of the real chances for spontaneous gene flow from cultivated plants to the wild flora of the Netherlands. *Gorteria supplement* 1: 1–100.

Dobson, A.P. and May, R.M. (1986) Patterns of invasions by pathogens and parasites. *In*: H.A. Mooney and G.A. Bernardi (eds): *Ecology of Biological Invasions of North America and Hawaii*. Springer-Verlag, New York, pp 58–77.

Doney, D.L., Whitney, E.D., Terry, J., Frese, L. and Fitzgerald P. (1990) The distribution and dispersal of *Beta vulgaris* L. ssp. *maritima* germplasm in England, Wales, and Ireland. *J. Sugar Beet Res.* 27: 29–37.

Faccioli, G. and Giunchedi, L. (1974) On the viruses involved on rhizomania diseases of sugar beet in Italy. *Phytopath. mediter.* 13: 28–35.

Fredshavn, J.R., Poulsen, G.S., Huybrechts, I. and Rudelsheim, P. (1995) Competitiveness of transgenic oilseed rape. *Transgenic Res.* 4: 142–148.

Frese, L. (1991) Sammlung, Erhaltung und Nutzbarmachung der genetischen Ressourcen von Beta-Rüben (*B. vulgaris* L.) und Wurzelzichorien (*Cichorium intybus* L.). *Landbauforschung Völkenrode* 41: 65–73.

Hornsey, K.G. and Arnold, M.H. (1979) The origins of weed beet. *Ann. Appl. Biol.* 92: 279–285.

Jones, D. (1995) *Proceedings of the 3rd International Symposium on the Biosafety Results of Field Test of Genetically Modified Plants and Microorganisms*, Monterey, California Nov. 1994.

Kareiva, P. (1993) Transgenic plants on trial. *Nature* 363: 580–581.

Kareiva, P., Parker, I.M. and Pascual, M. How useful are experiments and models in predicting the invasiveness of genetically engineered organisms? *Ecology*; *in press*.

Kjellsson, G. and Simonsen, V. (1994) *Methods for Risk Assessment of Transgenic Plants: I. Competition, Establishment and Ecosystem Effects*. Birkhäuser Verlag, Basel.

Klinger, T., Arriola, P. and Ellstrand, N. (1992) Crop-weed hybridisation in radish (*Raphanus sativus*): Effects of distance and population size. *Amer. J. Bot.* 79: 1431–1435.

Landsmann, J. and Casper, R. (1995) *Key Biosafety Aspects of Genetically Modified Organisms. Mitteilungen aus der Biologischen Bundesanstalt für Land- und Forstwirtschaft* 309, Braunschweig, Germany.

Letschert, J.P.W. (1993) *Beta* section *Beta*: Biogeographical patterns of variation, and taxonomy. *Diss. Universiteit Wageningen*.

Linder, C.R. (1994) *The Ecology of Population Persistence for Wild, Crop, and Crop-Wild Hybrid Brassica and its Implication for Transgenes Escaped from Canola*. Ph.D. thesis. Brown University.

Linder, C.R. and Schmitt, J. (1994) Assessing the risks of transgene escape through time and crop-wild hybrid persistence. *Molec. Ecol.* 3: 23–30.

Longden, P.C. (1974) Sugar beet as a weed. *Proc. 12th British Weed Control Conference* 301–308.

Longden, P.C. (1989) Effects of increasing weed-beet density on sugar-beet yield and quality. *Ann. Appl. Biol.* 114: 527–532.

Madsen, K.H. (1994) *Weed Management and Impact on Ecology of Growing Glyphosate Tolerant Sugarbeets*, Ph.D. thesis. The Royal Veterinary and Agricultural University, Weed Science, Denmark.

Manasse, R. (1992) Ecological risks of transgenic plants: Effects of spatial dispersion on gene flow. *Ecol. Appl.* 2: 431–438.

Mellon, M. and Rissler, J. (1995) Transgenic crops: USDA data on small-scale tests contribute little to commercial risk assessment. *Bio/Technology* 13: 96.

Monsanto (1993) *Petition for Determination of Nonregulated Status for Glyphosate-Tolerant Soybean*. United States Department of Agriculture, APHIS.

Monsanto (1994) *Petition for Determination of Nonregulated Status for Potatoes Producing the Colorado Potato Beetle Control Protein of Bacillus thuringiensis subsp. tenebrionis*. United States Department of Agriculture, APHIS.

Monsanto (1995) *Petition for Determination of Nonregulated Status: Bollgard™ Cotton Lines 757 and 1076 (Gossypium hirsutum L) with the Gene from Bacillus thuringiensis subsp. kurstaki*. United States Department of Agriculture, APHIS.

Mooney, H.A. and Bernardi, G. (1990) *Introduction of Genetically Modified Organisms into the Environment*. John Wiley and Sons, New York.

Morris, W., Kareiva, P. and Raymer, P. (1994) Do barren zones and pollen traps reduce gene escape from transgenic crops? *Ecol. Appl.* 4: 157–165.

Parker, I.M. and Kareiva, P. Assessing the risks of invasion for genetically engineered plants: Acceptable evidence and reasonable doubt. *Biol. Conserv.*; *in press*.

Pimentel, D. (1986) Biological invasions of plants and animals in agriculture and forestry. *In*: H.A. Mooney and J.A. Drake (eds): *Ecology of Biological Invasions of North America and Hawai*. Springer-Verlag, New York, pp 149–162.

Raybould, A.F. and Gray, A.J. (1993) Genetically modified crops and hybridization with wild relatives: A UK perspective. *J. Appl. Ecol.* 30: 199–219.

Regal, P.J. (1994) Scientific principles for ecologically based risk assessment of transgenic organisms. *Molec. Ecol.* 3: 5–13.

Rissler, J. and Mellon, M. (1993) *Perils Amidst the Promise: Ecological Risk of Transgenic Crops in a Global Market*. Union of Concerned Scientists, Cambridge.

Santoni, S. and Berville, A. (1992) Characterization of the nuclear ribosomal DNA units and phylogeny of *Beta* L. wild forms and cultivated beets. *Theor. Appl. Genet.* 83: 533–542.

Schierenbeck, K.A., Mack, R.N. and Scharitz, R.R. (1994) Herbivore effects on *Lonicera* growth and biomass allocation. *Ecology* 75: 1661–1672.
Stone, R. (1994) Large plots are next test for transgenic crop safety. *Science* 266: 1472–1473.
Tiedje, J.M.R., Colwell, R.L., Grossman, Y.I., Hodson, R.E., Lenski, R.E., Mack, R.N. and Regal, P.J. (1989) The planned introduction of genetically engineered organisms: Ecological considerations and recommendations. *Ecology* 70: 298–315.
Whitney, E.D. (1989) Identification, distribution, and testing for resistance to rhizomania in *Beta maritima. Plant Disease* 73: 287–290.
Williamson, M. (1992) Environmental risks from the release of genetically modified organisms (GMOs) – The need for molecular ecology. *Molec. Ecol.* 1: 3–8.
Williamson, M. (1993) Invaders, weeds, and the risks from GMOs. *Experientia* 49: 219–224.

Transgenic Organisms – Biological and Social Implications
J. Tomiuk, K. Wöhrmann & A. Sentker (eds)
© 1996 Birkhäuser Verlag Basel/Switzerland

Modern *versus* classical plant breeding methods – efficient synergism or competitive antagonism?

W. Friedt and F. Ordon

Institute of Crop Science and Plant Breeding, Justus-Liebig-University, Ludwigstraße 23, D-35390 Giessen, Germany

Summary. Besides classical plant breeding methods based on Mendel's laws of heredity, new technologies, e.g., cell and tissue culture techniques, molecular markers (RFLPs, RAPDs) and gene transfer, have gained evident importance in plant breeding in the past years. These techniques may enhance the efficiency of plant breeding procedures and new breeding goals can be achieved with the help of gene technology. However, as all these novel techniques are in general only covering very special aspects of plant breeding, they will not entirely replace classical breeding methods. Instead, a combination of new techniques with conventional methods will lead to more efficient and successful breeding procedures. In the frame of this chapter classical and new techniques as well as the combination of both are described with special consideration of the use of gene technology in breeding oilseed rape.

Introduction

It is the task of plant breeding to create high yielding and stable cultivars with the quality traits demanded by the market within the environmental conditions and the agricultural practices given. Methods to achieve these goals by systematic plant breeding are mainly based on the foundations of the Mendelian laws of heredity and their rediscovery by Correns, de Vries and von Tschermak at the beginning of the 20th century. Classical plant breeding methods are mainly based on applied genetics. However, during the last decades cellular and molecular techniques have found their way into practical plant breeding and offer the opportunity for creating new breeding strategies. Starting with cell and tissue culture techniques, followed by molecular methods of genome analysis – e.g., Restriction Fragment Length Polymorphism (RFLPs) or Random Amplified Polymorphic DNA (RAPDs) – and on to the isolation and transfer of genes, these techniques are very useful tools which may enhance the efficiency of breeding and enable the plant breeder to create new goals and achieve them by the application of gene technology. Classical breeding methods will first be introduced with a few examples and in the second part the application of modern breeding methods – especially the use and the future prospects of gene technology – will briefly be described.

winter wheat

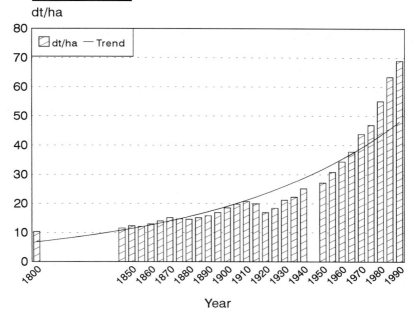

winter rye

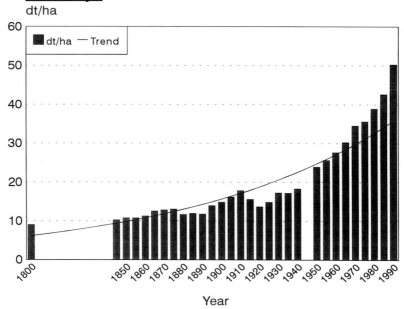

Figure 1. Yield increase in winter wheat and winter rye 1800–1993 (according to Bittermann, 1956; Anonymous, 1957–1994; Zschoche, personal communication).

Classical breeding methods

In simplified terms, breeding of cultivars mainly consists of the induction of genetic variation followed by its use in selection for the specific breeding goals relevant for the respective crop plant. Using these classical breeding methods, considerable yield increases (Fig. 1) as well as an improvement in resistance to biotic and abiotic stress factors resulting in enhanced yield stability were achieved. Furthermore, the quality of the plant products, e.g., the composition of seed protein or fat has been improved significantly.

Independently from botanical characteristics of a given crop and specific breeding methods resulting from these biological facts, the breeding procedure of a cultivar may be subdivided into three phases (Schnell, 1982): (i) Creation of genetic variation; (ii) selection of varietal parents; and (iii) testing and maintaining experimental varieties.

The resulting varieties can be grouped into four categories according to the propagational plant type and the mode of pollination, i.e., self- or cross-pollination: (i) Lines – generative propagation, self-pollination; (ii) populations – generative propagation, cross-pollination; (iii) hybrids – generative propagation, controlled crossing; and (iv) clones – vegetative propagation, self-/cross-pollination.

Induction of genetic variation

Genetic variation may be induced in different ways. Most frequently, intraspecific hybridizations i.e., crosses between individuals of the same species, are used in practical plant breeding, aiming at recombination of genes leading to novel genotypes which appear as new phenotypes in the offspring. However, in some cases there may be no or insufficient variation concerning traits of agronomical importance within species or genera. In this case variation may be created by hybridization between species (interspecific crosses) or even between genera (intergeneric crosses). But, it has to be taken into account that sexual crosses between species or genera are much more difficult to achieve than intraspecific crosses, due to different mechanisms inhibiting successful sexual hybridization between species and genera. These mechanisms may be circumvented by culturing immature embryos *in vitro* (embryo rescue) or – avoiding sexual crosses – "naked" somatic cells may be fused (protoplast fusion).

Besides recombination of genes in heterozygous individuals, induced mutations are another major source of genetic variation. In contrast to sexual crosses which result in new combinations of genes and alleles already existing, new genes or alleles (mutations) may be induced by different types of radiation and chemicals. Mutations are grouped into genome mutations (ploidy-mutations), chromosome mutations and gene mutations.

In all the cases mentioned above genetic variation required for selection is created by (i) joining together different sets of chromosomes; (ii) introducing single chromosomes from different alien species; or (iii) the change of individual genes at random, e.g., in the case of classical gene mutations. In contrast to these methods gene technology offers the opportunity for a selective transfer of single well characterized genes, which may not be present within the respective plant species or genus, in order to broaden the genetic variation. The use of gene technology in plant breeding will be reviewed later on. However, it has to be pointed out here, that gene technology is a very efficient and specific tool for the induction of novel genetic variation.

Selection of varietal parents and varieties

The phase of inducing genetic variation is followed by the selection of varietal parents and varieties, i.e., the restriction of the genetic variation in the direction of the respective breeding goals. Due to the characteristics of the different types of varieties mentioned above special selection procedures are needed which will be described in detail for line breeding and hybrid breeding in the following paragraphs.

Line breeding

In general genetic variation in self-pollinated crops, e.g., wheat, barley, oats, peas or flax, is induced by crosses between two parental lines in practical plant breeding. As these parental lines are in general homozygous varieties or advanced breeding lines, the F_1 consists of identical, heterozyous plants. Therefore, selection is not reasonable within crosses but may be applied on a limited scale between crosses. Starting with the first segregating generation (F_2) two completely different selection methods may be used (Fig. 2), i.e., the bulk population breeding method or the pedigree method.

The *pedigree method* normally starts with the selection of single plants, e.g., in the F_2 generation. Selection as early as in F_2 is only possible for simply inherited characters, e.g., monogenic traits. In subsequent generations selection is continued on between and within plant progenies (lines) as long as genetic variation is existing. In the F_5-F_6 generations when plants have reached a degree of homozygosity sufficient for yield tests, the offspring of a single plant, i.e., one row, is harvested completely and replicated yield tests are carried out in field plots at different locations and in subsequent years.

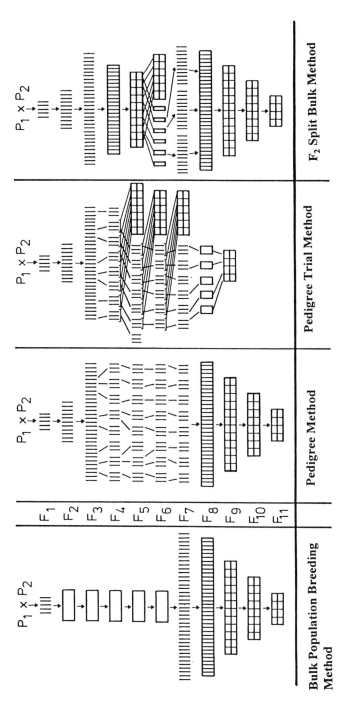

Figure 2. Different selection methods for line breeding.

In contrast to the pedigree method no selection in early generations is carried out when using the *bulk population breeding method*. In F_2 and the following generations the plants of a cross progeny are harvested completely (bulked) and are resown more or less without selection. The first selection step of single plants takes places at an increased level of homozygosity (F_5-F_6). The uniform offspring of these single plants are sown in single rows which are harvested completely and tested for yield in replicated trials as mentioned above for the pedigree method.

Both the pedigree as well as the bulk population breeding method have the disadvantage that it takes quite a long time until the breeder gets first information about the yielding ability of the selected lines. This shortcoming is partly overcome by modified methods such as the "pedigree trial method" and the "F_2 split bulk method". Here, first yield trials are carried out simultanously to selection – although a part of the plants may still be heterozygous to some extend. However, these trials give the breeder a first glance on the yielding ability of the selected progeny.

All these classical methods of line breeding have in common that it takes at least 6–7 years from the initial cross until the plants reach a level of homozygosity sufficient for reliable yield tests. Therefore, breeding of pure line varieties is a time-consuming process, which takes 10–12 years – including replicated yield tests and official seed tests. Today, this period may be abridged significantly by using doubled haploid lines (DHs, Foroughi-Wehr and Wenzel, 1990). Starting from F_1-plants obtained for example from crosses of two parental lines, haploid plantlets can be regenerated by anther or microspore culture, which after doubling of chromosomes, e.g., by using colchicin, give rise to homozygous doubled haploid plants as early as in the second generation, corresponding to an F_2 (A_1). As in classical line breeding these DH-lines have to be selected and tested in the following years. However, due to the fast return to homozygosity the breeding process is accelerated significantly (Fig. 3).

Furthermore, detailed RFLP-maps have been constructed in many inbreeding crops, e.g., barley (Graner et al., 1991, 1995), and RAPD-marker have been identified for important agronomic features (e.g., Ordon et al., 1995; Schweizer et al., 1995), facilitating efficient marker based selection procedures in order to accelerate the overall breeding schedule of line varieties.

Hybrid breeding

In population varieties, also referred to as panmictic varieties, of cross-pollinating species fertilization takes place at random. This type of variety and the respective breeding schemes will not be discussed in detail in this context. In contrast hybrid breeding in outbreeding species aims at a complete control of pollination (controlled crosses) in order to maximize heterotic effects. The magnitude of heterosis effects (hybrid vigor) is related to the degree of heterozygosity which is

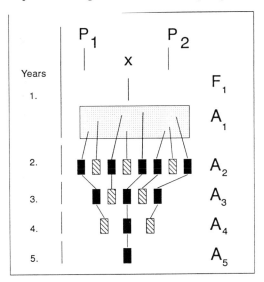

Figure 3. Use of doubled haploid lines (DHs) in line breeding (according to Foroughi-Wehr and Wenzel, 1990).

maximum in the F_1-generation. Consequently, heterosis effects are lost in subsequent generations (F_2 etc.) due to genetic segregation. Therefore, in contrast to line, population, and clone varieties hybrids cannot be reproduced identically by the farmer itself thereby ensuring seed exchange and warranting the breeders' rights. Hybrid breeding is generally grouped into three phases: (i) Selection of inbred lines with outstanding agronomic performance; (ii) identification of lines with maximum combining ability; and (iii) controlled crosses for the production of hybrid seeds.

(i) The development of inbred lines in cross-pollinating species normally has to face many difficulties. In general inbreeding depression is observed leading to lethality in a part of the inbred offspring due to the fact that recessive lethal factors (genes) become homozygous due to repeated selfing. Furthermore, the development of inbred lines is hampered by self incompatibility mechanisms in many cross pollinating plants.

(ii) In hybrid breeding the value of an inbred line is mainly due to its combining ability and only secondly to its own agronomic performance. Therefore, inbred lines have to be tested for their combining ability. In a first step they are tested for general combining ability (GCA), e.g., by using the polycross or the topcross test methods. These tests give hint to the average suitability of a line as a parent in hybrid breeding and are mainly used to restrict the number of lines which in a second step are tested for specific combining ability (SCA) in diallel or factorial crosses. Those lines expressing the highest SCA-values – i.e., the highest yield in individual cross combinations (F_1 hybrid) – will be used for hybrid seed production.

(iii) The seed production of a single cross hybrid (single) using cytoplasmic male sterility (cms) is shown in Figure 4. Cytoplasmic male sterility is a prerequisite for commercial hybrid seed production in androgynous crops like rye, rapeseed or sunflower.

Backcross

Backcrosses are in general used in order to incorporate mono- or oligogenically inherited traits, e.g., resistances to pathogens derived from primitive varieties or wild species into high yielding adapted cultivars. After several cycles of backcrossing out of the cultivar A the improved cultivar A' is developed which differs from A only because of the presence of a resistance gene derived from a primitive variety (Fig. 5). Concerning the procedure of backcrosses significant differences have to be noticed between incorporating dominant or recessive genes. In contrast to dominant genes the incorporation of recessive ones takes longer time due to the fact that each backcross has

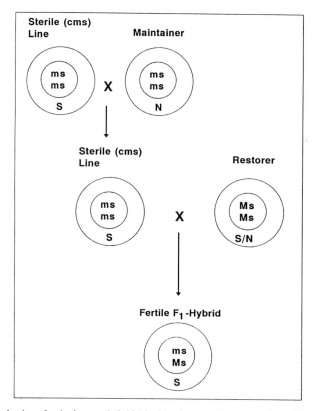

Figure 4. Seed production of a single cross hybrid (single) using cytoplasmatic male sterility (cms).

to be followed by a selfed generation in order to identify homozygous recessive genotypes which will be used for further backcrossing. As evident from Figure 5 even the improvement of a single trait by backcrossing takes a few years. In this respect gene technology offers opportunities to accelerate the process of incorporating individual genes coding for simply inherited, i.e., mono- or oligogenic traits.

Modern breeding methods

The classical breeding methods mentioned above are still the base of today's plant breeding, but more recently biotechnological methods are gaining more and more consideration and application. In this respect biotechnology is used as a term including all methods of cellular and molecular biology including gene technology. An overview on the use of biotechnology in plant breeding is given in Figure 6. Cell and tissue culture techniques for example permit sublethal embryos from interspecific or intergeneric crosses to develop *in vitro* (embryo rescue) and facilitate the

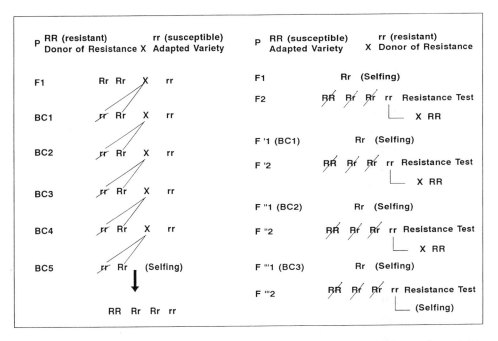

Figure 5. Backcross schemes for the incorporation of dominant (left) and recessive (right) genes from primitive varieties into adapted cultivars.

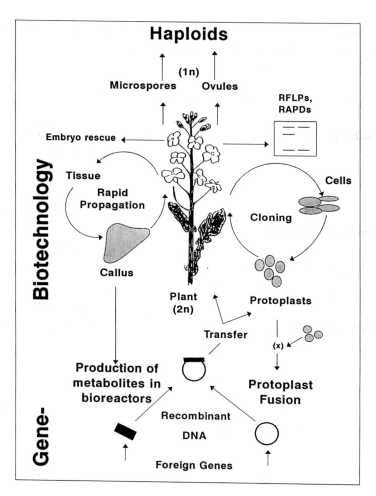

Figure 6. General overview on the use of biotechnology in plant breeding.

production of somatic hybrids (protoplast fusion). Furthermore, DH-lines produced by anther or microspore culture lead to a significant curtailment of long lasting breeding programmes and varieties of vegetatively propagated species can be maintained and propagated *in vitro* without the risk of getting infected by viruses and fungi. Today molecular methods like the RFLP- and the RAPD-technique are available which may be used for a better description of the genetic variation on the DNA level and for efficient marker based selection procedures. Last not least, foreign genes may be introduced into cultivated crops – independent from cross incompatibility mecha-nisms – by gene technology broadening the genetic variability in a way which cannot be achieved by other techniques. In the following paragraphs the use of gene technology will be illustrated on the example of oil crops – especially rapeseed (*Brassica napus*).

Present state and future prospects of gene technology in breeding of oil crops

As shown in Table 1 remarkable variation of the fatty acid composition of the seed oil is present in the major oil crops, i.e., rapeseed (*Brassica napus*), sunflower (*Helianthus annuus*), linseed (*Linum usitatissimum*) and soybean (*Glycine max*). Because of this variation, seed oil of these crops may be used as a source of human nutrition and also as a renewable source of raw materials for non-food purposes, e.g., in the oleochemical industries (Lühs and Friedt, 1994b). By classical plant breeding methods the fatty acid composition of rapeseed was drastically reduced from about 50% erucic acid (C22:1) to a content of almost zero percent C22:1, making rapeseed oil suitable for the production of margarine and other foodstuffs. Furthermore, plant breeding succeeded recently in creating rapeseed free of linolenic acid (C18:3), and high in oleic acid (C18:1) on the basis of induced mutations. Plants containing a large proportion of nervonic acid (C24:1), which normally is not present in rapeseed, were developed by using asymmetric protoplast fusion with *Thlaspi perfoliatum* (Murphy, 1994).

Table 1. Fatty acid composition (fatty acids %) of major crops (modified according to Lühs and Friedt, 1994a)

Crop	Palmitic	Stearic	Oleic	Linoleic	Linolenic	Eicosenoic	Erucic	Others
Soybean	8-13	2-5	18-26	50-57	5-10			
Breeding stocks	4-20	2-30	17-51	28-61	3-21			
Germplasm	3-28	2-30	7-70	8-64	2-33			0-1
Rapeseed	3	1	12	13	8	8	52	3
("high erucic")	2-6	1-3	8-60	11-23	5-13	3-15	5-60	0-3
Breeding stocks	1-5	1-3	7-50	9-23	4-13	3-16	5-61	0-3
Germplasm	2-11	0-5	6-50	7-48	2-21	2-23	5-63	0-5
Rapeseed	4	2	60	20	10	2	2	<1
("low erucic")	3-6	1-3	50-66	18-28	6-14	0-3	0-5	0-1
Breeding stocks	2-13	1-6	27-80	9-51	2-25	0-2	0-1	
Germplasm	0-13	1-7	27-87	4-51	2-25	0-1		
Sunflower	6-8	3-7	14-39	48-78				0-1
Breeding stocks	3-14	1-10	14-90	1-76				
Germplasm	3-30	1-14	9-95	1-84	0-1			0-1
Linseed	4-7	2-8	12-38	5-27	26-65			
Breeding stocks	4-13	2-10	12-44	5-70	2-70			
Germplasm	4-28	2-10	8-44	5-75	2-70			

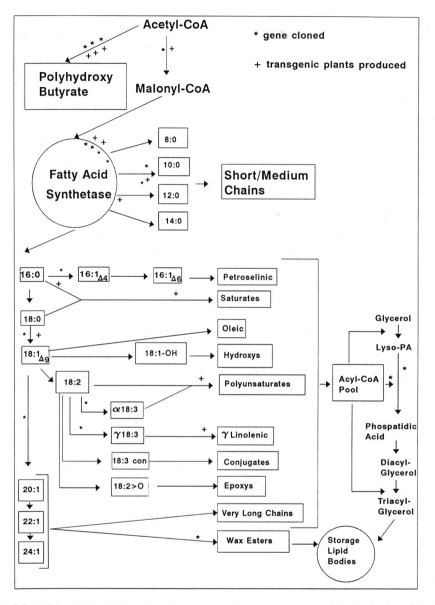

Figure 7. Potential pathways of fatty acid synthesis and modification in plants (according to Murphy, 1994).

While methods of classical plant breeding in combination with mutations and biotechnology are often useful in specific cases, the isolation and transfer of genes to create transgenic plants is probably the best all-purpose tool for increasing fatty acid diversity in oil crops. A first overview of the potential pathways of fatty acid synthesis and genes already cloned or transferred leading

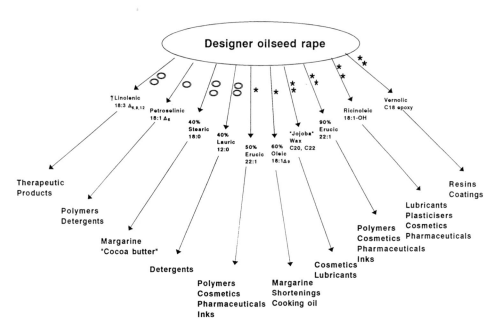

Figure 8. Summary of the "designer oilseed rape" varieties already produced or under development and their potential use for food and non-food purposes (according to Murphy, 1994).

to oil crops with designed fatty acid composition is given in Figure 7 (Murphy, 1994; Slabas et al., 1995). Concerning the transformation of rapeseed the close relationship between *B. napus* and the model plant species *Arabidopsis thaliana* (Brassicaceae) is advantageous, permitting the direct use of genes isolated from *Arabidopsis* in order to modify the seed oil composition of rapeseed. Furthermore, in comparison to other oil crops it is quite easy to transform and regenerate rapeseed. An overview on oil crops already transformed is given in Table 2; modifications of the fatty acid composition resistance to different herbicides and the *B.t.*-gene encoding for insect resistance are at the centre of transformation (Anonymous, 1995; see also Braun, this volume). Seed oil composition may also be modified by using "anti-sense" constructs (Fader et al., 1995). For example, the transformation of rapeseed and soybean with anti-sense *D12* and *D15* desaturase genes leads to a significant increase in oleic acid and a significant decrease in linolenic acid content, respectively (Tab. 3). Genotypes with modified seed oil composition can cover a wide range of use from food purposes to oleochemical and pharmaceutical products. In this respect a summary of the "designer oilseed rape" varieties already produced or under development and their different usabilities is given in Figure 8. For

Table 2. Guide to transgenic oilseeds (modified according to Anonymous, 1995)

Organisation	Crop	Altered trait	Target year for commercialization
AgrEvo Canada Inc.	Rapeseed	Herbicide tolerance	1995
	Soybean	Herbicide tolerance	1996 or 1997
Calgene	Rapeseed	Laurate	1995
		High-stearate	Latter 1990s
		Medium-chain fatty acids	Latter 1990s
		Cocoa butter alternative	Latter 1990s
	Cotton	Bromoxynil tolerance	1995
		Insect resistance (*B.t.*)	Latter 1990s
DuPont	Soybean	High-oleic, low-saturate	1998–1999
		High-lysine	1999–2000
		High-stearate, low-polystearate	1999–2000
	Corn	High-oil, high oleic	1999–2000
InterMountain Canola	Rapeseed	High-oleic	1996–1997
		High-oleic, low-saturate	1998
		High-stearate, low polystearate	1997–1998
Monsanto Company	Soybean	Roundup herbicide tolerance	1996
	Rapeseed	Roundup herbicide tolerance	1996
	Cotton	Insect resistance (*B.t.*)	1997–1998
Plant Genetics Systems	Rapeseed	Transgenic hybrid system	1996 or after
Crop Development Center,	Flax	Sulfonylurea tolerance	1995
University of Saskatchewan			

many purposes varieties developed by classical breeding methods or gene technology are already available today. But erucic acid and ricinoleic acid (both major industrial feedstocks) are still not found in transgenic plants showing maximum expression. However, for these traits transgenic rapeseed cultivars may be available at the end of this decade.

The examples presented show that gene technology is a very useful tool for the creation of genetic variability in a way which cannot be achieved by classical breeding methods. However, whether these "new" crops will gain any importance in practical agriculture depends on several factors. Farmers will ask for these varieties only if they will be economically superior in comparison to other cultivars, and the oleochemical industry will demand such oil as feedstock only if it is available in sufficient quantities and is cheaper in comparison to alternative feedstocks, e.g., mineral oil. Furthermore, these "new" varieties have to fulfill the requirements of Sortenschutz (registration for protection) and Sortenzulassung (variety release for distribution), considered in the last paragraphs.

Table 3. Fatty acid profile (fatty acids %) of oils from soybean and rapeseed breeding lines containing antisense delta 12 and delta 15 desaturase genes (Fader et al., 1995)

Crop	Lines	Palmitic	Stearic	Oleic	Linoleic	Linolenic
Soybean	Control-A2872	11,3	3,8	21,0	55,0	8,9
	Co-suppressed Delta 12-desaturase	8,0	4,3	76,1	2,5	8,1
Rapeseed	Weststar	3,9	1,8	68,0	18,9	6,9
	Co-suppressed Delta 12-desaturase	3,9	3,0	83,0	6,3	4,0
	Co-suppressed Delta 15-desaturase	3,8	1,6	69,0	21,3	1,4

Variety and variety law

"*Variety* means a plant grouping within a single botanical taxon of the lowest rank, which grouping, irrespective of whether the conditions for the grant of breeder's right are fully met, can be (i) defined by the expression of the characteristics resulting from a given genotype or combination of genotypes; (ii) distinguished from any other plant grouping by the expression of at least one of the said characteristics and (iii) considered as a unit with regard to its suitability for being propagated unchanged" (UPOV-Act, Article 1).

Because of this definition the phase of selection – regardless of dealing with line, population, clone or hybrid varieties – has to be followed by the phase of maintenance breeding, which aims at constantly maintaining the genetic constitution of a cultivar and thereby propagating it. In order to have an experimental variety licensed and released by the Bundessortenamt (Federal German Seed Board), a clear superiority in agronomical performance has to be demonstrated ("landeskultureller Wert", Saatgutverkehrgesetz), and it has to be new, uniform, distinct and stable. The latter requirements are asked for by the Sortenschutzgesetz (law of variety protection). In the case of fulfilling all these requirements an experimental variety will gain variety protection and can be released after being identified by an alphabetic name. In this case only the breeder of this variety is allowed to sell any seed of this cultivar for commercial purposes. But in contrast to protection by patent law, commercial seeds of released varieties can be used for breeding of new varieties (e.g., by using it for crosses) without permission of the respective breeder (Sortenschutzgesetz Abschnitt 1, §10). In contrast to patent law – which now is in the centre of discussion concerning gene technology – the law of variety protection explicitly permits the use of protected varieties for breeding new varieties without payment by simultaneously protecting the plant breeder's right (Kunhardt, 1993). Patent law and variety protection are contradictory in this case, because patents pending of single genes or DNA sequences, respectively, restrict the unlimited use of these genes in plant breeding. First attempts to solve this problem are included in Article 14 of the revised UPOV-Act which in summary says that anyone who is developing a new cultivar which is identical with the original variety in the major genotype infringes plant breeders' rights on a protected variety, since he is depending upon traits without permission of the breeder. For details of this complex problem in the field of gene technology (patent law) and classical plant breeding (variety protection) the reader is referred to the special literature (e.g., Straus, 1993).

Conclusions and future prospects

As described for oil crops plant species normally exhibit a wide range of genetic variability which can be used in practical plant breeding. In the past, this variability could only be exploited by using classical breeding methods. Today, modern techniques are very useful tools in plant breeding, and are expected to lead to an enhanced efficiency of breeding procedures. Cell and tissue culture techniques give new opportunities for creating genetic variability (e.g., by embryo rescue or protoplast fusion) and for accelerating the breeding process (e.g., by DH-lines). By using molecular techniques, e.g., RFLPs and RAPDs, the genetic variation can be described in more detail and efficient marker-based selection procedures may be carried out on the DNA-level.

Last but not least gene technology offers the opportunity to transfer single well defined traits, which are not present within the respective species or genus, in order to broaden the genetic variability. As described for rapeseed new fields of application may result from novel harvest products modified in this way. As the number and availability of defined genes is increasing, varieties with different quality traits may be created in shorter times in the near future. Gene technology is a very efficient tool for creating genetic variability which in combination with classical and various modern techniques gives new prospects to plant breeding. The only way for successful plant breeding in the future will be the efficient synergetic use of modern and classical plant breeding methods.

References

Anonymous (1957–1994) *Statistisches Jahrbuch über Ernährung Landwirtschaft und Forsten der Bundesrepublik Deutschland.* Bundesministerium für Ernährung, Landwirtschaft und Forsten (ed.), Landwirtschaftsverlag, Münster Hiltrup.

Anonymous (1995) Transgenic oilseed harvests to begin in May. *Inform* 6: 152–157.
Bittermann, E. (1956) Die landwirtschaftliche Produktion in Deutschland 1800–1950. *Kühn Archiv* 70: 1–149.
Fader, G.M., Kinney, A.J. and Hitz, W.D. (1995) Using biotechnology to reduce unwanted traits. *Inform* 6: 167–169.
Foroughi-Wehr, B. and Wenzel, G. (1990) Recurrent selection alternating with haploid steps – a rapid breeding procedure for combining agronomic traits in inbreeders. *Theor. Appl. Genet.* 80: 564–568.
Graner, A., Jahoor, A., Schondelmaier, J., Siedler, H., Pillen, K., Fischbeck, G., Wenzel, G. and Herrmann, R.G. (1991) Construction of an RFLP map of barley. *Theor. Appl. Genet.* 83: 250–256.
Graner, A., Bauer, E., Kellermann, A., Proeseler, G., Wenzel, G. and Ordon, F. (1995) RFLP analysis of resistance to the barley yellow mosaic virus complex. *Agronomie* 15: 475–479.
Kunhardt, H. (1993) *Sorten- und Saatgutrecht.* Fifth Edition. Verlag Alfred Strothe, Frankfurt a.M.
Lühs, W. and Friedt, W. (1994a) The major oil crops. *In*: D.J. Murphy (ed.): *Designer Oil Crops – Breeding Processing and Biotechnology.* VCH Verlagsgesellschaft, Weinheim, pp 5–71.
Lühs, W. and Friedt, W. (1994b) Non-food uses of vegetable oils and fatty acids. *In*: D.J. Murphy (ed.): *Designer Oil Crops – Breeding Processing and Biotechnology.* VCH Verlagsgesellschaft, Weinheim, pp 73–130.
Murphy, D.J. (1994) Transgenic plants – a future source of novel edible and industrial oils. *Lipid Technology* 6: 84–91.
Ordon, F., Bauer, E., Dehmer, K.J., Graner, A. and Friedt, W. (1995) Identification of a RAPD-marker linked to the BaMMV/BaYMV resistance gene *ym4*. *Barley Genet. Newsletter* 24: 123–126.
Schnell, F.W. (1982) A synoptic study of the methods and categories of plant breeding. *Z. Pflanzenzüchtg.* 89: 1–18.
Schweizer, G.F., Baumer, F., Daniel, G., Rugel, H. and Röder, M.S. (1995) RFLP markers linked to scald (*Rhynchosporium secalis*) resistance gene *Rh2* in barley. *Theor. Appl. Genet.* 90: 920–924.
Slabas, A.R., Simon, J.W. and Elborough, K.M. (1995) Information needed to create new oil crops. *Inform* 6: 159–166.
Straus, J. (1993) Pflanzenpatente und Sortenschutz – Friedliche Koexistenz. *GRUR* 10: 794–801.

Transgenic Organisms – Biological and Social Implications
J. Tomiuk, K. Wöhrmann & A. Sentker (eds)
© 1996 Birkhäuser Verlag Basel/Switzerland

Genetically modified food and its safety assessment

M. Teuber

Laboratory of Food Microbiology, Institute of Food Science, Swiss Federal Institute of Technology, ETH-Zentrum, CH-8092 Zürich, Switzerland

Summary. Genetic modification of microorganisms, plants and animals has been applied for the production of food additives (enzymes, amino acids, aroma compounds) and organisms which are directly used as food or food additives. After extensive biochemical, technological and toxicological evaluations, several products have gained legal acceptance and reached global markets. Fermentation-produced chymosin, microbial amylases and alpha-acetolactate decarboxylase are routinely used in Europe and America without the necessity of labelling the final products. Genetically modified tomatoes, potatoes, squash, soybeans, tobacco, maize and cotton are also on the market in different parts of the world. The strategies of genetic modification and safety evaluations are summarized and discussed. The labelling philosophy is presented here along with the European Union's legislation regarding novel food.

Background

Until a few decades ago, food had always been judged for its safety for human consumption by permanent trial and error, i.e., by its history of long and safe use. Over the millennia since the establishing of agriculture about 10000 years ago, humankind has by this "instinctive" strategy accumulated a wealth of experience regarding the production, properties and nutritional qualities of traditional food.

At first there was a permanent battle against the microbes which tended to spoil precious resources (Teuber, 1994). In the course of this struggle, microbes were detected which by spontaneous fermentation transformed specific raw materials (e.g., milk, meat, vegetables, fruits and fruit extracts) into microbiologically stable, agreeable products like sour milk, cheese, sauerkraut, soya sauce, wine, beer and others (Teuber et al., 1994).

These items originally spontaneously manufactured and now intentionally and on an industrial scale, were and still are consumed in part together with the live microbes used. In fact, the mean consumption of live microbes with fermented food, e.g., by the Swiss population, is estimated to be about 30000 million per person per day, based on the consumption statistics of the Swiss Nutrition Report (Teuber, unpublished data). This compares with only 100 million germs consumed with drinking water, air, and other mostly pasteurized or cooked food. In the human intestine, about 10000000 million living bacteria are already present (Drasar and Barrow, 1985).

Obviously, humans enjoy a natural beneficial cohabitation with these microbes which has developed during an evolution lasting millions of years.

The scientific assessment of food safety became necessary when traditional food was transformed by new technologies, leading to a new composition or new ingredients (e.g., UHT heating of milk, heat sterilization of canned food, chemical preservation, irradiation, and now genetic engineering). The traditional food was the necessary reference which in itself, however, had not been evaluated most of the time an the basis of the critical scientific investigation necessary in relation to novel food. To resolve this dilemma, several countries introduced into their food legislation the definition of traditional food items as generally recognized as safe (= GRAS) food if a long history of safe use was evident (Simon and Frommer, 1993).

However, we must realize that this GRAS definition can by no means be taken as an absolute guarantee of no risk. This applies to both traditional and novel food. For example, where the risk of a *Clostridium botulinum* food poisoning is involved, the raw material used has to be sterilized in a way which leads to a theoretical reduction of *C. botulinum* spores by 12 orders of magnitude. This is achieved by a 2.5 minute heating at 121°C. The batch of sterilized cans of food produced then bears a risk of 1 to one million of containing living *C. botulinum* spores if the original assumed theoretical content was 10^6 spores. This is an internationally accepted risk.

If we use traditional food as a reference, we can define different levels of novel food regarding its relatedness to the reference food. One recent proposal (ILSI, 1995) defines three levels of relatedness: (i) Substantial equivalence; (ii) sufficient similarity and (iii) insufficient similarity. However, novel food cannot so easily be described in its composition just by counting viable bacteria or by calculating sufficient thermal inactivation conditions.

It is paramount to any discussion and evaluation of safety to use the same scientific instruments for novel, e.g., genetically modified food as for its traditional counterparts or references. On this basis, it is clear that the final outcome of a safety assessment will therefore always have a certain uncertainty or – positively speaking – plausibility component. In other words, on the basis of the present state of scientific knowledge and technology, it is our task to arrive at the lowest possible risk attainable in order to obtain the highest possible probability of safety.

Genetically modified food

Microorganisms, plants and animals have been intensively investigated regarding their potential for improvement by genetic modification. In principle, we have two categories of products: (i) Genetically modified microorganisms, plants and animals which are directly used as food or food ingredients; (ii) products, e.g., enzymes, amino acids, vitamins, flavour compounds, sucrose,

Table 1. Examples of the application of genetic engineering in the production of milk

Stage of production	Genetic engineering on research level	Approaches on commercial level achieved
1. Fodder plants		1. Genetically modified plants with resistances to pesticides, insects, fungi and viruses
2. Silage	Development of starch degrading lactic acid bacteria	
3. Cow	1. Manipulation of rumen microflora to minimize carbon loss	1. Vaccines, hormones and antibiotics produced with genetically optimized microorganisms
	2. Modification of milk composition (new proteins, new fat, less lactose)	2. Feeding with fodder additives from genetically modified organisms (e.g., amino acids)
4. Dairy products	1. Genetically optimized starter cultures for cheese and sour milk production	1. Recombinant chymosin for milk coagulation in cheese making

starch, etc., manufactured with genetically modified microorganisms, plants and animals which are purified, i.e., separated from the genetically modified producer organisms and its recombinant DNA.

Since food production is a multibillion dollar business and of paramount importance for a dramatically increasing world population, it is not surprising that genetic engineering has a high priority in modern agricultural and food production research. It is the scope of this chapter to outline very briefly the main routes of research and application and to mention specific safety aspects and considerations regarding genetically modified food. The main consumer concerns are tabulated. The implication of genetic engineering techniques for the dairy industry is shown in Table 1.

Genetically modified microorganisms in food

Microorganisms traditionally used for the production of fermented food have been the subject of genetic modification for a series of purposes for the following reasons: (i) They are generally recognized as safe (GRAS) and provide a proper refence material; (ii) there is ample opportunity to handle them on a large industrial scale; (iii) their taxonomy, biochemistry, genetics and molecular biology are very well understood, at least for some genera like *Saccharomyces* or *Lacto-*

coccus/Lactobacillus; (iv) their behaviour in the environment (soil, water, food, intestine etc.) is also very well studied in many species, an important factor when it comes to the release of genetically modified microorganisms with food; (v) there is a substantial market potential.

Another advantage in the evaluation and assessment of the safety of genetically modified microorganisms in food is that only a limited number of microorganisms are used for food fermentations, as is evident from the following list of genera (Teuber, 1993): (i) Gram-negative bacteria: *Acetobacter* (vinegar), *Zymomonas* (ethanol); (ii) Gram-positive bacteria: *Brevibacterium* (cheese, amino acids), *Lactobacillus* (bread, cheese, yoghurt, pickles, sausages), *Lactococcus* (sour milk, butter, cheese), *Leuconostoc* (sour milk, butter, cheese, vegetables, wine), *Micrococcus* (sausages, amino acids), *Pediococcus* (soy sauce, silage), *Propionibacterium* (Swiss cheese), *Staphylococcus* (sausages), *Streptococcus* (yoghurt), *Streptomyces* (sausages, enzymes, antibiotics); (iii) yeasts: *Candida* (kefir), *Kluyveromyces* (kefir, enzymes), *Saccharomyces* (wine, beer, bread, bakers yeast, soya sauce), *Schizosaccharomyces* (alcoholic beverages, enzymes); (iv) moulds: *Aspergillus* (soy sauce, enzymes, citric acid), *Monascus* (colored rice), *Mucor* (cheese, enzymes), *Penicillium* (cheese, salami, antibiotics, enzymes).

Many economically important species among these genera are currently being closely investigated regarding genetics and genetic engineering potential. Methods of natural and artificial gene transfer (e.g., conjugation, electroporation) are available. The number of vectors developed for homologous and heterologous gene expression is ever increasing. *Lactococcus lactis* and *Saccharomyces cerevisiae* are almost as genetically managable as *Escherichia coli*, to name the two most important (Heinisch and Hollenberg, 1993; Gasson and de Vos, 1994).

Specific purposes of genetic modification of food microorganisms

These can be summarized as follows (Teuber, 1993): (i) Optimization of the function(s) of a single component culture; (ii) combination of properties from different biological systems in one microorganism.

In principle, genetic engineering is used to optimize the control of biochemical reactions already known and present in traditional food, which *per se* have a long history of safe use in their original genetic and food environment. Examples in *Saccharomyces cerevisiae* are constitutive maltose utilization, growth on starch with the aid of an enzyme system stemming from *Schwanniomyces occidentalis* or *S. diastaticus*, secretion of legume lipoxygenase, secretion of barley β-1,3-1,4-glucanase, reduction of diacetyl by an alpha-acetolactate dehydrogenase from *Acetobacter pasteurianus*, expression of malolactic enzymes and lactic dehydrogenase from lactic acid bacteria.

In lactic acid bacteria, examples include stabilization of technologically important functions by transfer of the corresponding genes from labile and conjugative plasmids into the chromosome (protease, lactose metabolism, citrate uptake), construction of bacteriophage-resistant starter cultures by recombination of different phage resistance mechanisms from different strains in one strain, establishment of starter cultures with high proteolytic activities to accelerate cheese ripening and aroma production, construction of starter cultures excreting bacteriocins and peptide antibiotics to inhibit pathogenic contaminant bacteria in food and fodder, and expression of amylase activity for better silage fermentations.

Due to the problems concerning the release into the environment of food containing genetically modified microorganisms and the difficulties in the safety assessment of such organisms when ingested by the consumer along with the food, only two genetically modified microorganisms have been legally accepted in one country, Great Britain: A genetically modified baker's yeast having a constitutive maltose uptake system and a brewer's yeast carrying the genetic information and expression capacity for glucoamylase derived from *Saccharomyces diastaticus*. In both cases, only genetic material from the genus *Saccharomyces* has been used, and all foreign genetic information has either been removed or, as in the case of introduced restriction and cloning sequences, has been kept to a minimum. The behaviour of these strains in the environment (water, food, soil, etc.) has been found to be similar to that of the unmodified parental organisms.

Food additives from genetically modified microorganisms

The main products of genetically modified microorganisms used in food are food grade enzymes like chymosin, proteases, amylases, β-galactosidase, lysozymes and glucanases. In addition, we have to assume that other food ingredients like amino acids, vitamins, citric acid, and nisin may already be manufactured with genetically optimized and modified cultures (see the tryptophan debacle; Swinbanks and Anderson, 1992). In this respect, it must be mentioned that at least 30 heterologous proteins have been expressed in *Lactococcus lactis* (Gasson and de Vos, 1994). A recent review (Heinisch and Hollenberg, 1993) lists in *Saccharomyces cerevisiae* 39 human proteins, 23 proteins from other mammals and higher eukaryotes, 11 viral proteins, 8 proteins from other fungi including food grade enzymes, and 5 bacterial proteins. If these products are sufficiently purified and free of the producing microbes, safety evaluation and assessment may be both easy and non-controversial.

Table 2. History of chymosin used for milk coagulation in cheese making

Year	Event
2000 B.C.	Use of rennet in high cultures of the Middle East
1875 A.C.	One calf stomach needed to curdle 2000 litres of milk for emmental cheese production
1900	Development of NaCl-extracts from calf stomach as chymosin source
1979	First description of amino acid sequence of bovine chymosin determined by classical Edmann degradation
1980	Isolation of m-RNA for bovine chymosin
1981	Cloning of chymosin cDNA in *E. coli*
1982	Description of nucleotide sequence of chymosin cDNA
1983	Cloning of chymosin in *Kluyveromyces lactis*
1984	Expression of bovine chymosin in *E. coli* and first cheese making experiments on pilot scale (US)
1985	Large scale preparation of recombinant chymosin from *E. coli* prochymosin inclusion bodies
1986	Complete intron/exon structure of bovine chymosin gene, first cheese making trials in Germany
1987	Expression of bovine chymosin in *Aspergillus niger*, proof of technological and biochemical identities of recombinant chymosins with calf stomach chymosin
1988	First official admission of recombinant *K. lactis* chymosin in the world (Switzerland)
1989	Recombinant chymosin regarded as suitable for cheese making by group of experts (B12) of the International Dairy Federation, Brussels
1990	*E. coli* chymosin accepted as GRAS by FDA
1995	60% of chymosin market in the US served by recombinant chymosin. More than 20 countries accepted worldwide recombinant chymosin (without specific labelling requirements for the produced cheese), estimated production level of cheese made with recombinant chymosin: 6 million tons.

Bovine chymosin produced with genetically modified microorganisms

The enzyme chymosin is traditionally used for the coagulation of milk in the process of cheese making. For a yearly world production of 14 million tonnes of cheese, the chymosin requirement is about 50 000 kg pure enzyme protein which would be extractable from the stomachs of 70 million calves. These are not available. Chymosin substitutes like pepsin, plant proteases or proteases from *Mucor miehei* and *Endotia parasitica* have different casein-splitting specifities and lead to differing products.

Therefore, chymosin has been one of the first food enzymes which is now produced with the aid of genetically modified microorganisms (Teuber, 1990; see also Tab. 2). It has been cloned into several apathogenic microorganisms (*Escherichia coli, Kluyveromyces lactis, Aspergillus niger*). The products developed with genetically modified microorganisms have been shown to be identical in their molecular, biochemical and technological properties to the traditional enzyme. No toxicological risk was detected (see Tab. 3). The products coming from *Escherichia coli, Kluyveromyces lactis* and *Aspergillus niger* are now accepted in more than 20 countries including the United States, Great Britain, Scandinavia, Australia, New Zealand and Switzerland. At the

Table 3. Components of safety assessment of recombinant chymosin preparations (*Escherichia coli, Kluyvero-myces lactis, Aspergillus niger*) as provided to the Swiss Health authorities prior to legal admission (BAG 1994)

---

1. Proof of technological functions
2. Proof of biological safety
    - no genetically modified producer organisms and recombinant DNA in products
    - identities of producer microorganisms
    - identities of vector DNAs
    - molecular and biochemical properties of recombinant chymosins
    - no pathogenicity of producer microorganisms as tested in experimental animals (mice) by intravenous, intraperitonal, nasal, cerebral and subcutaneous application
    - no short time toxicity in rats (5 g cheese daily for 3 weeks)
    - no acute toxicity in rats (5 g chymosin oral per kg)
    - no subchronic toxicity in rats (1000 mg chymosin per kg, 90-day test)
    - no allergenic sensibilization in Dunking-Hartley-Pirbright test in guinea pigs
    - no mutagenicity in Ames test for *Salmonella typhimurium*
    - no cytotoxicity for human cell cultures
3. Labelling
    - not necessary, according to new Swiss food law as of July 1, 1995, since identical with natural chymosin and free of producer microorganisms and recombinant DNA

---

moment, it is not accepted in Central Europe (Germany), due to overwhelming consumer concerns. As a consequence, cheese makers in the Netherlands, France, Denmark and Switzerland wanting to export their cheeses into Germany obviously do not use the recombinant enzyme. In contrast, at least 60% of the 50 million dollar chymosin market in the United States has been taken over by recombinant chymosins since their approval as a GRAS food supplement in cheese making by the Food and Drug Administration (FDA) in 1990.

Other food enzymes

Two other enzymes made with genetically modified bacteria have been accepted as GRAS by the FDA in 1990 as direct human food ingredients: A maltogenic amylase enzyme preparation derived from *Bacillus subtilis* and an α-amylase from *Bacillus stearothermophilus* (which is derived from *B. licheniformis*). Both these enzyme preparations were accepted in 1994 in France for use in starch hydrolysis, beer brewing, ethanol production, bread making and maltose syrup production. In addition, α-acetolactate decarboxylase of *Bacillus brevis* derived from *Bacillus subtilis* can be used in France to speed up beer ripening by the reduction of diacetyl. Neither these enzymes nor the chymosin need to be labelled on the final direct consumption products.

Genetically modified plants as food

As indicated in other contributions to this volume (Meyer, this volume; Friedt and Ordon, this volume), the genetic engineering of fodder plant varieties and food is very well developed. The scientific basis is provided by a series of methods to transfer and express heterologous and homologous genes into the desired plants (Watson et al., 1993; von Wettstein, 1993; Hines and Marx, 1995): (i) The use of the *Ti*-plasmid of *Agrobacterium tumefaciens* as a natural vector to transfer new genes into plant cells; (ii) the use of (attentuated) viruses as vectors; (iii) the use of protoplast fusion and transformation; and (iv) the use of ballistic transformation of plant cells with gold or platinum particles coated with the desired DNA.

In many plant varieties, the genetically modified transformed cells can be used to regenerate whole and fertile plants. A prerequisiste for all these techniques is the use of a selectable marker carried by the vector DNA in addition to the desired gene in order to select efficiently for the transformed cells. The kanamycin resistance gene (*nptII* gene = neomycin phosphotransferase II from *Salmonella typhimurium*) is just one but is a commonly applied system (Nap et al., 1992). The application of such resistance markers is an important point of concern in consumer circles (see below) and in the safety evaluation of genetically modified plants as food.

Some important functions incorporated into plants are summarized in Table 4. It is evident that the primary task is the protection of culturable plants against pests, viruses and fungi. Herbicide resistance and tolerance (glufosinate, bromoxynil, glyphosate, sulfonylurea) are also important as

Table 4. Important genetically modified agricultural plant varieties and their introduced properties (HT = herbicide tolerance; IP = insect protection; VR = virus resistance)

Species	Properties
Alfalfa	HT
Apple	IP
Rapeseed	HT, IP, alteration of the fatty acid composition, male sterility for improved breeding
Maize	HT, IP
Squash	VR
Melon	VR
Papaya	VR
Potato	·HT, VR, IP, composition of starch; lysozyme of egg white
Rice	IP, modified seed protein
Soybean	HT, modified seed protein
Strawberry	IP
Sunflower	modified seed protein
Tobacco	HT, IP, VR
Tomato	VR, HT, IP, slowdown of ripening
Walnut	IP

selection markers and as a selective protection of genetically modified plants when planted in the field. Virus resistance is achieved by incorporation of the genes for specific virus coat proteins. Insect protection is approached by the insertion and expression of the different *Bacillus thuringiensis (B.t.)* toxin genes. These genetic modifications do not normally influence the nutritional quality of the modified food and fodder plants. Nutritional improvements include changed fatty acid composition, additional seed proteins with more essential aminoacids (e.g., lysin, methionine) or delayed softening as in the case of the famous "Flavr Savr" tomato which contains antisense RNA to inhibit pectinase activities (Redenbaugh et al., 1992).

A number of such products have recently been cleared for the market by the Food and Drug Administration of the United States: Several tomatoes with delayed softening, cotton and soybeans with herbicide tolerance, squash with virus resistance, and a potato as well as maize with insect protection (*Bt*-gene). In Europe, genetically modified herbicide tolerant tobacco is planted in France and several other plants may be on the way to acceptance in the same country and most probably throughout the entire EU. In early 1996, transgenic rapeseeds for breeding purposes have been legalized by the EU, a prerequisite for a later acceptability as food and fodder source.

Genetically modified animals

The genetic modification of farm animals (cattle, pigs, sheep, goat, chicken) is being intensively investigated (Pühler, 1993). Traits include influenza virus resistance in pigs (*Mx* gene), milk with heterologous proteins (e.g., human lactoferrin in bovine milk), milk with expressed β-galactosidase to decrease lactose content (for lactose-intolerant consumers), and focus is especially on animals with engineered homologous or heterologous growth hormones to increase productivities. Transgenic swine with bovine growth hormone (GH) have been thoroughly investigated regarding carcass composition and nutritional qualities (Pursel and Solomon, 1993): "During the past decade, scientists have learned how to transfer recombinant genes into the genomes of livestock to produce "transgenic" animals. Microinjection of ova with copies of a gene is the primary method used, but the efficiency is low. About 1% of injected ova result in transgenic offspring. Initial research primarily involved genes encoded for growth hormone (GH). The GH transgenes that have thus far been used result in high concentrations of GH being produced throughout life. In general, GH (bovine) pigs did not grow larger than their sibs, but they gained weight up to 13% faster and they were 18% more efficient in utilizing feed. The excess GH dramatically altered carcass composition in comparison to sibs. At 92 kg, carcasses of GH pigs had 85% less total fat, which consisted of 85% less saturated fatty acids (SFA), 91% less monounsaturated fatty acids (MUFA), and 66% less polyunsaturated fatty acids (PUFA) than sibs. In meat cuts of

GH pigs, intramuscular fat was reduced by 43% in ham, 66% in loin, 64% in shoulder, and 69% in belly. No significant differences were detected in meat tenderness (shear force) for GH transgenics and sibs. Persistent excess GH in transgenic pigs was detrimental to their health. These problems were of such magnitude that these pigs could not be used for farming. When molecular biologists know more about gene regulation, transgenes can be constructed in which GH secretion might be tightly regulated. At that time, transgenic swine may be produced with positive attributes provided by a GH transgene, or other transgenes, with potential to improve carcass merit without adverely altering the health status of transgenic swine". This summary clearly indicates why transgenic animals are not likely to be available for use in agriculture in the near future.

The situation may be different with transgenic fish. The recent announcement of the FAO concerning world fish production (about 100 million tonnes) made it clear that the fish catch in the oceans is steadily declining due to overfishing. In contrast, aquaculture is on the rise, already comprising 15 million tons to balance the declining catches in the wild. In this respect, a recent paper was presented on the genetic modification of coho salmon with a gene construct (*pOnMTGH1*) completely derived from sockeye salmon (Devlin et al., 1994). This "all-salmon" construct consists of the metallothionein-B promotor fused to the full-length type-1 growth hormone gene. The linear *pOnMTGH1*-DNA was injected into the blastodisc region (animal pole) of coho salmon eggs that were developmentally arrested immediately after fertilization. Of more than 3000 injected eggs 6.2% developed into transgenic animals after one year. On average, the transgenic animals of the first (mosaic) generation were 11-fold heavier than non-transgenic controls. One individual was 37 times larger (41.8 cm) than controls. Seawater adaptability seemed to be normal in the transgenic animals.

The treatment of dairy cows with recombinant bovine growth hormone (BST) for improved milk production is another example of the dissemination of genetic engineering techniques in food production (see Tab. 1 and Bundesministerium für Ernährung, Landwirtschaft und Forsten, 1989). It is now a common and legal practice in the USA since 1994/1995; sales of BST in the first year of application have been reported to be around 100 million $.

Safety assessment of genetically modified food

Genetically modified organisms must be safe for the environment and the consumer (human or animal). The food scientist and nutritionist is mainly concerned with the nutrition and health aspects of genetically modified food. The release of genetically modified organims destined for food production is dealt with in several other chapters of this book and they should be consulted.

The food scientist must assume that the environmental problems (if any) have been solved before he approaches the different health issues.

Another prerequisite for the use of genetically modified organisms is the proof that the intended technology is functioning under production conditions. In addition, specific harmful effects on human health and the environment must be excluded on the basis of the state of science and the technology of detection. Whereas the safety assessment of isolated and purified products may be simple if the product is identical to a known compound (e.g., chymosin, see Tab. 3), the situation is quite complex if live genetically modified organisms are present in food. In this case, the organisms are released with the particular food item into the environment and are eventually consumed by humans (and animals).

The necessary basic elements of a safety evaluation exemplified for microorganisms are as follows (WHO, 1991): (i) Identity and knowledge of the recipient microorganism: species/strain identity, taxonomical position, pathogenicity for humans/animals/plants, function and behaviour in food including GRAS status, behaviour in human body and environment. This information is necessary to judge the genetic modification to be introduced. In that sense, the unmodified recipient organism is the reference material. There seems to be a consensus that only microorganisms in the lowest risk catagories (no risk or harmless) of the different systems should be used in food production. (ii) Identity and source of the introduced genetic material: Determination of the complete nucleotide sequence makes possible the exclusion of known protein toxins and pathogenicity factors and the identification of the genetic control elements and their identities with traditionally used elements and functions. It also allows the prediction of a possible migration of the introduced genetic material within the microbial community if it is released into the environment. (iii) Absence of pathogenicity and toxicity in the genetically modified microorganism. The absence of allergenicity may not be achievable if a molecule known to be allergenic in the first place is expressed. Antibiotic resistance markers with viable microorganisms should not be released into food and fodder. (iv) Fate of genetically modified organisms in the environment: Growth, proliferation and survival rates in the food, in the human/animal body, on plants, in air, water and soil should be known in order to compare it with the traditional recipient; (v) Nutritional properties of the produced food: It must be known whether the nutritional composition is significantly changed compared with the traditional product.

A very detailed discussion of different safety assessment systems (WHO, OECD, EBO, FDA) is given by Simon and Frommer (1993). Regarding the five different sets of data mentioned above, it becomes self-evident that at this stage only case-by-case evaluations of the use of genetically modified organisms and their products in food can be considered. As stated above, if the genetically modified microorganism functions as required and behaves otherwise like the traditional reference organism, a safety evaluation and assessment is possible.

If in the case of the evaluation the traditional organism turns out to represent hitherto unrecognized risk, this risk will have to be eliminated in accordance with the requirements concerning genetically modified organisms (International Life Science Institute, 1993, 1995).

Specific risk of antibiotic resistance genes as selectable marker in genetically modified organisms

Genetic engineering of microorganisms and plants has only been possible because powerful selection methods have been available. It should be mentioned that resistance transfer factors (plasmids) carrying more than one antibiotic resistance gene were discovered 40 years ago in a *Shigella* epidemic in Japan. These extrachromosomal elements provided the molecular basis for all the vectors used in early experiments in genetic engineering (Watson et al., 1993). Such plasmids, which may be conjugative and selftransmissible, occur in Gram-positive and Gram-negative bacteria. In antibiotic therapies in the treatment of infectious diseases of man and animals, bacteria used these genetic elements (and transposons in addition) to acquire and donate antibiotic resistance determinants. The introduction of a new antibiotic or class of antibiotics was always followed within a few years by the spread of antibiotic-resistant bacteria. This has led to the development of multiresistant bacterial strains which may no longer be sensitive to any useful antibiotic. Multiresistant bacteria are becoming an increasingly serious problem in nosocomial (=hospital acquired) infections (Neu, 1992).

The spread of antibiotic resistance is probably aggravated by the extensive use of antibiotics in animal farming, both as therapeutic and nutritive agents. Although antibiotics for human treatment may not be used as nutritive fodder additives, a clear separation is sometimes an illusion, since therapeutically necessary antibiotics (which are the same as those used in human treatment) can be added to the fodder of piglets, calves and chicken (Levy, 1978).

To further lessen the risk of antibiotic resistance it is vital to reduce the application of antibiotics to only severe cases in human as well as in veterinary medicine and animal nutrition. The release of genetically modified organisms together with antibiotic resistance markers and genes has to be considered in the light of this medical background. No release of genetically modified microorganisms containing antibiotic resistance genes can be accepted, so that the medically important antibiotics are retained as active life-saving drugs.

The release of genetically modified plants carrying antibiotic resistance genes (like *nptII*, e.g., kanamycin or neomycin resistane) is only acceptable if it can be shown that these genes are not transferred from plant cells into microorganisms.

Potential studies of toxicological aspects

The possibilities of assessing the toxicological risks of novel, i.e., genetically modified food, has been discussed recently on several occasions (Koschatzky and Massfeller, 1994; Lebensmittel-chemische Gesellschaft, 1994; Basel Forum on Biosafety, 1994).

Together with the consideration summarized by Simon and Frommer (1993), I would like to outline the basic toxicological procedure as follows: (i) In the particular instance of genetically

Table 5. Public concerns over biotechnology (Kemp, 1992)

Socioeconomic
  - Commercial exploitation
  - Role of multinationals
  - Patenting rights
  - Effects on Third World
  - Agriculture and profitability

Environmental
  - Threat to ecosystems and indigenous species
  - Competition
  - Predation
  - Parasitism
  - Food webs
  - Habitat destruction
  - Pollution
  - Loss of biodiversity
  - Transfer of DNA

Health
  - Resistance transfer
  - Long-term effects
  - Gene therapy
  - Genetic drift
  - Toxins in novel foods
  - Creation of uncontrolled organisms

Ethical
  - Man playing God
  - Human/animal rights
  - Links to biological warfare
  - Who decides
  - Effects on evolutionary process
  - Equality and the Third World
  - Secrecy/commercial confidentiality
  - Malthusianism

Trust in science
  - Human error
  - Commercial science
  - Mitigation measures
  - Dose-response relationships
  - Relevance of animal and tissue studies
  - Predictability, reliability

modified food microorganisms consumed as living organisms, the potential for colonization of the gastrointestinal tract (of humans and animals) and the transfer of genetic material in the gastrointestinal tract must be investigated. Acute and subacute toxicities should be studied in conventional rodents and/or in germ-free animals colonized with specific or total human gut microflora; (ii) the potential allergenicity must be addressed; (iii) subchronic toxicity should be tested in 90-day studies with rodents or other appropriate species (in combination with colonisation studies). Observations should include attention to indication of genotoxicity, neurotoxicity, immunotoxicity and reproductive function; (iv) safety for human consumption should be confirmed in human studies, including tolerance and examination of effects on faecal microflora (spectrum and content).

Such toxicological studies demanded by toxicologists and consumers for novel food are in principle not different from tests required for pharmaceutical products. This will open completely new perspectives for the food industry and will also influence the evaluation of traditional food items (ILSI, 1995).

I would like to emphasize that the safety assessment is only one step, but probably the most limiting in the development of genetically modified microorganisms for use in food production: (i) Realization of the genetic modification in the laboratory; (ii) proof of technological function and suitability; (iii) safety assessment for human consumption; (iv) safety assessment for the environment; (i) legal acceptance; and (vi) acceptance by the market and consumer.

This elaborate procedure will allow only economically strong, competitive, and well accepted products to gain a foothold in the market. As in other industries, increased legal constraints will work against small and medium-sized companies and will favour large enterprises. On the basis of a collection of concerns which have been and are being expressed (see Tab. 5), the extreme difficulties arising from public oppinion related to the field of genetic engineering have become evident (Sentker, this volume).

References

Basel Forum on Biosafety (1994) *Biosafety of Foods Derived of Modern Biotechnology.* BATS, Basel.

Bundesministerium für Ernährung, Landwirtschaft und Forsten (1989) *Folgen des Einsatzes von BST in der deutschen Milcherzeugung.* Landwirtschaftsverlag, Münster-Hiltrup.

Devlin, R.H., Yesaki, T.Y., Biagi, C.A., Donalson, E.M., Swanson, P. and Chan, W.-K. (1994) Extraordinary salmon growth. *Nature* 371: 209–210.

Drasar, B.S. and Barrow, P.A. (1985) *Intestinal Microbiology.* American Society for Microbiology, Washington.

Gasson, M.J. and de Vos, W.M. (1994) *Genetics and Biotechnology of Lactic Acid Bacteria.* Blackie Academic and Professional, London.

Heinisch, J.J. and Hollenberg, C.P. (1993) Yeasts. *In:* H.J. Rehm, G. Reed, A. Pühler and P. Stadler (eds): *Biotechnology,* Vol. 1, Second Edition. VCH, Weinheim, pp 469–514.

Hines, P.J. and Marx, J. (1995) The emerging world of plant science – Frontiers in biotechnology. *Science* 268: 653–691.

International Life Science Institute (1993) Nutritional appraisal of novel foods. Workshop organised by ILSI Europe, 28–30 September 1992, Brussels, Belgium. *International Journal of Food Sciences and Nutrition* 44, Supplement 1: 1–100.

International Life Science Institute (ILSI Europe) (1995) The safety assessment of novel foods. Guidelines prepared by ILSI Europe Novel Food Task Force. Brussels, Belgium, 16 pages.

Kemp, R. (1992) Social implications and public confidence: Risk perception and communication. *In*: D.E.S. Stewart-Tull and M. Sussman (eds): *The Release of Genetically Modified Microorganisms*. Plenum Press, New York, pp 99–114.

Koschatzky, K. and Massfeller, S. (1994) *Gentechnik für Lebensmittel? Möglichkeiten, Risiken und Akzeptanz gentechnischer Entwicklungen*. Verlag TüV Rheinland, Köln.

Lebensmittelchemische Gesellschaft – Fachgruppe in der GDCh (1994) *Gentechnologie – Stand und Perspektiven bei der Gewinnung von Rohstoffen für die Lebensmittelproduktion*. Behr's Verlag, Hamburg.

Levy, S.B. (1978) Emergence of antibiotic-resistant bacteria in the intestinal flora of farm inhabitants. *J. Infect. Diseases* 137: 688–690.

Nap, J.P., Bijfoet, J. and Stiekema, W.J. (1992) Biosafety of kanamycin-resistant transgenic plants. *Transgenic Res.* 1: 239–249.

Neu, H.C. (1992) The crisis in antibiotic resistance. *Science* 257: 1064–1073.

Pühler, A. (1993) *Genetic Engineering of Animals*. VCH, Weinheim.

Pursel, V.G. and Solomon, M.B. (1993) Alteration of carcass composition in transgenic swine. *Food Rev. Intern.* 9: 432–439.

Redenbaugh, K., Hiatt, W., Martineau, B., Kramer, M., Sheeky, R., Sanders, R., Houck, C. and Emlay, D. (1992) *Safety Assessment of Genetically-Engineered Fruits and Vegetables: A Case Study the FLAVR SAVR™ Tomatoes*. CRC Press, Boca Raton.

Simon, R. and Frommer, W. (1993) Safety aspects in biotechnology. *In*: H.J. Rehm, G. Reed, A. Pühler and P. Stadler (eds): *Biotechnology* Vol. 2, Second Edition. VCH-Weinheim, pp 825–853.

Swinbanks, D. and Anderson, C. (1992) Search for contaminant in EMS outbreak goes slowly. *Nature* 258: 96.

Teuber, M. (1990) Production of chymosin (E.C. 3.4.23.4) by microorganisms and its use for cheese making. *IDF Bulletin* 251: 3–15.

Teuber, M. (1993) Genetic engineering techniques in food microbiology and enzymology. *Food Revs. Intern.* 9: 389–409.

Teuber, M. (1994) Nahrungsmittelkonservierung – Mensch und Mikroben im Wettbewerb um die Nahrung. *Naturwiss. Rundsch.* 47: 59–63.

Teuber, M., Geis, A., Krusch, U. and Lembke, J. (1994) Biotechnologische Verfahren zur Herstellung von Lebensmitteln und Futtermitteln. *In*: P. Präve, U. Faust, W. Sittig and D.A. Sukatsch (eds): *Handbuch der Biotechnologie*, Fourth Edition. Oldenbourg, München, pp 479–540.

von Wettstein, D. (1993) Genetic engineering and plant breeding, especially cereals. *Food Revs. Intern.* 9: 411–422.

Watson, J.D., Gilman, M., Witkowski, J. and Zoller, M. (1993) *Rekombinierte DNA*, Second Edition. Spektrum Akademischer Verlag, Heidelberg, pp 255–273.

WHO (1991) *Strategies for Assessing the Safety of Foods Produced by Biotechnology*. WHO, Geneva.

Transgenic Organisms – Biological and Social Implications
J. Tomiuk, K. Wöhrmann & A. Sentker (eds)
© 1996 Birkhäuser Verlag Basel/Switzerland

Genetic intervention in human beings

M. Leipoldt

Department of Clinical Genetics, University of Tübingen, Wilhelmstraße 27, D-72074 Tübingen, Germany

Summary. Various aspects of human biology and medicine will be influenced by the information provided by the Human Genome Project regarding the entire human DNA sequence. In addition to scientific knowledge, the primary benefit for the community will lie in improvements in the diagnosis and therapy of monogenic as well as multifactorially inherited genetic disorders including many forms of cancer. An overview of the first five years is presented here and the achievements regarding DNA-based diagnosis, screening and therapy are discussed. The consequences for medical practice are outlined and placed in the context of individual, familial and social concerns. It is emphasized, however, that practicable interventions in human genomes leading to broad and long-term consequences are at present not a realistic possibility due to insufficient albeit rapidly improving technology. Negative developments could arise in the form of various kinds of genetic discrimination pending the revival of old and persisting eugenic concepts.

Status quo of the Human Genome Project

In 1988 the Human Genome Project (HUGO), i.e., the sequence determination of the total human genome across all autosomes and the sex-chromosomes with their three billion nucleotides was initiated, and began its 15-year realization schedule in 1990. From the outset it has been subject to enthusiastic praise as well as profound criticism (Committee on Mapping and Sequencing the Human Genome, 1988; Leipoldt, 1991). As has been put forward by advocates of the project, numerous questions and problems in human biology and medicine can be tackled on the basis of HUGO's achievements. And as has been discussed elsewhere (Leipoldt, 1991) HUGO did not represent an "absolute beginning". A number of quite specific parameters of the human genome, such as the average size of genes, the average number of genes per chromosome, as well as the number of genes mapped to specific chromosomal regions (more than 1500 in 1990) formed the project's basis (Morton, 1991). And undoubtedly, the financial support and the world-wide collaboration have been such that a number of intermediate goals will have been fulfilled within or even before the end of the 15-year period. The project can be divided into seven major areas: (1) Mapping and sequencing of the human genome (first five-year goal: creating a genetic map with one cM (Centimorgan) marker distances and a physical map with STS (sequence tagged sites) distances of 100 kb corresponding to about 0.1 cM); (2) mapping and sequencing the genomes of selected model organisms such as mouse, *Drosophila, Caenorhabditis elegans, Escherichia*

Table 1. Selected biological and clinical parameters of the human genome

Number of chromosomes per haploid genome	23
Number of nucleotides per haploid genome	3 billion
Number of nucleotides in the largest chromosome (# 1)	250 million
Number of nucleotides in the smallest chromosome (Y)	50 million
Estimated total number of coding genes	60 000–70 000
Average size of coding genes (range 800–2 000 000 nucleotides)	2500 nucleotides
Estimated total number of nucleotides in coding sequences	150 million
Estimated fraction of coding sequences	5%
Average distance between two coding genes	40 000 nucleotides
Total number of genetic traits	6680
Number of genes mapped to specific chromosomal regions	3 000
Number of heritable disorders	5 000
Number of heritable disorders mapped	900
Number of heritable disorders with known molecular defect	450
Total number of nucleotides sequenced	10 million
Number of anonymous DNA-markers	5300
Average distance between two markers (range 0.2–12 mbp)	1.6 mbp

Data taken from references cited in the text and results presented at the 17th ESHG congress, Berlin, May 1995

coli, etc. (first five-year goal: creating a physical and genetic map of the mouse genome, sequencing the genomes of *E. coli,* yeast, *Drosophila* and plants); (3) management and analysis of sequence data (first five-year goal: development of software and databases, development of new methods for the interpretation of genomic maps and DNA sequences); (4) ethical, legal and social considerations (ELSI), i.e., measures to maintain data confidentiality, to meet and make public mechanisms of individual discrimination as to a person's present or future health status; (5) training and education (first five-year goal: establishment of fellowships and workshops); (6) technological development (first five-year goal: support of new technologies for automatization); (7); technological transfer (first five-year goal: establishment and support of mutual relations between scientific institutions, industry and private economy, rapid and efficient transfer of technology into medical application).

The unbiased observer could easily get the impression that things are going well at present. Early in 1995 HUGO's results could be summarized as follows: (1) A total of 10 million nucleotides have been sequenced (comprising coding and non-coding sequences; this corresponds to 0.28% of the total genome); (2) 3000 genes have been successfully localized (this corresponds to about 45% of the 6680 known genetic traits); (3) the physical maps of chromosomes 22, 21 and Y are nearly completed (Foote et al., 1992; Chumakov et al., 1992; Bell et al., 1995); (4) 5300 anonymous DNA markers span the human genome with an average distance of 1.6 megabasepairs (the largest intervals span 10–12, the smallest not more than 0.2 megabasepairs (Weissenbach, 1995); the distance between two markers is probably on the average not more than 40 coding genes; (5) more than 1600 genes of the presumed total number of 60000–70000 (Fields et al., 1994) have been cloned (the average distance between two mapped genes narrows to 40000 basepairs (Tab. 1). Roughly 5000 genes are known to cause heritable disorders in humans (McKusick, 1994). More than 900 have already been chromosomally localized and for 450 the possible defects have been described, thus allowing direct genetic testing in most cases. As regards non-human model organisms, several viral genomes as well as the bacterial genomes of *Haemophilus influenzae* and *Mycobacterium genitalis* have been completely sequenced. Moreover, the first eukaryotic genomic sequence, i.e., that of *Saccharomyces cerevisiae*, with a total of 14 million basepairs (Williams, 1995) is almost completed and impressively demonstrates the success of collaborate sequencing efforts. With one third of the way into its schedule the aim of this contribution is to outline the implications of the project's achievements with regard to human genetic diseases. Special emphasis is placed on the molecular "management" of human diseases and its potential to shape a personal and social attitude towards human genetics.

Impact of the Human Genome Project on medical practice

The Human Genome Project will probably have consequences affecting almost every aspect of medical practice. The knowledge relating to genetic diseases, i.e., the localization, function, and malfunction of responsible genes is steadily increasing and will continue to do so over the next few years. New genetic traits will be discovered for diseases for which a genetic component has so far been unknown or merely speculative. Not a single discipline in medicine will remain uninfluenced by the data and evidence obtained by HUGO. HUGO's biological component attracts much less attention in view of the overwhelming medical significance of genome analysis, but will still be enriched by the new findings. Basic findings relating to gene regulation, genomic dynamics, genome organization, and human evolution may also be expected. Furthermore, when HUGO phases out in 2005, implications for medicine will continue to arise during the subsequent human sequence decodification efforts (elucidation of regulatory genomic interactions, especially with respect to complex traits resulting from multigene effects. New insights will also be gained in the field of human development). However, the gap between our knowledge of the molecular basis of genetic diseases and its therapeutical applicability will probably further increase (e.g., Huntington's disease, cystic fibrosis, neoplasias, etc.). With respect to clinical practice three topics are discussed below: Diagnosis, screening, and therapy. The Human Genome Project's achievements will not only bring about profound changes in the medical, clinical, and technological treatment of genetic diseases, but will also inevitably change the attitude towards genetic defects on the part of the clinician, the individual, and society as a whole (Nelkin, 1992).

The human genome as a target for diagnostic procedures

Compared with classical cytogenetic analyses there are some principal differences when molecular genetic analyses are performed. Whereas cells capable of undergoing mitotic division are needed in order to properly examine chromosomes, any nucleated cell type can be used for molecular studies. And whereas cytogeneticists consider the genome as a whole molecular geneticists often look at a single gene, parts of it, or even at a single nucleotide. However, recently developed techniques such as interphase-cytogenetics, and molecular scanning methods such as single-strand conformation polymorphism (SSCP) analysis provide broader applications for both approaches (Reiss et al., 1993). Since 1989 the powerful techniques of molecular gene analysis such as the use of restriction endonucleases, molecular hybridization and sequencing (Trent, 1993) have been increasingly supplemented and in part substituted by the polymerase chain reaction (PCR) (Saiki et al., 1988; for details in application see Newton and Graham, 1994). Without PCR, the field of molecular diagnostics would probably not be as advanced. Generally, a gene mutation or a mutated gene leading to genetic diseases can be detected by two different approaches: indirect genotyping is performed if the kind of mutation and its precise location is unknown or if a specific gene probe is not available. Diagnosis is based on linkage analyses using polymorphic extragenic markers (e.g., RFLP markers), not identifying the mutation itself but tracing the mutated gene within a family. If, however, a specific probe is at hand and/or the mutation is identified with respect to its localization, direct genotyping analysis will be preferred. Here diagnosis involves only a single individual; knowledge of the RFLP haplotypes of the relatives, especially that of the index patient is not necessary. This and the fact that indirect genotyping will statistically be less than one hundred percent correct (due to the possibility of meiotic recombination between the gene locus and the marker locus) renders the direct approach more attractive. An increasing number of genetic diseases to which linkage analysis has been applied in recent years can now be diagnosed by direct methods. This development is a direct consequence of HUGO's efforts as well as of the strong augmentation of the overall number that have been approached diagnostically since 1990 (McKusick, 1994). More than 400 genetic diseases can be diagnosed, a major fraction by direct methods. In Germany, routine diagnosis is offered for about 85 different diseases (Tab. 2). This method of molecular genetic analysis of human genes and their mutations not only helps to verify clinical diagnosis, more often it is applied in carrier detection, prenatal diagnosis, and increasingly in preclinical (presymptomatic) testing of hereditary late

Table 2. Molecular genetic diagnosis of genetic diseases offered in Germany

Achondrogenesis typeII	FraX E syndrome	Neurofibromatosis type 2
Achondroplasia	Friedreich's ataxia	Norrie's disease
Adrenoleucodystrophy	Hemophilia A	Ocular diseases
AED	Hemophilia B	Osteogenesis imperfecta
Agammaglobulinemia	Hb defects	OTC deficiency
AGS	Hunter syndrome	PKU
Alpha-1-Antitrypsin deficiency	Huntington's disease	Paramyotonia congenita
Alzheimer's disease	Hydrocephalus	Polycystic kidney disease
Androgen resistance	Hypercal. periodic paralysis	Polyposis coli
Angelman syndrome	Hypochondrogenesis	Prader-Willi syndrome
Anhidrotic ectodermal dysplasia	Ichthyosis	Premature hereditary
Aniridia-Wilms' Tumor	Kallmann syndrome	osteoarthrosis
ApolipoproteinB-deficiency	Kniest syndrome	Retinitis pigmentosa
CMT 1 A	Langer-Giedion syndrome	Retinoblastoma
CMT X	Lesch-Nyhan syndrome	Retinoschisis
Central core disease	Leukemias (Ph1)	Spinal muscle atrophy
Choroidemia	Lissencephaly	Spinobulb. muscle dystrophy
Chromosome aberrations	Lowe syndrome	Spinocerebellar ataxias
Congen. spond.epiph. dysplasia	MCAD deficiency	Stickler's syndrome
Cystic fibrosis	Machado-Joseph disease	Testicular feminization
Crouzon syndrome	Malignant hyperthermia	Thalassemias
Denis-Drash syndrome	Mental retardation (X)	Thyroid-hormone resistance
DMD/BMD	Metaphyseal dysplasia	Tuberous sclerosis
DiGeorge syndrome	Miller-Dieker syndrome	VCF (Shprintzen's syndrome)
Dopa-responsive dystony	Multiple endocrine neoplasia syndrome	WAGR syndrome
Dystonie-Parkinson-Syndrom (X)	Myoadenylate-desaminase deficiency	Waardenburg syndrome (I/III)
Ehlers-Danlos syndrome (IV,VII)	Myotonic dystrophy	Wiedemann-Beckwith syndrome
EMD	Nephrogenic diabetes insipidus	Williams syndrome
FraX A syndrome	Neurofibromatosis type 1	Wiskott-Aldrich syndrome

Data taken from Schmidtke, 1995.

onset conditions. The resulting consequences for the counselling of families with a genetic problem have been enormous. A number of frequent and rare genetic diseases can now be dealt with not only on the basis of formal genetics and probabilities according to Mendelian laws but also by precise diagnosis, specific carrier detection or exclusion and predictive testing.

Thus, the spectrum of prenatal diagnostic procedures which was still largely confined to cytogenetic and biochemical analyses ten years ago has increased considerably. Simultaneously, new developments in the isolation of fetal cells from maternal peripheral blood (Zeng et al., 1993; Simpson and Elias, 1994) and in preimplantation diagnosis (Delhanty, 1994) using sensitive molecular techniques offer the prospect that prenatal diagnosis might one day be a simple, reliable and risk-free test for numerous genetic diseases (see section "Has genetics in the 21th century the potential to shape humankind?"). Let us consider cystic fibrosis (*CF*), a common severe disorder of the exocrine glands (Boat et al., 1989), with an autosomal recessive mode of inheritance and an incidence of about one in 2000 to 2500 among caucasians, hence a frequency of heterozygous carriers of about one in 25 individuals. Since the identification in 1989 of the gene responsible, i.e., the cystic fibrosis transmembrane conductance regulator gene (CFTR) on chromosome 7q31, more than 500 different mutations have been identified as a consequence of the extensive molecular heterogeneity of the disease.

The CFTR gene encodes a transmembrane protein with 1480 amino acid residues with two membrane-spanning and two nucleotide-binding domains, as well as a regulatory domain that functions as regulated chloride channel. One mutation, i.e., the 3 bp deletion *dF508* in exon 10, accounts for 70–75% of all mutations on CF chromosomes of caucasians (Tab. 3). Mutations can affect every part of the peptide chain. However, a slight clustering of mutations is observed in exon 10 and 11 that code for the first nucleotide binding domain. Depending on the type of mutation (deletion, insertion, point mutation) and its environment mutation analysis is currently performed by directly sizing the PCR products, testing for gain or loss of restriction sites, allele-specific oligonucleotide hybridization, PCR-mediated site-directed mutagenesis, and DNA sequencing, to mention the most customary methods.

Table 3. The most common mutations in the CFTR gene on CF chromosomes in Germany

Mutation	Location	Nucleotide change*	Consequence	Frequency
dF508	exon 10	Deletion of 3 bp between 1652 and 1655	Deletion of Phe at codon 508	70.0%
R553X	exon 11	C to T transition at 1789	Arg > Stop at codon 553	2.3%
N1303K	exon 21	C to G transversion at 4041	Asn > Lys at codon 1303	2.3%
R347P	exon 7	G to C transversion at 1172	Arg > Pro at codon 347	1.6%
G542X	exon 11	G to T transversion at 1756	Gly > Stop at codon 542	1.4%
G551D	exon 11	G to A transition at 1784	Gly > Asp at codon 551	1.0%
3849 + 10 kb/ C-T	intron 19	C to T transition at 3849 + 10 kb	Aberrant splicing	1.0%
1717-1/G-A	intron 10	G to A transition at 1717-1	Splice mutation	0.9%
2789 + 5/G-A	intron 14b	G to A transition at 2789 + 5	Splice mutation	0.9%
3272-26/G-A	intron 17a	A to G transition at 3272-26	Splice mutation	0.9%

Modified from Tsui, 1992; Dörk et al., 1994.
*Numbers refer to cDNA sequence.

The mutation *R553X* (Tab. 3), a nonsense mutation in exon 11 is commonly identified as follows: (i) PCR amplification of a 425 bp fragment in exon 11 covering the mutation site (Zielenski et al., 1991); (ii) digestion of the PCR product with the restriction endonuclease *Hinc*II that recognizes a unique restriction site (5'-*GTPy/PuAC*-3') in wild-type DNA and thus cleaves the PCR product into two fragments of 239 and 189 bp, respectively. The result of this approach is confirmed by electrophoretic separation. In *R553* alleles however the *Hinc*II site is destroyed by the substitution of the 3'-C by T. Restriction enzymes with other specific recognition sites can be used to test for a number of different CFTR mutations. In the above-mentioned example, *Hinc*II is employed as a diagnostic tool, capable of discriminating not only between wild-type and mutant DNA but also of detecting heterozygous carriers of *R553X* alleles.

A variety of CFTR mutations occur with different frequencies in diverse populations (Tsui, 1992). Depending on the population studied and the detection method employed, a significant fraction of CF mutations remains unidentified. This is partially because the DNA sequence of the CF gene is far from being completely known, and partially because of possible mutations in the non-coding regions. A comprehensive survey of German CF patients (Dörk et al., 1994) has revealed more than fifty different mutations, thus leaving some 5% of the molecular defects on CF chromosomes undetected. For clinical purposes, this means that as well as patients who are homozygous for a single known mutation or compound heterozygous for two different known mutations, there are also patients with only one detected mutation. In the latter case the second mutation may remain unidentified for two reasons: (i) Diagnosis is limited to the most common mutations; (ii) the mutation is still unknown. This situation holds true for about 6–40% of all CF patients, depending on the study (Simon-Bouy et al., 1991; Devoto et al., 1991; Shrimpton et al., 1991; Reiss et al., 1993; Leipoldt and Stern, in preparation). Some patients clinically manifest CF but without any detectable known mutation (4–5% of all CF patients). The same situation applies to families where the carriers need to be detected among sibs or parents of (sometimes deceased) patients. In those cases clinical geneticists are left with the remaining option of indirect geno-typing by linkage analysis with polymorphic extragenic markers (Dörk et al., 1992).

The human genome as a target for screening programs

Testing for genetic diseases among particular large groups of the population was performed long before the beginning of the DNA era, e.g., PKU screening in newborns, screening for Tay–Sachs disease among adult Ashkenazi, and screening for sickle-cell anemia among adult blacks in the USA. The aim was, on the one hand, to identify PKU affected newborns in order to inititate immediate dietary therapy. On the other hand healthy heterozygous carriers of the autosomal recessive disorders, Tay–Sachs disease and sickle-cell anemia, were to be detected, with subsequent counselling and information about the increased risk for the disease among offspring. The examples show that screening programs may be initiated for a similar purpose, i.e., prevention, but may fundamentally differ in the way the goal is achieved. PKU screening can lead to an efficient therapy and is therefore considered successful and beneficial (Bierich, 1991). In the case of Tay–Sachs disease and sickle-cell anemia in particular (Caskey, 1992), the obligatory test usually means a mass of problems for the individuals involved and their families (Wexler, 1992). One of the main problems that has been confronted before the inauguration of a screening program (set up by the WHO) is how to deal with consequences in terms of therapy, early intervention, or reproductive choice.

This problem can be solved by renouncing or by the tedious procedure of prenatal diagnosis with its eventual – cynically termed – "therapeutic" abortion as the only option following a positive screening for Tay–Sachs disease or sickle-cell anemia. The human genome analysis provides a rapidly expanding spectrum of diseases that may be screened for genetically. Molecular genetic screening may indeed become simple, efficient, safe, and cheap for a number of early-onset as well as late-onset diseases, genetic predispositions of multifactorial diseases, and genetic risk factors in tumorigenesis (Tab. 4). Again, cystic fibrosis may serve as an example that illustrates well the implications of the new type of genetic screening. As mentioned in the section above, in Northern Europe about 75% of the CF chromosomes carry the mutation *dF508*, and an additional 10% display one of four or five other known mutations. Therefore, carrier screening for CF, in concentrating on the most common mutations, has a detection rate of 85%, and 72% for carrier couples. The proportion of carriers identified by this approach varies from one European country to another. It may reach up to 95% in Great Britain and Denmark, whereas in parts of Southern Europe the detection rate is much lower as the mutation *dF508* occurs at a frequency of only 30% and as no other major CF mutation has so far been observed (Williamson, 1993; Lucotte and Hazout, 1990). Screening proponents are thus left with the problem that apart from a dozen common mutations several hundreds of rare, very rare or so-called private mutations exist which easily escape large-scale screening. Scanning methods such as SSCP and DGGE (denaturing gradient gel electrophoresis) analyses may in a number of conditions with marked molecular heterogeneity help identify mutated regions in genes without precisely marking the site and kind of mutation. In instances of altered SSCP or DGGE tests a direct mutation analysis follows. Molecular CF screening methods differ from programs such as carrier-screening for hemoglobopathies or Tay–Sachs disease regarding the detected spectrum of mutations. If, for example, 85% of the CF carriers (i.e., one in 25 individuals) can be identified, adult CF screening will detect the expected 0.1% of "two-carrier" couples. Compared with untested couples with an *a priori* one in 2500 risk of having a CF child, the risk will increase to one in four if the couple is

Table 4. Present and future of screening for genetic diseases

Commonly tested diseases among newborns on the protein level

Disease	Incidence	Therapeutical options
Sickle-cell anemia	1: 400*	Hematological
Tay–Sachs disease	1: 4000*	None
Cystic fibrosis	1: 2500	Pulmonary, dietary
Duchenne muscular dystrophy (DMD)	1: 3500	Physical therapy
Phenylketonuria (PKU)	1: 10000	Dietary
Galactosemia	1: 50000	Dietary
Congenital hypothyroidism	1: 4000	Thyroid replacement

Diseases testable among newborns by molecular methods

Disease	Incidence	Number of allels tested	Therapeutical options
Hemoglobinopathies	1: 1000	6–8	Hematological
Gaucher's disease	1: 2500*	4	Hematological
Cystic fibrosis	1: 2500	50	Supportive
Type II hyperlipidemia	1: 2500	10	Dietary, medication
Neurofibromatosis	1: 3000	> 100	Supportive
DMD	1: 3500	> 100	Supportive
Adult polycystic kidney disease	1: 1000	5	Dietary, medication
Alpha 1-antitrypsin deficiency	1: 4000	2	Pulmonary
PKU	1: 10000	5	Dietary
Galactosemia	1: 50000	4	Dietary
Huntington's disease	1: 30000	1	Supportive

Diseases where carrier screening is possible by molecular methods

Disease	Number of alleles tested	Frequency	Carrier frequency	High-risk population
Hemoglobinopathy βS and βC alleles	2	1: 400	1: 12	Black
$\beta 39$ and $\beta 112$ alleles	2	1: 2000	1: 20	Mediterranean
Cystic fibrosis	50	1: 2500	1: 20	Broad
Gaucher's disease	3	1: 2500	1: 25	Ashkenazi Jewish
Tay-Sachs disease	2	1: 4000	1: 30	Ashkenazi Jewish
Alpha 1-Antitrypsin deficiency	2	1: 4000	1: 10	Broad
Phenylketonuria	4	1: 10000	1: 100	Broad

* Relevant for parts of the US population; modified from Caskey, 1992.

tested positive. If, however, one partner tests positive while the other tests negative, the residual risk will be one in 620 instead of zero because the negative test might have missed an unconsidered *CF* mutation in about 20% of the cases. The "one-carrier couple problem" is a strong argument against the large-scale screening of *CF* mutations as long as a carrier detection rate of 95% is difficult to obtain (Wexler, 1992). Genetic diseases are mostly not accessible by causal therapy. A couple who undergo screening for – let us suppose – an autosomal recessive monogenic disorder and test positive will face restricted options for family planning. The reproductive choices of renunciation, selective abortion following prenatal diagnosis, adoption, heterologous insemination (not practiced in Germany), and acceptance of handicapped children are excruciating decisions. New technologies in prenatal diagnosis such as preimplantative and prezygotic diagnosis or the isolation of fetal cells from maternal blood (Verlinski et al., 1992; Simpson and Elias, 1994; Delhanty, 1994) may be helpful in some instances but do not offer general solutions. Another matter of ongoing debate is the method of community screening (Brock, 1994, 1995). Should screening involve all newborns, members of families at risk (cascade screening), or all those who feel a personal necessity to have themselves screened? If not done at birth, should screening take place at school, on the occasion of marriage or pregnancy, or should it generally be offered by primary healthcare services? Perhaps even more important is the question as to "what" should be screened. Should screening be undertaken in accordance with technological achievements? If not, genetic testing will be restricted to those diseases that represent a "major health problem". This may include common diseases such as thalassemias in parts of India and Italy, severe diseases, which applies for most of the conditions considered as well as for both common and severe diseases. Finally, screening could be restricted to diseases that can be treated efficiently. This applies to the fewest genetic conditions. Time, scale, purpose, and target group of screening are thus discussed controversially, and it is evident that each alternative has advantages and disadvantages in a medical as well as in a social context. Most of the debate has dealt with *CF* screening and pilot studies have been carried out in Great Britain and the USA. It is not too speculative to predict that *CF* will be the first candidate for an "official" test by a DNA-based method. Clearly, therapeutical options are still restricted for *CF* although conventional efforts have considerably progressed in recent years and although genetic therapeutic approaches appear quite promising (see following section).

Apart from detecting genetic diseases that manifest at birth or show an early onset, DNA-based testing extends to late-onset diseases as well as to polygenic conditions. The aim of screening in these cases is different and far more complex for Huntington's disease, atherosclerosis, Alzheimer's disease, colon carcinoma, breast carcinoma, and possibly in the future, obesity, and schizophrenia.

The human genome as a target for therapeutic endeavours

Gene therapy is generally defined as the transfer of genetic material into the cells of an organism in order to treat a disease. According to the classification of Wivel and Walters (1993) gene therapy should be defined more comprehensively as fourfold genetic intervention: (i) Treatment or prevention of a disease by correcting a genetic defect in all somatic cell types except germ cells (somatic gene therapy). Correction may be achieved by adding a functional gene copy to the genome of the target cell or by the replacement of a non-functional gene through a functional copy; (ii) correction or prevention of genetic deficiencies by the transfer of functional gene copies into pre- or postzygotic "reproductive" cells (germ cell therapy); (iii) modification of physical or psychical characteristics in order to obtain their "improvement" by targeting somatic cells with an additional gene or by gene replacement; (iv) modification of physical or psychical characteristics in order to obtain their improvement by targeting reproductive cells with gene replacement. Whereas approaches (i) and (ii) can be regarded as gene therapies in the proper sense, approaches (iii) and (iv) must be considered as powerful but as yet only potential forms of genetic enhancement.

Consistently, the target group for approaches (i) and (ii) are individuals carrying a genetic defect that results in a clinically manifest phenotype, i.e., patients in the conventional sense, the target group for approaches (iii) and (iv), on the other hand, can be considered "healthy" individuals without any severe genetic deficiency. The definition above contains the dynamics and problematics of the molecular modification of cellular genomes in that it points to the fluid boundaries between genetic therapy and enhancement on the one hand, and, on the other hand, includes the possible application of germ-cell therapy, which in many recent publications has been rejected on methodological or ethical grounds (Caskey, 1992; Trent, 1993). Even if, as will be outlined below, current attempts in gene therapy obviously aim at correcting a clinically relevant defect in somatic cells, it should be kept in mind that present procedures are merely the onset of an expanding field of genetic modifications in humans, given that the technological requirements will be established and evaluated. In particular, the current reserved attitude towards germline therapy might well be a purely operational one, in view of the currently insufficient methods for safe and efficient gene targeting and the widespread apprehensive public attitude concerning the manipulation of human embryos.

Numerous reports in recent years on animal models, *in vitro* experiments, and preclinical trials have proved that somatic gene therapy is in a hot phase of experimentation heading for clinical testing (Dickinson et al., 1995; Miller and Boyce, 1995). Experiments focus on the search for a vector able to transfer the therapeutic gene in an optimal way with regard to efficiency, stability, and safety. They also focus on the search for appropriate target cells capable of incorporating and

Table 5. Vector-systems used in somatic gene therapy

Viral vectors	
Retroviral vectors	Recombinant RNA viruses (MMLV)
	Replicative deficiency
	Integration into host DNA (gene targeting)
Adenoviral vectors	Recombinant adeno-type DNA viruses
	Replicative deficiency, deficient expression
	No integration into host DNA (gene application)
Other viral vectors	Adeno-associated viruses (AAV)
	Incapable of replication
	Specific integration into host DNA (human chromosome 19)
	Herpes simplex virus type I (HSV I)
	Neurotropic
Non-viral vectors	
Liposomes	
Transferrin	
Polylysine	
Artificial mammalian chromosomes (MAC) capacity > 400 kbp	

expressing the therapeutic gene, and the optimal moment of the therapeutic intervention, which depends on the condition and the physiology of the target cell. A number of criteria should be met before implementing a particular human gene therapy protocol: (i) Knowledge of the structure and expression pattern of the therapeutic gene; (ii) knowledge of the physiology of the target cells; (iii) accessibility of the target cells for *in vivo* or *ex vivo* manipulation; (iv) transductive efficiency of the vector system applied; (v) stable integration of the therapeutic gene into the genome of the target cell or stability of non-integrating constructs as "episomal" particles; (vi) significant and coordinate (cell-specific regulation, external induction) expression of the therapeutic gene

Table 6. Characteristics of vector-systems used in somatic gene therapy

Parameter	Retroviral	Adenoviral	Adeno-associated	Liposomes biopolymers
Efficiency of transduction	+ +	+	+	+
Transduction of non-proliferating cells	–	+	?	+
In vivo transduction	–	+	?	+
Stable integration	+	–	+	–
Pathogenicity in case of recombination	+	+	–	–

following introduction into the host cell; (vii) safety of the approach with respect to possible recombination between the vector virus and the helper virus or endogenous (pro-) viruses of the host cell, respectively; (viii) safety with respect to immunological side-effects; (ix) safety with respect to arbitrary, nonhomologous recombination and integration of the vector in genomic regions of functional importance; and (x) adequate timing of the therapeutic approach with respect to the differentiation of the target cell and the etiology of the disease. The most common approaches used at present (Dickson, 1995) involve recombinant defective viral vectors (Tab. 5) carrying the therapeutic gene, replication-competent helper viruses, and isolated target cells in an *ex vivo* protocol with subsequent reincorporation of the manipulated cells. As can be inferred from Table 6 none of the existing systems comprise enough favorable characteristics to be regarded as optimal.

On the basis of today's technology the ten main criteria forementioned, especially those dealing with stability of integration, efficiency of transduction, coordinate expression, and possible side-

Table 7. Candidates for somatic gene therapy

Monogenic disorders

Disease	Deficient gene product	Target cells
Severe combined immunodeficiency (SCID)	Adenosine desaminase	T-/B-lymphocytes
Sickle cell anemia	Beta-globin	Erythrocytes
Thalassemias	Alpha- and Beta-globin	Erythrocytes
Duchenne muscular dystrophy	Dystrophine	Myoblasts
Cystic fibrosis	CFTR protein	Lung, pancreas
Phenylketonuria	Phenylalanin-hydroxylase	Hepatocytes
Hypercholesterinemia	LDL-receptor	Hepatocytes
Hemophilia A, B	Factor VIII, IX	Platelets, keratinocytes
Lesch-Nyhan syndrome	Hypoxanthinephospho-ribosyltransferase	Basal ganglia

Tumors

Tumor	Target	Therapeutic gene	Strategy
Brain tumor	tumor cells	*HSV tk*	GCV-suicide
	tumor cells	*MDR 1*	cell protection
Malignant melanoma	tumor cells	*IL - 2*	immune stimulation
	fibroblasts	*IL - 4*	immune stimulation
Colon carcinoma	tumor cells	*TNF*	induction of tumor necrosis
Lung carcinoma	tumor cells	*antisense k-ras*	inhibition of expression

(tk) thymidine kinase; (GCV) ganciclovir; (MDR) multidrug resistance; (IL) interleukin;
(TNF) tumor necrosis factor

effects can only be partially fulfilled. As a consequence, for several candidates of DNA-based therapy, where the molecular genetics, the biochemistry, and the pathophysiology of the underlying defect are well understood, the experimental stage of exploitation is still pending the beginning of preclinical trials (Tab. 7). Since 1990 and despite a number of shortcomings, about one hundred somatic gene therapies have been initiated (Tab. 8). In cancer therapy several remarkable results have been obtained (Moorman and Culver, 1995; Bridgewater and Collins, 1995).

It should be emphasized that in most cases cancer therapy trials are not gene therapies *in sensu strictu*; they rather aim at the destruction of tumor cells at the immunostimulatory level or at the introduction of toxin genes in either cancer or tumor-infiltrating cells. Undoubtedly, DNA-based therapies for human tumors are much more frequent than therapies for monogenic disorders, if one considers both the number of treatment centers and the number of patients undergoing therapy. Furthermore, humans suffering from severe infectious diseases have become good candidates for gene therapy. In 1990 the first experimental models were set up for the treatment of AIDS. Meanwhile, studies have been extended to hepatitis B, influenza A, rabies, and malaria using DNA-based immunizations. The latter ones are still restricted to animal models (Davis and Whalen, 1995). In the future human cancer and infectious diseases will be the predominant targets for gene therapy, as a result of the large number of patients and of the equally broad appli-

Table 8. Somatic gene therapy in humans – current and approved protocols

Disease	Therapeutic gene	Strategy	Target cells	Success reported
SCID	*ADA*	Restoration of enzyme activity	lymphocytes	yes[1]
Malignant melanoma	*TNF*	Induce tumor necrosis	tumor infiltrating lymphoytes	yes[2]
Hypercholesterinemia	*LDL*-receptor	Synthesis of functional receptor	hepatocytes	yes[3]
Cystic fibrosis	*CFTR*	Restore chloride-channel	airway epithelia	yes[4]
Ovarian cancer and brain tumor	*HSV tk*	Induce GCV-mediated tumor destruction	cancer cells	no
Factor IX deficiency	*f9-YAC*	Synthesis of factor 9	keratinocytes	no
HIV infection	*sCD4*	Inhibition of HIV-binding to cellular *CD4* by competition	lymphocytes	no[5]
HIV infection	viral *CD4*-receptor antisense	Inhibition of HIV-binding to *CD4* by blocking receptor synth.	lymphocytes	no[6]

[1]Kinnon, 1995); [2]Moorman and Culver, 1995; [3]Prentice and Webster, 1995; [4]Wadsworth and Smith, 1995; transient expression only, inflammatory complication; [5]approved but first clinical phase not yet started; [6]Heidenreich et al., 1995; clinical trial, phase II

cability of immunostimulation and cellular immunization methods which can be applied to various types of cancer (genetic diseases, on the contrary afford individual and highly specific protocols). Therefore, the discussion on genetic interventions in humans should also include infectious diseases. Most clinical trials for the genetic treatment of monogenic disorders involve a very small number of patients (e.g., 50–60 patients world-wide in the case of cystic fibrosis). Even if one hypothesizes that world-wide all CF sufferers were to undergo therapy the total of roughly 60 000 still appears insignificant compared with about 20 million cancer patients and 200–300 million individuals suffering from pandemic infectious diseases every year.

What will the achievements of the Human Genome Project be as far as future developments in gene therapy are concerned? As well as the discovery of new and clinically significant genes, a better understanding of genomic interactions will help in the construction of gene therapeutics that can be reliably and efficiently regulated. A steadily increasing knowledge of the human DNA sequence will facilitate the construction of vector molecules capable of exact, sequence-specific recombination, thus rendering the gene-to-host genome transfer a precise event.

Social and ethical considerations

The Human Genome project is sometimes compared to the Apollo 11 Project. This is appropriate if the enormous financial support is considered. The overall beneficial effect for society and the individual, however, is certainly debatable. For most humans the venture played no immediate part in their lives. Does this also hold true for the Human Genome Project? According to human geneticists every single human carries half a dozen recessive lethal alleles. Theoretically, over the next years everybody should be able to be informed about his or her negative traits. Will this information be at everybody's disposal? The potential abuse of gene manipulation and individual genetic data has been repeatedly stressed and discussions began long before HUGO came into being (Kaufmann, 1964; Beckwith, 1976; Leipoldt and Schempp, 1983a, 1983b). Undoubtedly, HUGO will have a tremendous scientific impact on the diagnosis and treatment of genetic disorders, but the project's findings also carry the inherent risk of promoting new eugenic concepts. These basically resemble the classical notions but gain a completely new dimension in view of modern technologies and the concerted power of HUGO's research (Garver and Garver, 1994). New eugenic concepts will not appear as rudimentary as those held, e.g., in the USA of 1924 or in the Germany of 1933 when they were directed against "inferiors", ethnic groups, and races. Today these notions will probably invade our community in a more subtle way, i.e., in terms of genetic discrimination. This of course presumes the existence of the appropriate technology, the readiness to realize discriminative concepts, and the susceptibility of our society to eugenic theories. The technological advances made in the course of HUGO's duration will surely refine the existing repertoire of predictive and presymptomatic tests, the identification of predisposing factors, and the establishment of genetic concepts explaining variant or "aberrant" human behavior (Hamer et al., 1993; Brunner et al., 1993). Genetic discrimination is already being practiced in the fields of insurance, employment, education and even leisure activities (Billings et al., 1992). Simultaneously, we face the phenomenon of geneticizing various fields of medical practice. As has been outlined in the preceding sections, genetic, i.e., molecular concepts and methods will increasingly stamp clinical diagnosis, prenatal diagnosis, reproductive medicine, newborn and adult screening, health care, as well as disease treatment and management. Molecular biologist Sidney Brenner from Cambridge University in the UK, recently stated in a BBC interview: "Genetics will no longer exist as a biological discipline by the end of our millennium, rather will every field in biology and medicine be a sub-discipline of a universal genetic concept." In view of such predictions it seems vitally important that the American Human Genome Project established an accompanying "Ethical, Legal and Social Implications Program" (ELSI) dealing with the community's apprehensions relating to the new genetics and the monitoring of fatal developments. Other aspects of biomedicine such as embryo-research, gene technology, and gene therapy are embedded in national regulations and legislations (e.g., the strict regulation of gene therapy in the US since 1980, the establishment of the "Human Fertilization and Embryology Authority" in the UK in 1991; a comprehensive French bioethic legislation launched in 1994). The European Council is preparing a bioethic convention that will deal with topics such as gene therapy, predictive testing, embryo research, and privacy of genetic data (Zell and Leipoldt, 1995). In Germany, we have the "Gentechnikgesetz" and the "Embryonenschutz-gesetz"; somatic gene therapy is presently regulated by the German "Arzneimittelrecht" and a "Gentherapiegesetz" is in preparation. Bioethics has already been established in the U.S.A. for some years and has become a popular discipline in Europe with respect to national programs as well as supra-national network projects. Bioethic research programs are expanding but should remain a multi-disciplinary and independent area. Otherwise they might degenerate to either fundamentalistic round tables of unmitigated criticism or unquestioned acceptance and application.

Has genetics in the 21st century the potential to shape humankind?

The Human Genome Project illustrates well that the community and its individual members are largely influenced by new molecular genetic technologies. As far as the scientific, technological, and applied aspect is concerned we have already achieved and expect in the near future an even greater wealth of new insights into genome organization, gene regulation, and developmental genetics. Technological improvements will permit the analysis of complex genomic segments with a degree of precision, speed, and economics hitherto unanticipated. Genetic defects in monogenic, multifactorial, and familial diseases, as well as genetic factors in tumorigenesis will subsequently be discovered and will lead to exact diagnosing and screening methods, finally opening up possibilities for genetic therapy. Theoretically, in Germany more than a thousand families every year could benefit from these achievements. This figure must, however, not obscure the fact that most genetic diseases are rare, hence from the population's point of view only few individuals would be involved. In particular, the number of genetic treatments of diseases is by far much lower than the number of organ transplantations carried out. As long as gene therapy only involves somatic cells there are of course no consequences for our gene pool (Müller, 1993). Any revolution brought about by the Human Genome Project thus will obviously not be a genetic or biological one. The revolution – if the term is correct at all – may perhaps take place in our way of thinking. We will learn through the various projects of the human genome analysis that genetic components can be detected in almost all conditions of human diseases, behavior, predilections and sexual proclivities. From time to time we may get the impression that our genetic constitution is the undisputed basis for health or illness, acceptable or deviant behavior, social adaptation and integration or the inability to integrate. If the concepts of a genetically determined human life gains a secure foothold in the next century we might experience a community more backward than it is at present. In his discussion on the genetic contribution to the phenotype, Wolf (1995) concludes: "… the genotype–phenotype relationship cannot be expected to follow strict laws, but rather to be irregular. Indeed, this relationship is highly prone to variation. The phenotype is not deducible from the genotype. The fact that variation is restricted to the extent that, within certain limits, individual phenotypes can be assigned to individual mutations and *vice versa* can be ascribed to the extensive similarity, if not identity, of the majority of conditions for the development and functioning of the organism, so that the respective mutation is the main variable within the system."

References

Beckwith, J. (1976) Social and political uses of genetics in the United States: Past and present. *Ann. N.Y. Acad. Sci.* 265, 46–58.

Bell, C.J., Budarf, M.L., Nieuwenhuijsen, B.W., Barnoski, B.L., Buetow, K.H., Campbell, K., Colbert, M.A.E., Collins, J., Daly, M., Desjardins, P.R., DeZwaan, T., Eckman, B., Foote, S., Hart, K., Hiester, K., van Het Hoog, M.J., Hopper, E., Kaufman, A., McDermid, H.E., Overton, G.C., Reeve, M.P., Searls, D.B., Stein, L., Valmiki, V.H., Watson, E., Williams, S., Winston, R., Nussbaum, R.L., Lander, E.S., Fischbeck, K.H., Emanuel, B.S. and Hudson, T.J. (1995) Integration of physical breakpoint and genetic maps of chromosome 22. Localization of 587 yeast artificial chromosomes with 238 mapped markers. *Hum. Mol. Genet.* 4, 59–69.

Bierich, J.R. (1991) Postnatale Diagnostik – Screening. *In*: D. Beckmann, K. Istel, M. Leipoldt and H. Reichert (eds): *Humangenetik.* P. Lang, Frankfurt/M., pp 175–179.

Billings, P.R., Kohn, M.A., de Cuevas, M., Beckwith, J., Alper, J.S. and Natowicz, M.R. (1992) Discrimination as a consequence of genetic testing. *Am. J. Hum. Genet.* 50: 476–482.

Boat, T., Welsh, M. and Beaudet, A. (1989) Cystic fibrosis. *In*: C. Scriver, A. Beaudet, W. Sly and D. Valle (eds): *The Metabolic Base of Inherited Disease.* McGraw-Hill, New York, pp 2649–2860.

Bridgewater, J. and Collins, M. (1995) Vaccine immunotherapy for cancer. *In*: G. Dickson (ed.): *Molecular and Cell Biology of Human Gene Therapeutics.* Chapman and Hall, London, pp 140–156.

Brock, D.J.H. (1994) Carrier screening for cystic fibrosis. *Prenat. Diagn.* 14: 1243–1252.

Brock, D.J.H. (1995) Heterozygote screening for cystic fibrosis. *Eur. J. Hum. Genet.* 3: 2–13.

Brunner, H.G., Nelen, M., Breakefield, X.O., Ropers, H.H. and van Oost, B.A. (1993) Abnormal behavior associated with a point mutation in the structural gene for monoamine oxidase A. *Science* 262: 578–580.

Caskey, C.T. (1992) DNA-based medicine: Prevention and therapy. *In*: D.J. Kevles and L. Hood (eds): *The Code of Codes.* Harvard Univ. Press, Cambridge/MA., pp 112–135.

Chumakov, I., Rigault, P., Guillou, S., Ougen, P., Billaut, A., Guasconi, G., Gervy, P., LeGall, J., Soularue, P., Grinas, L., Bougueleret, L., Bellanne-Chantelot, C., Lacroix, B., Barillot, E., Gesnouin, P., Pook, S., Vaysseix, G., Frelat, G., Schmitz, A., Sambucy, J.-L., Bosch, A., Estivill, X., Weissenbach, J., Vignal, A., Riethman, H., Cox, D., Patterson, D., Gardiner, K., Hattori, M., Sakaki, Y., Ichikawa, H., Ohki, M., LePaslier, D., Heilig, R., Antonarakis, S. and Cohen, D. (1992) Continuum of overlapping clones spanning the entire chromosome 21q. *Nature* 359: 380–387.

Comittee on Mapping and Sequencing the Human Genome (1988) *Mapping and Sequencing the Human Genome.* Natl. Acad. Press, Washington, D.C.

Davis, H.L. and Whalen, R.G. (1995) DNA-based immunization. *In*: G. Dickson (ed.): *Molecular and Cell Biology of Human Gene Therapeutics.* Chapman and Hall, London, pp 368–387.

Delhanty, J.D.A. (1994) Preimplantation diagnosis. *Prenat. Diagn.* 14: 1217–1227.

Devoto, M., Ronchetto, P., Fanen, P., Orriols, J.J.T., Romeo, G., Goossens, M., Ferrari, M., Magnani, C., Seia, M. and Cremonesi, L. (1991) Screening for *non-dF508* mutations in five exons of the cystic fibrosis (CFTR) gene in Italy. *Am. J. Hum. Genet.* 48: 1127–1132.

Dickinson, P., Dorin, J.R. and Porteous, D.J. (1995) Modelling cystic fibrosis in the mouse. *Mol. Med. Today* 1: 140–148.

Dickson, G. (1995) *Molecular and Cell Biology of Human Gene Therapeutics.* Chapman and Hall, London.

Dörk, T., Neumann, T., Wulbrand, K., Wolf, B., Kühn, N., Maaß, G., Krawczak, M., Guillermit, H., Ferec, C., Horn, G., Klinger, K., Kerem, B.S., Zielenski, J., Tsui, L.-C. and Tümmler, B. (1992) Intra- and extragenic marker haplotypes of CFTR mutations in cystic fibrosis families. *Hum. Genet.* 88: 417–425.

Dörk, T., Mekus, F., Schmidt, K., Boßhammer, J., Fislage, R., Heuer, T., Dziadek, V., Neumann, T., Kälin, N., Wulbrand, K., Wolf, B., v.d. Hardt, H., Maaß, G. and Tümmler, B. (1994) Detection of more than 50 different CFTR mutations in a large group of German cystic fibrosis patients. *Hum. Genet.* 94: 533–542.

Fields, C., Adams, M.D., White, O. and Venter, J.C. (1994) How many genes in the human genome? *Nature Genet.* 7: 345–346.

Foote, S., Vollrath, D., Hilton, A. and Page, D.C. (1992) The human Y chromosome: Over-lapping DNA clones spanning the eucromatic region. *Science* 258: 60–66.

Garver, K.L. and Garver, B. (1994) The human genome project and eugenic concerns. *Am. J. Hum. Genet.* 54: 148–158.

Hamer, D.H., Hu, S., Magnuson, V.L., Hu, N. and Pattatucci, A.M. (1993) A linkage between DNA markers on the X chromosome and male sexual orientation. *Science* 261: 321–327.

Heidenreich, O., Shin-Heh Kang, X.X. and Nerenberg, M. (1995) Application of antisense technology to therapeutics. *Mol. Med. Today* 1: 128–133.

Kaufmann, R. (1964) *Die Menschenmacher. Die Zukunft des Menschen in einer biologisch gesteuerten Welt.* S. Fischer-Verlag, Frankfurt/M.

Kinnon, C. (1995) Inherited immuno-deficiencies. *In*: G. Dickson (ed.): *Molecular and Cell Biology of Human Gene Therapeutics.* Chapman and Hall, London, pp 157–174.

Leipoldt, M. (1991) Prädiktive Medizin: Analyse des menschlichen Genoms. *In*: D. Beckmann, K. Istel, M. Leipoldt and H. Reichert (eds): *Humangenetik.* P. Lang-Verlag, Frankfurt/M., pp 155–164.

Leipoldt, M. and Schempp, W. (1983a) Genetische Manipulation, Part I. *Notabene Medici* 5: 373–377.

Leipoldt M. and Schempp, W. (1983b) Genetische Manipulation, Part II. *Notabene Medici* 6: 464–469.

Lucotte, G. and Hazout, S. (1990) A NW-SE decreasing gradient of the *dF508* frequencies in Europe. *Newslett. EC Cystic Fibrosis Consort.*, 1.

McKusick, V.A. (1994) *Mendelian Inheritance in Man*, 11th Edition. J. Hopkins Univ. Press, Baltimore.

Miller, J.B. and Boyce, F.M. (1995) Gene therapy by and for muscle cells. *Trends Genet.* 11: 163–165.

Moorman, D.W. and Culver, K.W. (1995) Gene transfer and cancer chemotherapy. *In*: G. Dickson (ed.): *Molecular and Cell Biology of Human Gene Therapeutics.* Chapman and Hall, London, pp 125–139.

Morton, N.E. (1991) Parameters of the human genome. *Proc. Natl. Acad. Sci. USA* 88: 7474–7476.

Müller, H.J. (1993) Treatment of human disorders with gene therapy and its consequences for the human gene pool. *In*: K. Wöhrmann and J. Tomiuk (eds): *Transgenic Organisms: Risk Assessment of Deliberate Release.* Birkhäuser Verlag, Basel, pp 229–244.

Nelkin, D. (1992) The social power of genetic information. *In*: D.J. Kevles and L. Hood (eds): *The Code of Codes.* Harvard Univ. Press, Cambridge/MA., pp 177–190.

Newton, C.R. and Graham, A. (1994) *PCR.* Bios Scientific Publ., Oxford.

Prentice, H. and Webster, K.A. (1995) Cardiovascular disease. *In*: G. Dickson (ed.): *Molecular and Cell Biology of Human Gene Therapeutics.* Chapman and Hall, London, pp 281–300.

Reiss, J., Ellermeyer, U., Rininsland, F., Ballhausen, P., Lenz, U., Wagner, S. and Schlösser, M. (1993) A comprehensive CFTR mutation analysis of German cystic fibrosis patients. *Hum. Mol. Genet.* 2: 809–811.

Saiki, R.K., Gelfand, D.H., Stoffel, S., Scharf, S.J., Higuchi, R., Horn, G.T., Mullis, B. and Erlich, H.A. (1988) Primer-directed enzymatic amplification of DNA with a thermo-stable DNA polymerase. *Science* 239: 487–491.

Schmidtke, J. (1995) Molekulare Diagnostik in der BRD und Nachbarländern. *Med. Genetik* 1/95: 40–46.

Shrimpton, A.E., McIntosh, I. and Brock, D.J.H. (1991) The incidence of different cystic fibrosis mutations in the Scottish population: Effects on prenatal diagnosis and genetic counselling. *J. Med. Genet.* 28: 317–321.

Simon-Bouy, B., Mornell, E., Serre, J.L., Taillandier, A., Boue, J. and Boue, A. (1991) Nine mutations in the cystic fibrosis (*CF*) gene account for 80% of the CF chromosomes in French patients. *Clin. Genet.* 40: 218–224.

Simpson, J.L. and Elias, S. (1994) Isolating fetal cells in maternal circulation for prenatal diagnosis. *Prenat. Diagn.* 14: 1229–1242.

Trent, R.J. (1993) *Molecular Medicine.* Churchill Livingstone, Edinburgh.

Tsui, L.-C. (1992) The spectrum of cystis fibrosis mutations. *Trends Genet.* 8: 392–398.

Verlinsky, Y., Rechitsky, S., Evsikov, E., White, M., Cieslak, J., Lifchez, A., Valle, J., Moise, J. and Strom, C.M. (1992) Preconception and preimplantation diagnosis for cystic fibrosis. *Prenat. Diagn.* 12: 103–110.

Wadsworth, S.C. and Smith, A.E. (1995) Cystic fibrosis and lung disease. *In*: G. Dickson (ed.): *Molecular and Cell Biology of Human Gene Therapeutics.* Chapman and Hall, London, pp 237–251.

Weissenbach, J. (1995) Genethon report 1994/95. Oral contribution, *17. ESHG congress,* Berlin, May 1995.

Wexler, N. (1992) Clairvoyance and caution: repercussions from the human genome project. *In*: D.J. Kevles and L. Hood (eds): *The Code of Codes.* Harvard Univ. Press, Cambridge/MA., pp 211–243.

Williams, N. (1995) Closing in on the complete yeast genome sequence. *Science* 268: 1560–1561.

Williamson, R. (1993) Universal community carrier screening for cystic fibrosis? *Nature Genet.* 3: 195–201.

Wivel, N.A. and Walters, L. (1993) Germ-line gene modification and disease prevention: Some medical and ethical perspectives. *Science* 262: 533–538.

Wolf, U. (1995) The genetic contribution to the phenotype. *Hum. Genet.* 95: 127–148.

Zell, R. and Leipoldt, M. (1995) Entwurf einer Konvention zum Schutz der Menschenrechte und der Menschenwürde im Hinblick auf die Anwendung von Biologie und Medizin. Stellungnahme des AK Biomedizin / VdBiol. *BIUZ* 2: 17–18.

Zeng, Y., Carter, N.P., Price, C.M., Colman, S.M., Milton, P.J., Hackett, G.A., Greaves, M.F. and Ferguson-Smith, M.A. (1993) Prenatal diagnosis from maternal blood: Simultaneous immunophenotyping and FISH of fetal nucleated erythrocytes isolated by negative magnetic cell sorting. *J. Med. Genet.* 30: 1051–1056.

Zielenski, J., Rozmahel, R., Bozon, D., Kerem, B.-S., Grzelczak, Z., Riordan, J.R., Rommens, J. and Tsui, L.-C. (1991) Genomic DNA sequence of the cystic fibrosis transmembrane conductance regulator (CFTR) gene. *Genomics* 10: 214–228.

History of and progress in risk assessment

J. Tomiuk and A. Sentker[1]

Department of Population Genetics, University of Tübingen, Auf der Morgenstelle 28, D-72076 Tübingen, Germany
[1]Eduard-Spranger-Straße 41, D-72076 Tübingen, Germany

Summary. The history of and the progress in risk assessment for the deliberate release of transgenic organisms is described in the light of three cornerstones: The conference at Asilomar (1975), at Kiawah (1990) and at Goslar (1992). Since Asilomar, one of the major goals has been to determine the probability that genes or organisms can escape human control and invade ecosystems. Studies related to risk assessment have provided more precise information on gene flow and the spread of genes or organisms. The conclusion consistently drawn at all meetings, however, was that although our understanding of the risk potential of transgenic organisms has increased considerably, more information is needed. The "creation" of new organisms and their release into the environment always bears some unpredictable hazard potential. The ecologically-based risk of organisms, however, can be analyzed by means of analogy-based studies of species' invasions or population biology, but the evolutionary-based risks are unpredictable. We are now at a stage where we can define more precisely the risk potential but can still not exclude any one kind of risk.

First considerations of risk

The first intensive considerations of a risk involving the application of new biological technologies and the release of genetically modified organisms were raised in the early seventies. In 1971 Paul Berg announced plans to clone the adenovirus SV40 in *Escherichia coli*. The plans, however, were aborted because the animal tumor virus SV40 can combine with virus S2 which is often present in humans, and the scientists were discouraged by the unknown hazard potential due to the possible transfer of cancer gene characters into humans. Berg and some of his colleagues demanded that laboratory research on the recombination of DNA from different species be deferred as long the risk potential remained uncertain (Berg et al., 1974). "There is a serious concern that some of these artifical recombinant DNA molecules could prove biologically hazardous" (Berg et al., 1974).

Asilomar

The discussion concerning the possible inherent risk in gene-technological experiments finally led to the Asilomar conference in California in 1975. Here the participants considered the potential biohazards of recombinant DNA molecules and the ways in which scientific work could be undertaken with minimal risks for laboratory workers, for the public at large, and for the animal and plant species sharing our ecosystems. High security standards were established to avoid any hazard potential due to inadequate laboratory working conditions (Berg et al., 1975). Different risk levels were defined and laboratory procedures were proposed in order to avoid the related biohazards. It was agreed that biological and physical barriers had to be implemented in order to ensure the containment of newly created organisms. It was proposed that research concentrate on the development of safe vectors, hosts and laboratory procedures, and that the safety hazard potential be reconsidered as new information became available.

It was clearly recognized that the future state of knowledge might indicate that the possible safety risks were in fact lower than formerly had been anticipated. But as long as the risk potential remained unknown, the experimental design should focus on optimizing the containment conditions of recombinant DNA material, meticulously and on an increasing scale. In the longer term, however, serious problems were expected with regard to large-scale applications of genetically modified organisms. The limited information concerning the survival of laboratory strains in different ecological niches under natural conditions demanded that more research be undertaken before the strains could be handled on a larger scale (Berg et al., 1975).

After the Asilomar conference a series of publications, national meetings, and state regulations in the USA dealt with the complex of problems connected with the handling of recombinant DNA material or genetically modified organisms (in 1976 the US National Institute of Health, NIH, set out guidelines for their physical and biological containment; since 1978 the NIH guidelines have continuously downgraded the containment levels). Knowledge concerning the molecular genetics of microorganisms had considerably increased and the risk potential of many of them had been assumed to be relatively low. Consequently, in 1980 the NIH guidelines for the handling of recombinant DNA material were further downgraded, and in 1981 experiments with *Escherichia coli, Bacillus subtilis*, and yeast no longer had to be registered.

New progress in research permitted the design of organisms which could survive in nature. As a result, this also led to requests for the release of transgenic organisms. In the USA, the third request for the release of GMOs in 1982 from the University of California at Berkley raised public protest. The former safety supplements that had mainly regulated the containment in laboratories were now extended to containment in field tests. Because of concern about the possible damage to ecosystems, the advisory boards dealing with gene-technological experiments were supplemented by ecologists and population geneticists. Step-by-step procedures were proposed to test the biological characteristics of organisms, i.e., first excluding risks then increasing the potential risk levels – from the laboratory to the greenhouse *via* small-scale field tests to large-scale field tests and commercial applications.

In microorganisms the rate and nature of horizontal (infectious) genetic transmission and the stability of the engineered genetic changes (role of transposable genetic elements) were considered to be relevant genetic properties for risk assessment studies. The likelihood and nature of host range shifts, unregulated propagation, and changes in virulence (parasites and pathogens), the effects on competitors, on prey, hosts and symbionts, on predators, parasites and pathogens, the role of introduced microorganisms as vectors, pathogens and, finally, the effects on ecosystem processes and on habitats were seen to be important evolutionary and ecological characteristics (Fiksel and Covello, 1986). For only a few species are many of these characteristics known (e.g., *E. coli*) which allow a relatively precise prediction of the behavior of the released organisms. But important for the analysis of releases is an understanding of the characteristics of the original, nonmodified, parental organism. Analogy-based approaches to the risk assessment of released organisms have limited applicability when little is known about the parental organism, related indigenous organisms, or even whether known organisms are released into a foreign environment (Fiksel and Covello, 1986, pp 1–34).

Most publications dealing with microorganisms considered optimistically the available methodology of risk assessment (Klingmüller, 1988). The detailed information relating to various microorganisms suggested the risk potential could largely be recognized in advance and kept at a minimum if organisms were released into the environment. Nevertheless, the consistent demand was always that more information and an improvement in the available test procedures were needed to define more exactly the risk potential (Fiksel and Covello, 1986, pp. 1–34).

The 1st international meeting in Kiawah

In November 1990 the first international symposium on "The Biosafety Results of Field Tests of Genetically Modified Plants and Microorganisms" was organized in Kiawah, South Carolina (MacKenzie and Henry, 1990).

"We are clearly at a significant juncture in the devlopment of agricultural and environmental biotechnology" (Hess, 1990, published in MacKenzie and Henry, "Preface", p. XV).

Fifteen years after Asilomar and following a steady growth of knowledge in the field of molecular genetics, the initial cautious view with respect to the release of genetically modified organisms had more or less disappeared. For example, an audacious argument was that the "background microbial populations are complex, thus the release of genetically modified microorganims (GMM) may be insignificant by comparison" (MacKenzie and Henry, 1990, "Towards a Consensus", p. 277). It was further argued by some that on the basis of their own extended experience with gene technology they could detect no obvious risk. Consequently, the representation of scientific research and of the available scientific data should now forge ahead so that the application of the new technology on an industrial scale can be accepted by the public. It was put forward that one should look carefully at the economic costs, thus implying that case-by-case procedures were too expensive and therefore not generally practicable. Furthermore, policy should also consider the benefits of gene-technological research and not focus exclusively on its problems. The main arguments in favor of gene technology and the release of GMOs dealt with their potential benefit to humankind and, of course, to breeders and industry (Bridges, 1990, published in MacKenzie and Henry, 1990, "Towards a Consensus", p. 273). Scientists had left their academic ivory tower. In the papers contributed to the symposium, the biosafety of genetically modified organisms was considered controversially. On the one hand, continued caution was recommended where no information was available concerning, e.g., the interrelationship between genotype, crop type, outcrossing potential, disposal of final products, site security and the social–political climate (Muench, 1990, published in MacKenzie and Henry). Kareiva et al. (1990, published in MacKenzie and Henry) stated that the statistical analyses of gene flow and spread must be improved. Rees et al. (1990, published in MacKenzie and Henry) pointed out the importance of the experimental design. It was argued that we needed test procedures, more precise estimates for the isolation distances in field tests, more practical experience in field tests, better use of laboratory dates for the release, and better and more rapid measures of gene flow and gene expression. It was also suggested that studies should be conducted into the differences between small and large-scale releases. Lacy and Stromberg (1990, published in MacKenzie and Henry) again explained that little was known about the biology of microbes. We must study the environmental destiny of some gene products or secondary products, the competition between GMOs and populations, the significance of population densities and the phenomenon of viable but non-culturable bacteria cells, and the relationships between microcosm results and field test performance of microbes. On the other hand, many authors found no risk at all in the release of genetically modified organisms (e.g., Bakker et al.; Cook and Weller; Deshayes; Fuchs and Serdy; Klingmüller; Kluepfel and Tonkyn; Kluepfel et al., 1990, published in MacKenzie and Henry). The message for future work in gene technology was to identify "useful" genes, to map genomes and to develop efficient methods of transformation. Research should focus on the improvement of detection methods, sampling methodologies, monitoring protocols, and modelling techniques (MacKenzie and Henry, 1990, "Towards a Consensus", p. 274). A better understanding of the risk potential should be further supported by the representation of information about negative research results of the kind indicating that no gene transfer had been observed. The use of information from small-scale tests combined with literature and laboratory data was considered to be the starting point for large-scale testing and the commercial use of genetically modified organisms (Anderson and Betz, 1990, published in MacKenzie and Henry).

The precision of the new technology made possible genetic modifications in microbial strains that could be characterized more fully (NSC, 1989, p. 123). It was argued that breeders could better improve traits by the transfer of single genes excluding the noise of genomic backgrounds. But in analyzing the population consequences of genetic design in sexually-reproducing organisms, only a low value of gene technology for practical purposes was found, but a high potential for the long-term understanding of the processes in animal and plant breeding, and therefore also for the understanding of evolutionary processes. The advantage of handling single genes was levelled out by the uncertainties involving the various interactions between a gene and its genomic background (Christiansen, 1990).

Only a few mathematical or statistical models were provided to consider biotechnology risk assessment (but see for example Ginzburg, 1991; Levin and Strauss, 1991). The procedures for estimating relevant parameters, e.g., gene flow, gene spread and selection coefficients, could be further improved and estimates could be more accurate (Lenski; Manasse, 1991, published in Ginzburg). However, the theoretical approaches showed that there was no generally valid solution to the problem. General models were not suitable in the instance where considerable effects on the environment were expected. If environmental effects were the very sensitive product of the input disturbance, then individual studies were needed to determine the extent of risk (Condit, 1991, published in Ginzburg). Finally, engineering could never completely prevent disasters, but the probability that a disaster would occur could be tested (Kim et al., 1991, published in Ginzburg).

The 2nd international meeting in Goslar

"The promised benefits for producers and consumers still lay ahead. On the 2nd International Meeting in Goslar (Germany, 1992) great progress was reported compared to the first symposium" (Casper and Landsmann, 1992, p. XI).

There was a consensus among scientists that risk is a function of the traits of new organisms rather than the consequence of a specific technique to modify them (Tiedje et al., 1989). Furthermore, biological laws are valid, whether the varieties are produced by traditional breeding or by modern techniques. Crawley (1992, published in Casper and Landsmann) argued that release does not pose substantial new threats to the environment. In the case of herbicide-tolerant oilseed rape and maize, results suggested that further ecological work was not necessary before the decision to allow large-scale use. Other results contradicted Crawley's view. Darmency and Renard (1992, published in Casper and Landsmann) found that recombination genes survived longer in the soil if interspecific hybridization occurred. Dale et al. (1992, published in Casper and Landsmann) proposed that for potatoes at least 3–4 years of monitoring is necessary to obtain a reasonable risk assessment. But Crawley (1992, published in Casper and Landsmann) also argued that the results obtained for some organisms could not be simply extrapolated to the behavior of other organisms or to other transgenic constructs. The conclusion of the symposium was again as before – more studies were needed to understand the biology of organisms and, therefore, the destiny of transgenic organisms in the field.

In agricultural science, we have a long history of breeding new traits in species, with no serious damage to the environment, e.g., no problems have been observed during the cultivation of potatoes in Europe. Based on the results of numerous field tests where no events harmful to our environment were described, it was concluded that we could now concentrate on the large-scale application of GMOs. Likewise, no risk was assumed in relation to other organisms carrying antibiotic resistances, because these are already widespread in the natural environment and thus exchange is commonplace.

In the same year, 1992, a symposium was organized in Basel (Switzerland). Perhaps the ecological preponderance fostered a more cautious point of view. Although no negative ecological consequences following the introduction of transgenic crop plants had been reported to date, this had to be regarded only as a provisional statement on the present situation (Altmann, 1993). The evolutionary behavior of organisms was unpredictable and thus considerations about their impact on the target area or the ecosystems always bore a trace of risk (Regal, 1993; Sukopp and Sukopp, 1993; Williamson, 1993). The use of transgenic microorganisms promised a benefit for the repair of environmental damage (Lenski, 1993) but their autonomous replication made it impossible to remove them completely from the habitat if negative effects were observed.

In 1992 a symposium on the risk assessment for the deliberate release of transgenic organisms was held in Blaubeuren (Ulm, Germany; proceedings published by Wöhrmann and Tomiuk, 1993). The hazards were not considered as lightly as in Goslar, and opinions were similar to those held in Basel. Many examples of the risk of introduction of plants or animals into a new habitat were known from studies of species' invasions (see, e. g., Wöhrmann and Tomiuk, 1993). The consensus among ecologists and molecular biologists was again that little is known about the biology of species and we need more research to quantify the risk that can attend the release of alien organisms. For example, a study of the chances of spontaneous gene flow from 42 cultivated plants of the wild flora of the Netherlands indicated that gene transfer is expected to cause no ecological effects in 50% of species, there might be a small effect in 15%, a considerable effect in 25% and there is an unknown risk for 10% of the plants (de Vries et al., 1992).

In 1993 and 1994 a series of workshops and symposia were held, during which the risk of released GMOs was considered. The workshop on "Application of Agricultural biotechnology and Safety Considerations" (Casper, 1993) in China focused on the large-scale use of engineered plants in agriculture. This workshop report conveys the impression that there were two interest groups present: Scientists from industrial countries wanting to know the Chinese experience with their manifold large-scale field releases, and Chinese scientists interested in the technical improvements in genetic engineering. The 3rd International Meeting in Monterey (USA) can only be mentioned briefly (the final report of the symposium was not available). Stone (1994) reviewed the contributions to the meeting where researchers again discussed controversially the risk that can occur through the release of GMOs. There was agreement, that carefully managed experimental plots had given insufficient information on transgenic hazards. One ecological message was that gene flow among crops and their wild relatives has to be studied before transgenic crops were used in a large scale. But "biotech proponents" found that most crops pose little risk to the environment, and many scientists in developing countries found environmental safety issues to be secondary to the demand for increased crop production.

Two European workshops are finally mentioned, where controversial points of view again became obvious. Risk assessment for the release of transgenic organisms appeared no longer to be the major topic at the biosafety meeting in Mainz (Germany, 1994). It was proposed that large-scale field tests and the commercial use of transgenic organisms, however, now be accompanied by concomitant biological research in order to obtain some preventive risk management and to avoid damage to the environment. A few months later, a European workshop on the "Regulation of Releases of Genetically Modified Organisms" was held in Brussels (Belgium). Ecological facts, of course, were considered with respect to the consequences of the release of transgenic organisms into the environment but the major interest here concerned the state regulations in this regard.

Conclusions

Based on our experiences in population biology and from studies of invasion, we are able to group biological attributes of organisms into harmful and harmless characters, related to the potential of genes or organisms to escape human observance and to invade ecosystems. Various test procedures have been applied in ecology, population biology, and population genetics. A comprehensive list of such methods that can be used for the risk assessment of transgenic plants is given by Kjellsson and Simonsen (1994).

Van Elsas et al. (1990, published in MacKenzie and Henry) and many other studies provide lists of data related to the biology of microbes, plants and animals. But the problem remains that even accurate knowledge about the biology of one species does not enbable us to draw conclusions about the behavior of related species (for review see, Wöhrmann and Tomiuk, 1993).

Ecologists know very well that even closely related species can interact differently with different environments. We can draw some basic conclusions about the ecological behavior of transgenic organisms from studies of their natural relatives, but to get an idea of the "evolutionary risk" of organisms we have only statistical approaches. For instance, to predict the weed potential of plants in relation to a specific character is often impossible. It is only statistically possible to quantify the probability that species will become pests. The "ten-to-ten" rule states that 10% of species introduced into a new habitat will become permanent populations of which again 10% will become pests (Holdgate, 1986; Williamson and Brown, 1986). Nevertheless, Köhler and Braun (1995) stated recently that it is highly important to develop and to apply appropriate tests to the different experimental designs in order to obtain a reliable risk assessment. Furthermore, information regarding negative results must also be provided, not only in the sense of preserving and sharing data that indicate no survival, no dispersal, and no genetic transfer (MacKenzie and Henry, 1990, "Towards a Consensus", p. 279), but also uncommon and unexpected results that may indicate potential risks and not simply artefacts. Therefore, it seems imperative that case-by-case studies are performed.

Many studies demonstrate the uncertainties that are still present when the risk potential of some organisms is considered. Crawley's statement (1992, published in Casper and Landsmann) that further research on oilseed rape is not necessary before its large-scale field testing or commercial use appears hasty in the light of other results. The effect of a transgene in a novel background such as an interspecific hybrid depends on the DNA construct and has to be assessed on an individual basis (Leckie et al., 1993; see also Christiansen, 1990). Risk assessment thus becomes a question of the possible effects of genes. The results of Leckie et al.'s (1993) intraspecific hybridization experiments suggest that if cross-pollen is available to a feral rape plant it is more likely to produce hybrids than selfs. Thus the probability can increase that a gene conferring herbicide resistance is able to escape from cultivated rape to its wild relatives and consequently cause economic damage. Moreover, recently Manasse (1992) and Kaveira et al. (1994) analyzed gene flow between transgenic crops and their wild relatives, and their results demonstrate that we need estimates of spatial distribution rates of pollen rather than simple counts of numbers of transgenic pollen at different distances, when we are interested in the probability that transgenes escape into the wild.

Asilomar was characterized by fears that the new biotechnological possibilities could result in unpredictable harm to humans and to the environment. The present study of the history and the development of risk assessment for the deliberate release of transgenic organisms obviously shows that our understanding of the biology of some species, microbes, crops and their wild relatives has made progress. Since the Asilomar conference, however, the argument has consistently been made that although our knowledge regarding the biology of species is increasing rapidly it is still too limited to exclude any one risk. The arguments applied to conventional breeding decades ago have resurfaced – namely, that any progress in science contributes to human well-being (Casper and Landsmann, 1992). The controversy will surely continue, but it seems that economic interests will finally dominate the debate and the process of GMO releases.

References

Altmann, M. (1993) Introduction: Gene technology and biodiversity. *Experientia* 49: 187–189.

Berg, P., Baltimore, D. and Boyer, H.W. (1974) NAS ban on plasmid engineering. *Nature* 250: 175.

Berg, P., Baltimore, D., Brenner, S., Roblin, R.O. III and Singer, M.F. (1975) Summary statement of the Asilomar conference on recombinant DNA molecules. *Proc. Natl. Acad. Sci. USA* 72: 1981–1984.

Casper, R. (1993) *Report – Workshop on "Application of Agricultural Biotechnology and Safety Considerations"*. China-E.C. Biotechnology Center, Hainan Island, China.

Casper, R. and Landsmann, J. (1992) Proceedings of the 2nd International Symposium on "The Biosafety Results of Field Tests of Genetically Modified Plants and Microorganisms". Goslar, Germany May 11–14, 1992. Biologische Bundesanstalt für Land- und Forstwirtschaft, Braunschweig.

Christiansen, F.B. (1990) Population consequences of genetic design in sexually reproducing organisms. *In*: H.A. Mooney and G. Bernardi (eds): *Introduction of Genetically Modified Organisms into the Environment - Scope 44*. John Wiley and Sons, New York, pp 43–55.

de Vries, F.T., van der Meijden, R. and Brandenburg, W.A. (1992) *Botanical Files - A Study of the Real Chances for Spontaneous Gene Flow from Cultivated Plants to the Wild Flora of the Netherlands*. Gorteria Supplements, W.J. Holverda, R. van der Meijden, and C.M. Werker (eds), State University Leiden, Netherlands.

Fiksel, J. and Covello, V.T. (1986) *Biotechnology Risk Assessment - Issues and Methods for Environmental Introductions*. Pergamon Press, New York.

Ginzburg, L.R. (1991) *Assessing Ecological Risks of Biotechnology*. Butterworth-Heinemann, Boston.

Holdgate, M.W. (1986) Summary and conclusions: Characteristics and consequences of biological invasions. *Phil. Trans. Royal Soc.* B314: 733–742.

Kareiva, P., Morris, W. and Jacobi, C.M. (1994) Studying and managing the risk of cross-fertilization between transgenic crops and wild relatives. *Molec. Ecol.* 3: 15–21.

Kjellsson, G. and Simonsen, V. (1994) *Methods for Risk Assessment of Transgenic Plants - 1. Competition, Establishment and Ecosystem Effects*. Birkhäuser Verlag, Basel.

Klingmüller, W. (1988) *Risk Assessment for Deliberate Release*. Springer-Verlag, Berlin.

Köhler, W. and Braun, P.W. (1995) Populationsgenetische und statistische Aspekte der Begleitforschung. *In*: S. Albrecht und V. Beusmann (eds): *Zur Ökologie transgener Nutzpflanzen*. Campus Verlag, Frankfurt, pp 163–181.

Leckie, D., Smithson, S. and Crute, I.R. (1993) Gene movement from oilseed rape to weedy populations – A component of risk assessment for transgenic cultivars. *Aspects Appl. Biol.* 35: 61–66.

Lenski, R.E. (1993) Evaluating the fate of genetically modified microorganisms in the environment: Are they inherently less fit? *Experientia* 49: 201–209.

Levin, M. and Strauss, H. (1991) *Risk Assessment in Genetic Engineering - Environmental Release of Organisms*. McGraw-Hill, Inc., New York.

MacKenzie, D.R. and Henry, S.C. (1990) *Biological Monitoring of Genetically Engineered Plants and Microbes - Proceedings of the Kiawah Island Conference*. November 27–30, 1990. Agricultural Research Institute, Bethesda, Maryland.

Manasse, R.S. (1992) Ecological risks of transgenic plants: Effects of spatial dispersion of gene flow. *Ecol. Appl.* 2: 431–438.

National Research Council (NSC) (1989) *Field Testing Genetically Modified Organisms – Framework for Decisions*. National Academic Press, Washington.

Regal, P.J. (1993) The true meaning of "exotic species" as a model for genetically engineered organisms. *Experientia* 49: 225–234.

Stone, R. (1994) Large plots are next test for transgenic crop safety. *Science* 266: 1472–1473.

Sukopp, H. and Sukopp, U. (1993) Ecological long-term effects of cultigens becoming feral and of naturalization of non-native species. *Experientia* 49: 210–218.

Tiedje, J.M., Colwell, R.K., Grossman, Y.L., Hodson, R.E., Lenski, R.E., Mack, R.N. and Regal, P.J. (1989) The planned introduction of genetically engineered organisms: Ecological considerations and recommendations. *Ecology* 70: 298–315.

Williamson, M. (1993) Invaders, weeds and the risk from genetically manipulated organisms. *Experientia* 49: 219–224.
Williamson, M.H. and Brown, K.C. (1986) The analysis and modelling of British invasions. *Phil. Trans. Royal Soc.* B314: 505–522.
Wöhrmann, K. and Tomiuk, J. (1993) *Transgenic Organisms: Risk Assessment of Deliberate Release*. Birkhäuser Verlag, Basel.

Transgenic organisms and evolution: Ethical implications

T. Potthast

Center for Ethics in the Sciences and Humanities, University of Tübingen, Keplerstraße 17, D-72074 Tübingen, Germany

Summary. The issue under discussion is the link between evolutionary biology and the ethical aspects of the risks of releasing transgenic organisms. References to "evolution" give rise to contradictory positions: (i) Because humans are free to act and free to decide, and because of the potential threats which have been identified until now, humankind is responsible for protecting diversity in nature and its evolutionary potential. However, there is disagreement concerning the grounds for this evolutionary responsibility and its range; (ii) in the recent past there have been efforts to postulate the significance of interspecific gene exchange processes (horizontal gene transfer) in "natural" populations. This stipulated significance cannot serve to legitimize genetic engineering or releases of transgenic organisms in particular since that would be a classic naturalistic fallacy; (iii) for risk assessment of releases, the openness in principle of evolutionary processes is a fundamental problem. Statements concerning long-term effects of transgenic organisms are determined by fundamental aspects of views of nature and of history as well as by the metaphorical load and the linguistic usage of the concept "evolution". As a conclusion, reflective explanation of basic ethical assumptions is demanded, also and in particular within the framework of scientific debates about the problems of release.

Introduction

There are several levels in public discussions as well as in the scientific community, on which questions in evolutionary biology about the release of transgenic organisms are influenced by ethical issues:

(i) Since the beginning of the 1970s both biologists and environmental ethicists have been emphasizing that natural evolutionary processes are not just a given fact; rather, safeguarding their natural progression is a moral obligation which goes beyond an immediate or potential practical application by humans (see section on *Evolution as a good to be protected or as a value in itself*, below).

(ii) It is controversial to what extent genetic engineering can be compared to evolutionary genetic transfer mechanisms and what the ethical implications are for human action (see section on *Evolution, genetic engineering and the naturalistic fallacy*, below).

(iii) Within the debate about the risks of release there are considerable differences of opinion with respect to the impacts of transgenic organisms on natural evolution. The question is to what extent the processes of species formation and phylogenesis as well as the long-term dynamics of ecological systems – in short: "evolution" – can be modified by transgenic organisms and whether this happens in a way which is problematic for humankind and nature (see sections on *Prognoses of evolutionary biology und conceptual framework* and *Horizontal gene transfer and transformation of risk assessments*, below). This question is directly connected with the discussion whether the expected effects or a decision with uncertainty about possible effects are acceptable and thus can be considered as responsible action (see section on *Discussion and conclusions*, below).

The two last-mentioned levels also mark the two fundamental points of controversy in the scientific debate on the release of transgenic organisms. Assessments of possible effects and consequences are based on different receptions of concepts from evolutionary biology and ecology as well as on the (also divergent) attitudes towards the question of a fundamentally new quality of transgenic organisms. Thus, positions concerning "evolution" necessarily consist of a combination of arguments from science, philosophy of science and normative ethics.

Evolution as a good to be protected or as a value in itself

Regarding the genetic impoverishment among cultivated plant varieties and the overall endangered species-richness, the plant geneticist Frankel (1970, 1974) recognized that humankind has become the decisive factor in the biosphere. He was the first to draw conclusions in a way to make human "evolutionary responsibility" a subject matter. Against the background of a global threat to biological diversity, the maintenance of those processes on which humans have as little influence as possible ("natural" processes) is seen as something which humans have to protect actively (Frankel and Soulé, 1981; Wilson, 1988; Blab et al., 1995). In the meantime, there seems to be a consensus that natural evolutionary processes are a good to be protected. Note that "consensus" designates free approval by everyone who takes part in the discussion – not to be confused with a compromise in which diverging positions remain but have to be moderated in order to reach an agreement (Habermas, 1983, p. 82).

Highly significant for the ethical dimension of the release discussion, however, is the fact that since its beginning there have been different "ways of substantiating" the claim that evolutionary processes are worth protecting. A classic anthropocentric argument is as follows: For economic reasons and for reasons of breeding it is necessary to protect genetic richness since it forms the basis for the development of cultivated plant varieties and for the chemistry of natural substances (basis for new drugs). In this respect, natural genetic diversity is certainly a resource for genetic engineering as well, and for some decades there have been attempts to preserve it in sperm and gene banks. Genetic engineering as a mean of protecting genetic or even species diversity (Moore et al., 1992) has to face a critical objection: *ex situ*-preservation might prove to be a medium- and long-term impasse because in case of isolated genetic material or seeds in fact coevolutionary mechanisms significant for adaptation are excluded (Frankel and Soulé, 1981; Weizsäcker and Weizsäcker, 1987; Wilson, 1988).

Other ways of substantiating the protection of evolutionary processes go beyond the immediate usefulness for the survival of humankind. Species-richness is to be protected also if it does not obviously serve this purpose. The lines of argument ramify even further: For some authors evolution resembles a value in its own right. The fact of phylogenesis is an inherent value which humans are not to destroy or endanger in terms of its potentiality (Rolston, 1988). This argument makes explicit or implicit use of religious or secularized motives (right to life, sacredness of nature, responsibility for the creation; Altner, 1987). Other authors reject this type of argument and relate worthiness of protection back to humankind in so far as we should strive for a different understanding of the relationship between man and nature. This new conception would depend on an appreciation of evolutionary potentials as beautiful, edifying, and valuable beyond their practical usability. This is to assume an intrinsic value of evolutionary processes which, however, is clearly distinguished from biocentric argumentation (Passmore, 1980). This is the point where the interests of future generations of humans come in: we must not deny to them the opportunities to decide whether and how they want to influence, impair, control evolution (Frankel, 1970).

Thus, natural processes and potentials of evolution have gained essentially undisputed normative relevance. However, the respective reasons given for the claim to protection are very different: First, aspects of usability concerning raw materials, breeding and pharmaceutical chemistry; second, aesthetic and moral value for humankind (here, the claim to satisfaction of scientific curiosity is often mentioned as well); and third, an inherent value of the processes of evolution beyond all human interests (but humans have to recognize, accept, and operationalize this inherent value in the context of their activities).

The role of evolutionary biology must not be underestimated. As a science, it draws for us the picture of evolution, its causes, mechanisms and course of events. Therefore, biological concepts are crucial factors in forming and altering assessments about the scope and aims of nature protection. This becomes obvious, e.g., in the ongoing replacement of the static characterization of natural processes (in ecological concepts such as "equilibrium", "climax", "cycles") by dynamic concepts (Peters, 1991; Breckling, 1993; Jax, 1994). As a consequence, less emphasis is put on the protection of certain transient stages of ecological systems, but more on the dynamical aspect of change within habitats. At the same time, the future-oriented emphasis on an "evolutionary potential" has become a topic also because science has established the present-day threat to this resource. Thus, the above-mentioned normative decision is necessarily based on a scientific description and conceptualization of what the characteristic properties of the good to be protected actually are. Therefore, scientific analysis forms an integral part of ethical reflection

(Mieth, 1993), with scientific concepts being subject to change and thereby influencing normative decisions.

Evolution, genetic engineering and the naturalistic fallacy

The question of the relationship between evolution and genetic engineering is crucial, not least because of the positive connotation of "evolution" sketched above. In what follows, the ethical implications resulting from different positions will be discussed. As a second stage, these positions themselves will be evaluated.

In its "classic form", the theory of evolution comprises four causal factors driving evolutionary change: mutation, selection, isolation and genetic drift (chance). According to Schmitt (1993), humans have played the role of selection by breeding; now genetic engineering offers procedures of more or less well-aimed mutation because natural coherence in genetic regulation and combination can be manipulated specifically. The latter differs from conventional methods of non-specific mutagenesis (colchicine as mitosis inhibitor for polyploidization, radiation treatment, etc.). It is generally emphasized that, naturally, humankind as a biological species has an influence on evolution. Since the end of the Middle Ages, however, this influence has reached new and unthought-of dimensions. This last point is largely uncontroversial (e.g., Markl, 1986; Chadarevian et al., 1991).

What conclusions for the assessment of genetic engineering can be drawn from such considerations? Weber (1994) points out that Charles Darwin (1988, orig. 1859) already differentiated clearly between man-made and natural selection – "natural selection" being a metaphor since there can be no conscious selection without humans. Thus, Darwin distinguished between two ways of looking at things: a metaphorical and a scientific-analytic way. If this is right, natural processes and conscious human actions can be described analogously for illustrative reasons, but they cannot be equated.

An opposing position holds that humankind can use only mechanisms which already exist in nature and that, e.g., genetic engineering is nothing but an application of identical natural processes of gene transfer. A formulation is: Evolution is natural genetic engineering (Markl, 1986, p. 19). This amounts to a "naturalistic justification" (Rottschaefer, 1991) of human activity by equating it to analogous processes in nature. A second step is the insinuation that genetic engineering and its consequences do not have to follow specific criteria in assessing potential risks.

The inadvertent or decided transition from the factual "is" (in a description) to the moral "ought" (directing or justifying action) has been criticized since David Hume (1978, orig. 1739). Inferring what ought to be the case from what is the case is to commit the "naturalistic fallacy"

(Moore, 1978; Ball, 1988). The above described naturalistic justification of genetic engineering is a paradigm case of the naturalistic fallacy – regardless of whether it seems scientifically plausible to equate methods of genetic engineering to evolutionary processes of natural gene transfer (Chadarevian et al., 1991; Potthast, 1996; please see also section on *Horizontal gene transfer and transformation of risk assessments*, below). For if not only an analogy but also an equation (identity) could be established, in contrast to the dynamics of natural evolution humans must always account for their actions. Logically, this goes also for their technological interventions in the dynamics of nature – at least in so far as these actions affect other humans and/or structures in nature. This holds true for applied genetic engineering in general as well as for releases of transgenic organisms in particular. The existence of gene-dynamics in natural evolution cannot justify the admissibility of methods and applications of genetic engineering.

Prognoses of evolutionary biology and conceptual frameworks

There is a fairly general agreement that evolution does not simply consist of the modification of nucleotide sequences, with everything else being a mere epiphenomenon of these processes. Rather, evolution proceeds in the form of many and diverse epigenetic interactions on the levels of individuals, populations, ecological communities and of (geo-) ecosystems. Interconnected reciprocal actions conform only rarely to linear–causal laws (Futuyma, 1990). For quite some time, research has been done on the connection between ecological dynamics and individual factors of evolution; however, generalizations have seemed hardly possible (Orians, 1962; Cooper, 1993).

The crucial point affecting the release of transgenic organisms, is that population dynamics and ecological long-term dynamics resulting from the richness of natural genetic and ecological processes, are unpredictable in principle (Gabriel, 1993; Wöhrmann et al., 1993; Weber, 1994). In this branch of the discussion, "evolutionary consequences" seem to be taken as almost synonymous with ecological long-term consequences: the time scale for evolutionary processes spans 10 000 years and more ("for ever time-scale of concern"; Frankel, 1970).

Statements on long-term consequences of the release of transgenic organism, on nature's stability and dynamics in general, are not least determined by ideas about the historicity of nature (on historicizing nature see Lepenies, 1976; Foucault, 1988; on evolutionary biology see Mayr, 1991). Not only with respect to genetic engineering, molecular biology seems to be looking through two different time windows: One shows the so-called "laboratory evolution" which is concerned with characteristics changing within a few weeks up to some years; the other opens – almost supertemporally – on to the "general evolution" with its fundamental mechanisms since the origin of life and even further to general cosmologies (e.g., the principle of self-organization;

Prigogine and Stengers, 1980; Eigen, 1992). In ecology there is – like it or not – less generalization: The focus is on the medium period with specific landscape modifications brought about by humans within several dozens up to hundreds of years. Meanwhile, ecology emphasizes for conceptual reasons the singularity of ecological events. This amounts to a warning against simple and rash generalizations and prognoses (Breckling, 1992; Jax, 1994).

Theoretical references to evolutionary changes (history of organisms) in nature and in its mechanisms cut across the dividing line between the *pros* and *cons* of genetic engineering: On the one hand, a "dynamic nature" plays an important role in the risk debate, especially in arguments concerning little-known gene transfer processes or unexpected effects. On the other hand, in the course of the discussions about a natural and global equilibrium, great importance has been attached to the stability of certain states of nature, not least in order to be able to criticize anthropogenic disturbances. Today, a more differentiated formulation would be: There are evolutionary grown states of interconnectedness in nature which have to be protected against drastic interventions; otherwise, the inherent – natural – dynamics will get into a mess. Transgenic organisms intended for release are especially problematic because of their necessarily dominantly exprimed transgenes and a high competitive potential. Thus they do not meet the criteria of a possible withdrawal of transgenic organisms from the environment and of error-friendliness (in sum: Reversibility of effects; Weizsäcker and Weizsäcker, 1987; Weizsäcker, 1991; Weber, 1994).

Meanwhile, those who support release of transgenic organisms have discovered the dynamics of natural gene transfer processes for their arguments. They hold that nature is so "strict" in terms of selection – this is where "orthodox" concepts of selection are used – and treats genes and resources so "economically" that introducing transgenic organisms even on a large scale (observing some safety measures) would not particularly threaten its stability. There are no distinct statements on the past and future development of nature or evolution; rather, it is merely asserted that there are law-governed mechanisms of this development which humans cannot disturb (Markl, 1986).

Linguistic usage – the term "evolution"

The use of the term "evolution" is problematic even within biology (Mayr, 1991, 1993) since it cuts across various fields of research and does not always mean the same thing. Schmitt (1993) lists four different definitions: Descent with modification; shift of allele frequencies in populations; development from simplicity to complexity; and history of the biosphere. However, "evolution" is also generally understood as the development of any kind of object or process over time, not restricted to organisms or molecules. Such differences in content can be described as the results of, among others, various phenomena in the history of a language. Here, the use of English as the dominant language of the scientific community in the western industrialized countries is significant. In German, the concept "evolution" underwent a considerable extension of meaning – largely due to the life sciences. In English usage, on the other hand, "evolution" has always meant development over time in a more general sense. Microbiology and molecular biology with their almost exclusively Anglo-Saxon discourse are shaped by this interpretation, and this can be a source of communication problems or actual misunderstanding.

The concept "evolution" has spread into other sciences and everyday usage because of the increasing technological success of the life sciences and because evolution theory is of central importance to humans' conceptions of themselves. As a consequence, "evolution" is used more and more to describe vastly diverse natural, technological and social developmental processes with very different causes or effective factors. This leads to shifting connotations and ultimately "evolution" might degenerate into a "plastic word" (Pörksen, 1988) with no explanatory value. Whether such trends should be welcomed or not may be left open; in any case, they represent a well-known phenomenon of change in meaning – by your leave, evolution of language. In certain respects, "evolution" has thus returned to its starting point in linguistics: In the middle of the 19th century, August Schleicher had drawn up family trees indicating the relationships between different languages, and these served as the model for the phylogenetic tree of the groups of organisms popularized mainly by Ernst Haeckel (Krauße, 1987).

The use of metaphors – evolution as a cipher

A metaphor for a biological theory or scientific statements about nature helps the theory gain acceptance among its contemporaries. Perhaps existing metaphors even foster the theory's initiation itself. A paradigm case is the significance of "natural selection" and of the economic term "competition" for the work of Charles Darwin (Young, 1985; Trepl, 1994; in this chapter, I do not want to work out the differences between metaphor, analogy, etc.). The presentation of the genetic code as analogous to human language and the conception of genes as the blueprint, or later on the computing program for organisms (Jacob, 1972; Mayr, 1991) are necessary prerequisites for the "construction of new forms of life" in genetic engineering: Experimental practice is inextricably linked with learning to conceive the living world of nature in this form (Herbig and Hohlfeld, 1990; Fleck, 1993). In this respect, "evolution" in the resulting debate has two opposed implications. On the one hand, it is considered to be a cipher for the significance of a changeability, a dynamics over time which is at work also without humans being involved. In this view, changes brought about by transgenic organisms appear to be less important elements of a general change. On the other hand, "evolution" can function as a symbol for the opinion that, against the background of global problems of nature conservation and in view of the value of processes uninfluenced by humans, transgenic organisms as yet another strain should not be added. These are alternatives for preliminary decisions on which all scientific evaluations and prognoses of long-term natural changes are based.

Horizontal gene transfer and transformation of risk assessments

For some years, the "naturalness of genetic engineering" has referred to the scientific hypotheses and findings on horizontal gene transfer, which was discovered in bacteria. It is defined as DNA transfer onto other microorganisms not belonging to the same species (possibly independently of reproduction). At the beginning of the 1980s, some scientists warned about the possibility that transgenes might not only remain in the manipulated organisms but might also spread *via* horizontal gene transfer through bacteria or on other ways yet unknown. Such considerations were doubted, and this hypothesis was mostly rejected. In its 1987 report, the German Enquete-Commission still assumed explicitly that transfer of plant or animal genes through viruses or bacteria could practically be ruled out (Deutscher Bundestag, 1987). This served as an argument for the "safety" of the transgenes which were said to be restricted to their original carriers and therefore controllable with little risk involved. By the end of the 1980s, however, the early hypotheses were confirmed by experimental data which indicated gene transfer between different species of bacteria much more often than initially assumed. Now, horizontal gene transfer came to be a knock-down argument against the "safety" of genetically manipulated organisms because of the uncontrollable spreading of the transgenes. Unwanted, unwelcome, and possibly dangerous side-effects seemed much more likely to occur (Potthast, 1990; Bernhardt et al., 1991), particularly since horizontal gene transfer was detected also between plants or animals, mediated by viruses, mites and the like. But the more frequently occurrence and significance of horizontal gene transfer was described (as "cornerstone of evolution"; Amábile-Cuevas and Chicurel, 1993), the more it was discovered as an argument for genetic engineering as being "safe". Various authors conclude from their work that methods of genetic engineering bear a strong resemblance to the processes of natural evolution, at least with regard to microorganisms. Thus, they maintain that the use of transgenic microorganisms involves no particular risk (among others Lorenz and Wackernagel, 1993; see also this volume).

This discussion can be divided into four stages (following Potthast and Weber, 1995): (i) Critical questions about whether processes of gene modification and gene transfer hitherto thought of as rare or purely hypothetical should be assumed also for transgenic organisms and whether this might have unexpected consequences; (ii) rejection of such assumptions as unsound, unscientific speculations (Bartsch and Sukopp, 1993); (iii) at least partial confirmation of the assumptions and therefore also of the postulated potentials of risk; (iv) processes initially considered to be at least rare are redefined as "important mechanisms of natural evolution". This leads to a legitimation of interventions by means of genetic engineering which are taken to be comparable with the natural mechanisms.

This sequence is presently repeated in the discussion on the genetically engineered production of virus-resistant cultivated plants and possibilies of recombination with natural viruses. Recently, it has reached stage 4 (Falk and Bruening, 1994; Greene and Allison, 1994; on heterologous transcapsidation). Another example is the intake of "naked" tumour-DNA through the injured skin of mammals (Bernhardt et al., 1991; Burns et al., 1991).

On another level, the question on ecological risks of release must be whether the fact of more and more frequently detected natural horizontal gene transfer can be at all relevant to assessing risks of transgenic organisms (Tiedje et al., 1989). First, the phenomenon remains that horizontal gene transfer does not appear everywhere but is restricted in the course of evolution to particular cases and has a limited frequency. Methods of genetic engineering and release of transgenic organism certainly open up new possibilities of regulation and spreading, respectively. Secondly, we do not know to what extent natural horizontal gene transfer already provides paths for the spreading of pathogenic properties. Thirdly, on the conditions just sketched, the conception of a so-called field-containment to prevent the – desired or undesired – spreading of transgenic organisms and of the transgenes is no longer tenable: This ought to have consequences for the safety concepts and for the relevance of accompanying ecological research of deliberate release.

Discussion and conclusions

In the debate on deliberate release and in the entire discussion about genetic engineering there is no standard use of the terms "evolution", "theory of evolution". However, all participants in the discussion refer to "evolution" in order to legitimize their respective attitudes towards the central controversies about genetic engineering (see above) by moral–ethical appeals. The questions cannot be answered independently of ethical–normative preliminary decisions. At the same time, however, all normatively oriented answers are shaped by the state of the art in the respective science. Ideas from biology about what nature is and how it functions have a decisive influence on the positions concerning genetic engineering. Of course, this is also true *vice versa*: world views and values set the conditions for the way in which science makes "nature" its research topic (Young, 1987; Trepl, 1987, 1994). Anyone who does not acknowledge such a dilemma has a blinkered attitude to their work. We have to resolve the mutual influence in the individual case since this is the only way to accept scientific statements within their theoretical and normative contexts as intersubjectively binding insights.

A synopsis of the discussion about risks of transgenic organisms reveals an argument whose structure is paradoxical in more than one respect: On the one hand, critics refer to the surprising and in detail little-known possibilities of natural processes of gene exchange. Thus, nature is supposed to bear dangerous potentials which humans might activate by genetic engineering and which ultimately would turn against humankind. Without human intervention, on the other hand, in ecosystems there is an environmental impact (*"Umweltverträglichkeitsprüfung"*; quote from a paper in Deutscher Bundestag, 1987, p. 317) of new genotypes in nature. I would like to sharpen the position to a provocative point: Without humans, nature is in a state of perfect harmony and modifications are tested carefully. As soon as humankind intervenes too strongly in nature, e.g., *via* genetic engineering, nature in all its dynamics becomes the threatening and unpredictable "other" again. On the supporters' side the same inconsistencies are evident – but in mirror-inverted form: A dangerous nature with disease-causing agents (cancer, AIDS, hereditary diseases, parasites) as well as the struggle for food (scarce resources, "over-population", etc.) has to be fought in an enlighted manner and this serves as argument for applied genetic engineering (Markl, 1986). Simultaneously, however, insights in natural gene exchange processes are used for the naturalistic legitimation of genetic engineering methods. On the consequences of anthropogenic intervention, confidence in a nature which regulates itself – mainly *via* mechanisms of selection – seems to be boundless, but this strictly contradicts the concept of dangerous elements in nature.

From a sociological point of view, Beck (1986) has discussed the connection of the risk debate with ethical aspects. He points out that the risk debate about genetic engineering is an implicit debate about ethics since the question about risk acceptability directly implies the question about social and economic values on which support and application of any technology are based. Risk assessments must necessarily commit themselves to scientific as well as ethical value-judgements. It should be demanded that this has to be done in a reflective manner. Nobody should call his position neutral in terms of values ("objective") and at the same time reject other positions as affected by values.

The general conclusion I want to draw, not only with respect to the topic of (evolution and) deliberate release, is as follows: Biologists should point out, more clearly than in the past and also in public, that "exact" long-term prognoses on the consequences of genetic engineering are largely impossible. Following Aristotle, the claim should be "for every field of research to demand [and to assert! the author] only so much precision as the nature of the object permits" (Aristotle, 1991, p. 107). The positions within biology are too disparate about the evaluation of certain evolutionary factors such as selection, competition, isolation, self-organization and the dynamics and stability of ecological systems. There are obviously big differences which can be traced back to, e.g., "selectionists" (including sociobiology; Wilson, 1975) and to theories in molecular biology accentuating processes of self-organization in a broad sense or internal developmental constraints. I doubt whether a synthesis of these positions is possible or useful. A much-desired uniform/unified theory (van der Weele, 1993) would not fit the phenomena. Moreover, I agree with Ahrendholz' (1993) proposal to appreciate a pluralism of theories, methods, and – my addition – metaphors as the strong point of (evolutionary) biology and at the same time be clear about the limited scope of the respective concept.

Biology needs "pictures" not only to popularize its concepts; rather, images, metaphors and analogies form an indispensable basis for the formation process of scientific theory itself. New hypotheses and images can open up new research approaches and insights into the functional structure of nature. At the same time they offer the opportunity of describing and defending scientific positions in a way which brings them to public attention. In the case of dispute, however, this could be a problem if the scope of validity of the images is extended (too far) unthinkingly or deliberately: Speculations in the philosophy of nature and in metaphysics concerning forbidden human interventions in the evolution of interrelated and harmonious systems should be identified as metaphors, just like talk about nature as the "first genetic engineer practising for billions of years". The validity of particular statements can be checked by comparing them to the scientific state of the art. Definite "true or false decisions" are not to be expected, but at least in the individual case the images lose their seeming arbitrariness.

In the public discussion preceding political decisions when all participants try to legitimize their positions by using "evolution" in the various above-described ways, arguments with reference to the authority of evolution(ary biology) are highly significant. A detailed social scientific analysis of this phenomenon would be very helpful (Krimsky, 1982). Regarding the debate within the scientific community the situation seems to be ambiguous: On the one hand, the theoretical relevance of evolutionary biology is taken seriously and permeates into expert opinions and advisory statements for politicians – not to mention that it is a fascinating field of biology. On the other hand, more detailed interest and financial support dwindle quickly when it cannot be integrated into projects of accompanying research concerning genetic engineering, and/or when precise statements are untenable from the very beginning or at least dubious for short term research.

The above-mentioned underlying levels of argument come in again when concepts from evolutionary biology are operationalized for the release risk debate. Depending on whether the accent is on the aspect of utilisation, the inherent value, or the broadly aesthetic significance of a nature evolving without human intervention, there will be a different willingness to accept release of transgenic organisms, the uncertainty of prognoses, and the estimated dimensions of the project.

The criteria for and the judgement about the risk assessment itself also depend on this (Levidov, 1995).

The conclusions and demands in summary: First, to spell out implicit assumptions and premises, i.e., to state clearly one's attitude towards evolutionary responsibility and its substantiation when evolutionary risks of transgenic organisms are discussed. Secondly, it is necessary to specify the respective scientific version of the concept "evolution" because its great bandwidth reflects not least vastly divergent scales and contexts of evolutionary processes. We should strive for a critical evaluation of the validity of statements and prognoses.

Acknowledgements
I would like to thank Uta Eser, Dietmar Mieth and Barbara Skorupinski (Tübingen) for helpful comments on an earlier draft of the paper.

References

Ahrendholz, W.-R. (1993) Das Ende der synthetischen Evolutionstheorie? *Ethik und Sozialwissenschaften* 4: 16–18.

Altner, G. (1987) *Die Überlebenskrise in der Gegenwart – Ansätze zum Dialog mit der Natur in Naturwissenschaft und Theologie*. Wissenschaftliche Buchgesellschaft, Darmstadt.

Amábile-Cuevas, C.F. and Chicurel, M.E. (1993) Horizontal gene transfer. *Amer. Sci.* 81: 332–341.

Aristoteles (1991) *Die Nikomachische Ethik*. Artemis, Zürich and München.

Ball, S.W. (1988) Evolution, explanation and the fact/value distinction. *Biol. Philos.* 3: 317–348.

Bartsch, D and Sukopp, H. (1993) *Ermittlung und Bewertung des ökologischen Risikos beim Umgang mit gentechnisch veränderten Organismen. Dokumentation eines Fachgesprächs am 14./15.10.1991*. Texte 20/93, Umweltbundesamt, Berlin.

Beck, U. (1986) *Risikogesellschaft – Auf dem Weg in eine andere Moderne*. Suhrkamp, Frankfurt/M.

Bernhardt, M., Weber, B. and Tappeser, B. (1991) *Gutachten zur "biologischen" Sicherheit bei der Nutzung der Gentechnik* (Auftraggeber: Büro für Technikfolgen-Abschätzung des Deutschen Bundestages). Öko-Institut e.V., Freiburg im Breisgau.

Blab, J., Klein, M. and Ssymank, A. (1995) Biodiversität und ihre Bedeutung in der Naturschutzarbeit. *Natur und Landschaft* 70: 11–18.

Breckling, B. (1992) Uniqueness of ecosystems *versus* generalizability and predictability in ecology. *Ecol. Model.* 63: 13–27.

Breckling, B. (1993) *Naturkonzepte und Paradigmen in der Ökologie – einige Entwicklungen*. Discussion Paper FS II 93304, Wissenschaftszentrum für Sozialforschung, Berlin.

Burns, P.A., Jack, A., Nielson, F., Haddow, S. and Balmain, A. (1991) Transformation of mouse skin endothelial cells *in vivo* by direct application of plasmids DNA encoding the human T24 H-*ras* oncogene. *Oncogene* 6: 1973–1978.

Chadarevian, S. de, Dally, A. and Kollek, R. (1991) *Experimente mit der Evolution. Zum Verhältnis von Evolution, Züchtung und Gentechnologie*. Werkstattreihe 76 des Öko-Instituts e.V., Freiburg i.B.

Cooper, G. (1993) The competition controversy in community ecology. *Biol. Philos.* 8: 359–384.

Darwin, C.R. (1988) *Über die Entstehung der Arten durch natürliche Zuchtwahl*. Wissenschaftliche Buchgesellschaft, Darmstadt, (Reprint der 9. A. der dt. Übs. von JV Carus (1920) Schweizerbart'sche Verl.buchh., Stuttgart; Orig. 1859).

Deutscher Bundestag (1987) *Chancen und Risiken der Gentechnologie*. Der Bericht der Enquete-Kommission "Chancen und Risiken der Gentechnologie" des 10. Deutschen Bundestages. – Themen parlamentarischer Beratung 1/87, Bonn.

Eigen, M. (1992) *Stufen zum Leben – Die frühe Evolution im Visier der Molekularbiologie*, Second Edition. R. Piper, München and Zürich.

Falk, B.W. and Bruening, G. (1994) Will transgenic crops generate new viruses and new diseases? *Science* 263: 1395–1396.

Fleck, L. (1993) *Entstehung und Entwicklung einer wissenschaftlichen Tatsache. Einführung in die Lehre vom Denkstil und Denkkollektiv*, Second Edition. Suhrkamp, Frankfurt/M. (Orig. 1935) .

Foucault, M. (1988) *Die Ordnung der Dinge*. Suhrkamp, Frankfurt/M. (Orig. 1966) .

Frankel, O.H. (1970) Variation – The essence of life. *Proc. Linn. Soc. New South Wales* 95: 158–169.

Frankel, O.H. (1974) Genetic conservation: Our evolutionary resposibility. *Genetics* 78: 53–65.

Frankel, O.H. and Soulé, M.E. (1981) *Conservation and Evolution*. Cambridge University Press, Cambridge.

Futuyma, D.J. (1990) *Evolutionsbiologie*. Birkhäuser Verlag, Basel.

Gabriel, W. (1993) Technologically modified genes in natural populations: some sceptical remarks on risk assessment from the view of population genetics. *In*: K. Wöhrmann and J. Tomiuk (eds): *Transgenic Organisms: Risk Assessment of Deliberate Release*. Birkhäuser Verlag, Basel, pp 109–116.

Greene, A.E. and Allison, R.F. (1994) Recombination between viral RNA and transgenic plant transcripts. *Science* 263: 1423–1425.

Habermas, J. (1983) Diskursethik – Notizen zu einem Begründungsprogramm. *In*: J. Habermas (ed.): *Moralbewußtsein und kommunikatives Handeln*. Suhrkamp, Frankfurt/M., pp 53–125.

Herbig, J. and Hohlfeld, R. (1990) *Die zweite Schöpfung – Geist und Ungeist in der Biologie des 20. Jahrhunderts*. Hanser, München and Wien.

Hume, D. (1978) *A Treatise of Human Nature*, Second Edition. Clarendon, Oxford (Orig. 1739/40).

Jacob, F. (1972) *Die Logik des Lebenden – Von der Urzeugung zum genetischen Code*. S. Fischer, Frankfurt/M.

Jax, K. (1994) Mosaik-Zyklus und Patch-dynamics: Synonyme oder verschiedene Konzepte? Eine Einladung zur Diskussion. *Z. Ökol. Natursch.* 3: 107–112.

Krauße, E. (1987) *Ernst Haeckel*. Biographien hervorragender Naturwissenschaftler, Techniker und Mediziner Bd. 70, Teubner, Leipzig.

Krimsky, S. (1982) *Genetic Alchemy. The Social History of the Recombinant DNA Controversy*. MIT Oral History Programm, Massachusetts Insitute of Technology Press, Cambridge (MA).

Lepenies, W. (1976) *Das Ende der Naturgeschichte – Wandel kultureller Selbstverständlichkeiten in den Wissenschaften des 18. und 19. Jahrhunderts*. Hanser, München and Zürich.

Levidov, L. (1995) Whose ethics for agricultural biotechnology? *In*: V. Shiva and I. Moser (eds): *Biopolitics: Biotechnology, Feminism and Ecology*. London, Zed Books, pp 175–190.

Lorenz, M.G. and Wackernagel, W. (1993) Horizontal gene transfer in the environment. *In*: K. Wöhrmann and J. Tomiuk (eds): *Transgenic Organisms: Risk Assessment of Deliberate Release*. Birkhäuser Verlag, Basel, pp 43–64.

Markl, H. (1986) Evolution und Gentechnik. *In*: H. Markl (ed.): *Evolution, Genetik und menschliches Verhalten. Zur Frage wissenschaftlicher Verantwortung*. Piper, München, pp 12–37.

Mayr, E. (1991) *Eine neue Philosophie der Biologie*. R. Piper, München.

Mayr, E. (1993) Proximate and ultimate causations. *Biol. Philos.* 8: 93–94.

Mieth, D. (1993) The release of microorganisms – Ethical criteria. *In*: K. Wöhrmann and J. Tomiuk (eds): *Transgenic Organisms: Risk Assessment of Deliberate Release*. Birkhäuser Verlag, Basel, pp 245–256.

Moore, G.E. (1978) *Principia Ethica,* Revised Edition. Cambridge University Press, Cambridge (Orig. 1903) .

Moore, H.D.M., Holt, W.V. and Mace, G.M. (1992) *Biotechnology and the Conservation of Genetic Diversity*. Symposia of the Zoological Society of London No. 64, Clarendon Press, Oxford.

Orians, G.H. (1962) Natural selection and ecological theory. *Amer. Natur.* 96: 257–264.

Passmore, J. (1980) *Man's Responsibility for Nature – Ecological Problems and Western Tradition*, Second Edition. Duckworth, Gloucester.

Peters, R.H. (1991) *A Critique for Ecology*. Cambridge University Press, Cambridge.

Pörksen, U. (1988) *Plastikwörter – Die Sprache einer internationalen Diktatur*. Klett-Cotta, Stuttgart.

Potthast, T. (1990) Freisetzung gentechnologisch veränderter Organismen. *In*: AK BuSiB e.V. (ed.): *Gentechnologie*, Göttingen, pp 189–210.

Potthast, T. (1996) Die Debatte um Gentechnik und Evolution – vorläufige Bestandsaufnahme und kritische Bemerkungen. *In*: J. Ries (ed.): *Darwin und Darwinismus – Materialien zu einem Symposion (14–15.4.1994)*, Wissenschaft im Deutschen Hygiene Museum, Dresden, pp 182–210.

Potthast, T. and Weber, B. (1995) *Der Einsatz gentechnisch veränderter Mikroorganismen zur Reduzierung von Schadstoffbelastungen*. Werkstattreihe 92 des Öko-Instituts e.V., Freiburg im Breisgau.

Prigogine, I. and Stengers, I. (1980) *Dialog mit der Natur – Neue Wege naturwissenschaftlichen Denkens*. R. Piper, München and Zürich.

Rolston III, H. (1988) *Environmental Ethics – Duties and Values in The Natural World*. Temple University Press, Philadelphia.

Rottschaefer, W.A. (1991) Evolutionary naturalistic justifications of morality: A matter of faith and works. *Biol. Philos.* 6: 341–349.

Schmitt, M. (1993) Für ein ganzheitliches Evolutionsverständnis. *In*: H. Kaiser, B. Matejovski and J. Fedrowitz (eds): *Kultur und Technik im 21. Jahrhundert*. Campus, Frankfurt/M. and New York, pp 285–287.

Tiedje, J.M., Colwell, R.K., Grossman, Y.K., Hodson, R.E., Lenski, R.E., Mack, R.N. and Regal, P.J. (1989) The planned introduction of genetically engineered organisms. Ecological considerations and recommendations. *Ecology* 70: 298–315.

Trepl, L. (1987) *Geschichte der Ökologie. Vom 17. Jahrhundert bis zur Gegenwart – 10 Vorlesungen.* Athenäum, Frankfurt/M.

Trepl, L. (1994) Competition and coexistence: On the historical background in ecology and the influence of economy and social sciences. *Ecol. Modell.* 75/76: 99–110.

van der Weele, C. (1993) Explaining embryological development. Should integration be the goal? *Biol. Philos.* 8: 385–397.

Weber, B. (1994) Evolutionsbiologische Argumente in der Risikodiskussion am Beispiel der transgenen herbizidresistenten Pflanzen. *In:* W. van den Daele, A. Pühler and H. Sukopp (eds): *Verfahren zur Technikfolgenabschätzung des Anbaus von Kulturpflanzen mit gentechnisch erzeugter Herbizidresistenz,* Heft 5. FS II 94-305, Wissenschaftszentrum für Sozialforschung, Berlin, pp 1–146.

Weizsäcker, E.-U. von and Weizsäcker, C. von (1987) How to live with errors? On the evolutionary power of errors. *World Futures – The Journal of General Evolution* 23: 225–235.

Weizsäcker, C. von (1991) *Fehlerfreundlichkeit und Gentechnologie. Fragen zu Evolution und Risiko.* – Vortrag auf der Fachtagung "Freisetzung von genmanipulierten Mikroorganismen und Pflanzen – Anwendungsmöglichkeiten und Risiken", veranstaltet vom OekologInnenverband der Schweiz (OeVS) und dem Schweiz. Bund für Naturschutz (SBN) am 25.10.1991 an der Eidgenössischen Technischen Hochschule, Zürich.

Wilson, E.O. (1975) *Sociobiology – The New Synthesis.* Belknap (Harvard University Press), Cambridge (MA).

Wilson, E.O. (1988) *Biodiversity.* National Academy Press, Washington D.C.

Wöhrmann, K. and Tomiuk J. (1993) *Transgenic Organisms: Risk Assessment of Deliberate Release.* Birkhäuser Verlag, Basel.

Wöhrmann, K., Tomiuk, J., Pollex, C. and Grimm, A. (1993) *Evolutionsbiologische Risiken bei Freisetzungen gentechnisch veränderter Organismen in die Umwelt.* – Studie in Auftrag des Umweltbundesamtes, Forschungsvorhaben 108 02 099, Tübingen.

Young R.A. (1985) *Darwin's Metaphor – Nature's Place in Victorian Culture.* Cambridge University Press, Cambridge.

Transgenic Organisms – Biological and Social Implications
J. Tomiuk, K. Wöhrmann & A. Sentker (eds)
© 1996 Birkhäuser Verlag Basel/Switzerland

Genetic engineering and the press – Public opinion *versus* published opinion

A. Sentker

Eduard-Spranger-Straße 41, D-72076 Tübingen, Germany

Summary. The hostility towards technology and particularly the attitude of rejection towards genetic engineering are frequently repeated arguments in discussions about scientific and economic development in Germany. The validity of these arguments, however, is only rarely tested. Extensive studies from recent years have shown that German hostility towards technology is out of the question. Public opinion is even increasingly positive about genetic engineering. Above all certain fields of application such as medicine and pharmacy are met with approval. Scientific and technological reports still bear considerable deficiencies. Many of these deficiencies are of a structural nature. Reports on technology in general and on genetic engineering in particular are, however, on an average balanced and show as well as the public opinion a distinctive valuation with a mainly positive tendency. Better reporting will only be possible, if scientists and journalists cooperate intensively.

Introduction

Science and technology are affecting more and more the economic, political and social development in the industrial states (Ruß-Mohl, published in Flöhl and Fricke, 1987). Since research and technology are shaping the future of society (and increasingly, everyday life) they are very much in the focus of public interest (Depenbrock, 1976). Examples of this development are nuclear energy, genetic engineering and – catchwords like information highway and multimedia are showing the trend – information technology.

The interest of the public in technological development is growing with its increasing influence upon people's lives. However, a deficit (which has already been predetermined historically) becomes more and more apparent: The public is only inadequately and partly even wrongly informed about the work of the universities and research institutes, about industrial developments and aims, and about the questions of fundamental research (Depenbrock, 1976). As early as in 1980 the Gesellschaft Deutscher Naturforscher und Ärzte (Society of German Nature Researchers and Physicians) published a declaration in which it criticized the lack of agreement and communication between scientists and the public. The society particularly called upon scientists to stimulate the public dialogue (Thorbrietz, 1986, p. 50). Even twelve years later, at the conference of the Gesellschaft Deutscher Naturforscher und Ärzte the problems of agreement between scientists and public were the center of discussions (Schnabel, 1992). It is still an emotive issue.

The media's role in the dialogue on new technologies and scientific developments is constantly emphasized by communication experts, politicians and also by journalists and publishers themselves (Depenbrock, 1976). There is, however, wide dissension on the duties of journalists (Lohmar, 1972). The advocates of new technologies detect growing hostility by the public and attribute responsibility for this trend to the media. In the evaluation of technology, attention is now more and more drawn to the so-called acceptance dilemma. Without research into the acceptability of technological progress, estimation of technical development remains incomplete and often defective (Jaufman et al., 1989; Kliment et al., 1994). Public acceptance has become an important, but no longer unquestionable condition for successful research and technology politics (Hennen, 1994).

Acceptance and hostility towards technology

In the public discussion the Federal German population is often reproached for being particularly sceptical towards technology. Moreover, this hostility towards technology is said to be a typically German phenomenon. In view of the seemingly more technical- and progress-minded opinion in foreign countries, experts are warning that a hostile mood to innovation could endanger the economic and industrial position of Germany (Kliment et al., 1994)

At the Akademie für Technikfolgenabschätzung (Academy for the Valuation of Technological Issues) in Stuttgart, Kliment et al. (1994), after an extensive analysis of different surveys and studies, come to the surprising conclusion that a Federal German hostility towards technology is out of the question. In spite of all the differences of the questions and inquiry methods, the available public opinion polls demonstrate that the majority of the German population perceives technology in a mostly positive way (Kliment et al., 1994, p. 6; comp. also Hennen, 1994). The studies of Jaufman et al. (1989) and Kistler (1991) had already reached a similar conclusion. The attitudes towards technology have changed, perhaps in that the distinctive technological euphoria of the sixties has given way to a more ambivalent attitude (Hennen, 1994). The Stuttgart authors see this ambivalence as a partial phenomenon of an attitude towards modern industrial society, which is altogether divided. This attitude is, however, not a typically German phenomenon, as it can be found in all industrial nations (Kistler, 1991; Hennen, 1994). There is an increasing distancing from technology and a decline of the population's knowledge of technology to be found in nearly all industrial countries. The above quoted warnings that a country's ability to compete will be jeopardized are evident in all highly developed societies (Kistler and Pfaff, 1990).

Kliment et al. (1994) distinguish the acceptance of technology in the different spheres of public life. Acceptance in the field of consumption is indicated by the willingness to buy a certain product – for instance a transgenic tomato. Acceptance on the level of production is indicated by the willingness to use a new technique at a job and also by tolerance to a technical installation in one's own surroundings.

How useful is this distinction can be demonstrated by the results of an up-to-date investigation: Consumption technology is mainly positively rated. Technology at the place of work is received positively for the most part, if it is gradually introduced and does not endanger jobs. Technology close at hand is more often rejected in large technical applications – an observation, which is confirmed in the field of genetic engineering in laboratories and production buildings as well as in experiments of deliberate release.

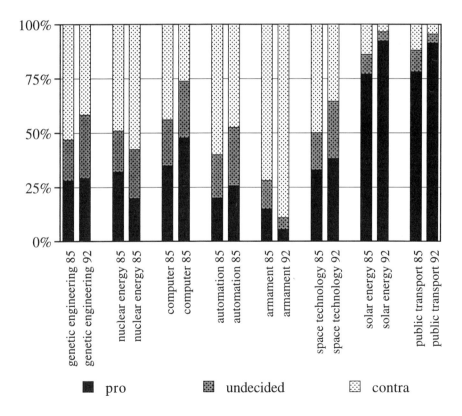

Figure 1. Valuation of different fields of technology. In 1985 (Frederichs, 1986) and 1992 (Hennen and Stöckle, 1992) about 1000 people older than 18 years were asked how they valued the aspects of technology (modified after Hennen and Stöckle, 1992).

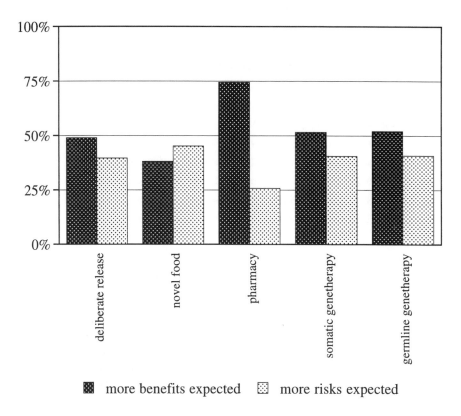

Figure 2. Public valuation of different fields of application in genetic engineering. 1049 people older than 18 years were asked in 1992 (modified after Hennen and Stöckle, 1992).

None of the hitherto existing inquiries, however, show a general scepticism towards technology. Uses of technology are, on the contrary, very much differentiated by the public (Fig. 1). Technology as a necessary condition of a modern society and its influence upon the competitive ability of a country are no longer disputed. The fears are just as unanimous, however, that technologies are being applied before their negative effects have really been examined (Kliment et al., 1994). Exactly these fears also characterize the critical attitude towards applications of genetic engineering.

Genetic engineering and the public

The evaluation of genetic engineering in the population corresponds mainly to the above trend of growing scepticism towards technology (Hennen and Stöckle, 1992). In Germany one can observe a particular sensitivity to the risks of genetic engineering. In comparison to the rest of the EC-states, the fear of the risks of genetic engineering is in general more distinct. The Danish and Dutch are, however, clearly still more sceptical in this matter (Kliment et al., 1994; Jaufmann et al., 1989). The attitude towards genetic engineering generally is still changing. The trend being more in the direction of more positive thinking (Kliment et al., 1994).

Particularly interesting are the results of studies on applications in genetic engineering. In a recently conducted inquiry (Hennen and Stöckle, 1992) people were asked which fields they would associate with genetic engineering. First named were treatments with human genes (19%) as well as the manipulation of non-human genes (15%). Only 7–8% of the people asked named applications in animal husbandry and cultivation of plants. Despite the importance of the subject, in the discussion upon the amendment of the genetic engineering law, the deliberate release of transgenic organisms was only named by 0.5% of the people asked.

The studies demonstrate that the different fields of application in genetic engineering are evaluated very differently (Fig. 2). Treatments of the genetic substance of man for health reasons as well as the production of pharmaceutics are mostly considered as useful and of low risk. The reverse, however, is found, regarding the application of biotechnological knowledge in the field of crops and foodstuffs (Kliment et al., 1994, p. 35). The population is more willing to use methods of genetic engineering if their application is exactly described. The more certain applications are questioned in detail, the greater the general increase in acceptance of genetic engineering. Similarly to the evaluation of genetic engineering as a whole, the attribution of advantages and risks to individual fields of application of genetic engineering is still extremely variable and may change according to the rate of information and the course of the public debate (Kliment et al., 1994, p. 24).

Public opinion and published opinion

The public must clearly be better informed upon the aims, results and possible consequences of scientific research and technological development than up to now. This demand is shared by most experts today (Fig. 3). Equally, there are repeated complaints about considerable deficiencies in the communication of information (Flöhl and Fricke, 1987). The need for scientific reporting is therefore unquestionable. Scientific reporting in the media has not yet become commonplace, but

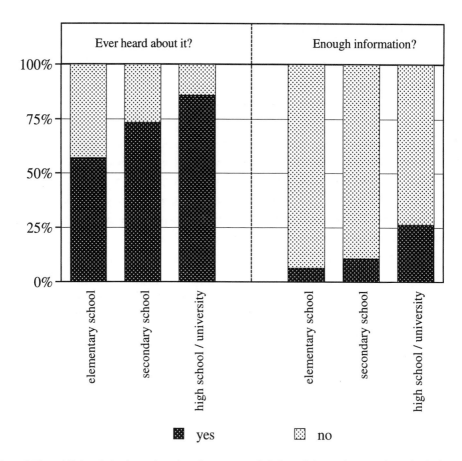

Figure 3. The public is only inadequately and partly even wrongly informed about science, as shown by the knowledge about genetic engineering, e.g., the human genome project. 1049 people older than 18 years were asked in 1992 (modified after Hennen and Stöckle, 1992).

the quality of scientific journalism has been recognized as a problem (Ruß-Mohl, published in Flöhl and Fricke, 1987).

Particularly interesting in this connection is the observation that there is no lack of information, but indeed a lack of essential qualities in the communication of information (Reschenberg, published in Bamme et al., 1989). Above all four reasons are made responsible for this: The relatively small number of scientific–technologically trained journalists; the orientation of reports towards the interests and knowledge of the readers; specific rules for the choice of news; and the orientation of the weighting and evaluation of information towards that of non-science journalistic colleagues (Reschenberg, published in Bamme et al., 1989). Even on the part of science, especially in universities and research institutes there are still considerable deficits in the presentation and communication of information (Neidhardt, 1994).

Position and function of the media in reporting

The duty of the media is to create an informed public and thus to make social processes controllable. This task consequently has to be extended to the complex field of science and technology (Guha, published in Bamme et al., 1989). Furthermore journalism should make transparent the complexity of a society of technology and bring about a communication beyond the limits of the individual specialities. It is insufficient, however, to translate science simply into a popular language (Beste, published in Bamme et al., 1989).

Information on new scientific–technical facts and their connections, and also upon processes, procedures and methods have to be relayed in an intelligible way. The effects of science and technology upon the individual and society as well as the peculiarities of scientific work and scientific reflection should become visible. If the scientific journalist takes his tasks seriously he, ideally, contributes directly or indirectly to the scientific progress (Ronneburger, published in Bamme et al., 1989). This ideal is, however, rarely achieved.

Science reporting

Publicity in science journalism is an interesting topic (Böhme-Dürr and Grube, 1989). Thorbrietz (1986) compiled the data of the situation of science journalism:

The proportion of reporting upon scientific themes is very low in relation to the contents offered in total by the media and amounts on an average to less than five percent. There is hardly a regularity in reporting. Only a few newspapers have a separate science editorial staff. Among the daily newspapers with full editorial staff in the Federal Republic of Germany scarcely two thirds have their own section for science reporting. Among the small newspapers, only one in six employs a science journalist, and among the bigger ones it is nearly one in two. However, even the larger newspapers with high output rarely have more than two journalists concentrating on the field of science and research (Hömberg, published in Bamme et al., 1989; Ruß-Mohl, published in Flöhl and Fricke, 1987). The science page appears either once a week (29%) or every two weeks (14%). Most of the newspapers of course publish the section irregularly (34%). That means: Science pages are often shifting or changing pages; they only come into the paper if the volume of advertizing allows it (Hömberg, published in Bamme et al., 1989). Science is often, depending on current events, a stopgap and page filler.

Objectivity and orientation to facts as journalistic ideals lead to a kind of reporting which is strongly dependent upon announcements, statements, declarations of the press and current events, and it therefore reacts. The perception of current events in journalism makes a problem interesting only if it has actually occurred with reported consequences. Science journalists overestimate their readers' ability, mostly assuming a clearly too advanced level of education and thus switch off their readers (Hömberg, published in Flöhl and Fricke, 1987). The splitting of science reports into a specialist branch closely connected to science itself and into a popular branch rather remote from science has still not been perceived in publication research (Flöhl, published in Flöhl and Fricke, 1987). Popularly oriented journalists often practise a very literal and even-treated science reporting; they only rarely communicate long term background information permitting an individual judgement of the facts to the reader (Thorbrietz, 1986). Journalists lack above all the faculty of exactly observing rapid change and of recognizing controversial questions early.

Focal points of reporting

Even in an ordinary news agency there are more than 1000 announcements of agencies coming in daily. Press communications of institutes and associations augment these as well as offers from freelance contributors and internal offers. The editors have to make a selection. Different factors contribute to this selection: the taste of the public, judgements of the editorial staff, the news matter itself or the subjective and spontaneous decisions of the editor (Emmerich, 1984).

In science reporting the focal points are technology, medicine, biology, geosciences and, to a lesser extent, physics. Since the eighties the focal points of reporting have lain primarily in the fields of chemical technology, medical technology and environment technology. The spread of computers into the daily life of man has made data processing an important theme. Genetic engineering has been added as a new application-oriented field of technology and has a high profile in reporting (Kepplinger et al., 1991).

In reporting, the reluctance to evaluate and interpret results and developments becomes evident (Hömberg, published in Flöhl and Fricke, 1987). Necessary for such evaluation is of course the professional objectivity which cannot be provided by most of the journalists. Kepplinger (1989) notes, however, an increasing change in the presentation of technology, for which he gives two possible reasons: A changing reality and a changing perception of reality. The first assumption suggests that the increasingly negative presentation of technology results from the growing threat to man and his environment as a consequence of technical development. The second reason indicates a growing sensitivity of man, which may be influenced by rational as well as by irrational factors. Kepplinger proves (in an exemplary manner) that reality and presentation often diverge. The intensity of the presentation of technological damage says little or nothing about its *actual* intensity (Kepplinger, 1989, p. 138). The presentation of technology and its transformation is therefore dependent on a changing perception of reality.

The attitude of journalists towards technology has changed, too. This is shown in their comments. They always draw upon the opinion of experts, whose (predominantly negative) statements prevail. The differential selection of expert opinions, depending on the editorial attitude, reinforces the impression that scientists are at odds or disagreeing (Kepplinger, 1989).

The lack of training of journalists is for Kepplinger an important reason why some developments are wrongly evaluated. Publicists criticize above all the small part played by active investigation, causing many interesting and exciting themes to remain undiscovered. Individual results of research are generally reported, especially spectacular successes, but not research in process, stages in research, wrong directions and digressions of research. It is also striking how often deficiencies in journalistic training are criticized; deficiencies which are particularly observed when treating complex themes and connections (Ruß-Mohl, published by Flöhl and Fricke, 1987). The problem is that the presentation of technology in the media influences public opinion. But Kepplinger indicates that reporting in the media is not the only reason for an altered valuation of technology by the public.

Deficient infrastructures

Technology has become a favourite topic of general and political reporting. The fact that not only science journalists report upon genetic engineering is to Kepplinger significant, because not professionally trained journalists but humanistically and sociologically trained editors will be responsible for reporting on genetic engineering. The structural transformations in the reporting of the media may change public conceptions all the more, as political reporting usually has a greater resonance within the population than economics and science reporting, biased as they are towards technology and being received mainly by interested minorities (Kliment et al., 1994, p. 26). Thorbrietz (1986) criticizes that the scientific nature of many questions is given less attention than the political aspects.

As the editorial staffs lack a science journalistic infrastructure, redoubled efforts on the part of science are necessary (Thorbrietz, 1986). Appropriate information centres and places of reference for investigation are needed as well as qualifications and further training. Thirdly there is a need for critical institutions such as in science journalism research (Ruß-Mohl, published in Flöhl and Fricke, 1987).

Language barriers

The publicists see the technical language and the flood of information in science as well as the lack of cooperation between scientists and the press as particular obstacles in the way of a "public science" (Depenbrock, 1976). Language barriers not only exist between different social levels, between experts and laymen, but also amongst the specialists themselves. Professional language as an artificially formed, precise terminology serves as means of communication among colleagues. It loses its value if the results of science and technology have to be made intelligible to a wider public (Depenbrock, 1976; Reschenberg, published in Bamme et al., 1989; Pörksen, 1986). The gap between natural and scientific language and thus the gap between science and society threatens communication and transfer of knowledge to an ever greater extent. Bungarten (1981) distinguishes "professional barriers" from such language barriers, which are raised consciously and intentionally by the speaker. Linguists confirm this diagnosis. A learned vocabulary has two functions: firstly the function of communication in giving expression to thoughts – and among those the important, difficult and enigmatic thoughts; secondly the social function to endow its users with reputation and to evoke respect and reverence in those who do not understand it (Hayakawa, 1976). The language barriers exist, however, in both directions: On the one hand the question arises, how the scientific knowledge may be communicated to the public, but also how knowledge which is socially demanded, may be transmitted to the producers of scientific knowledge, the scientists themselves.

Science happens largely behind anxiously closed doors, in carefully protected fortresses, in a labyrinth without signpost, opaque and immense. Language becomes a secret defensive weapon (Mohl, published in Flöhl and Fricke 1987, p. 27). In this field journalists have to mediate, translate results of research, present their context, their conditions, their authenticity, validity and possibly their practical significance. For this technical competence is needed, knowledge of scientific production and communication-conditions, knowledge of science not only as an instrument of growing recognition, but also as working process and operating system (Hömberg, published by Flöhl and Fricke, 1987, p. 36). Many science reporters, however, lack such knowledge. Correspondingly scientists often complain about the lack of competence and the sensationalism of the journalists. These in turn deplore the minimal understanding by scientists of journalists and the reluctance to face the public (Krüger, published by Flöhl and Fricke, 1987) – a dilemma, which above all in the domain of genetic engineering has been exacerbated in recent years. On complex issues, the individual is often hardly capable of taking decisions and unable to support decisions taken for him by specialists. A further elementary problem of laymen is that critical voices from within their ranks, particularly in the discussion on genetic engineering, are often rightly condemned as being unqualified. This criticism is sometimes premature. Scientists seem to have forgotten how to communicate with laymen. Particularly interesting in this connection is the assumption that journalism does not rectify existing information and communication deficiencies, but on the contrary intensifies them. It is assumed that educated readers use the media more quickly and effectively than those without adequate qualification.

Genetic engineering and the media

As the discussion on genetic engineering becomes more and more politicized, the press gets into the crossfire of criticism. For the adversaries of technology the press is not critical enough, for scientists it is not competent enough and in the eyes of the advocates of technology it propagates an unreasonable panic. The accusations, however, have hardly been checked for accuracy

Kepplinger et al. (1991) as a reaction upon the present discussion, have made an extensive empirical study on the presentation of genetic engineering in the print media. In the first part of their analysis the authors compare the opinions of scientists and science journalists about genetic engineering, and in confronting these views reach some very revealing results. The science journalists themselves demonstrate a mainly positive attitude towards genetic engineering; however, non-science journalists mostly display an attitude of rejection. Science journalists see more potential risks in genetic engineering than do scientists and accept a development or a further application only under the condition of restrictive safety measures. The scientists see the public discussion as well as the tendencies of reporting as a clear handicap to their work and are afraid of a migration of research and production into foreign countries which are science-friendly. Most scientists find the image of genetic engineering in the media wrongly presented and the risks dramatically exaggerated. They often doubt the professional competence of journalists.

Science journalists in return, however, consider scientists working in industry as being competent, but as not very trustworthy. They mistrust the established body of science as well as its scientific institutions. The formal training of most of the science journalists is on a high level; however, many lack a scientific education. Journalists who have not been trained scientifically see more potential risks in genetic engineering than their colleagues with a scientific education. Scientifically trained journalists in the daily press hardly have a chance to speak. Up-to-date concerns of genetic engineering, which are of general interest, are often taken off their hands by journalists of other areas, especially by political editors.

Kepplinger et al. (1991) in a second step analyzed the contents of different journals extensively. They examined extra-regional and regional daily papers, weeklies and public magazines as well as some popular science magazines.

Only a third of the contributions on genetic engineering had been written by science journalists. Nearly twelve percent of the information comes from news agencies. Genetic engineering and its applications are generally presented in a relatively balanced manner, the conditions of genetic engineering, however, are presented very negatively. Reports about genetic engineering are mainly placed in the political sections of the published organs. The valuation of technology there is slightly negative, on the science pages or in the economy section however, it is more positive. News articles about genetic engineering dominate and convey a mostly positive image. In the commentaries (which are rarer) the impression is, however, contradictory.

General statements on genetic engineering have a rather negative tendency, characteristics and consequences of technology are, however, mostly presented in a positive manner. The greatest part of reporting is of a general nature and does not refer to particular fields of application of genetic engineering. In statements about specific fields of application, genetic diseases and human gene therapy dominate other fields. Ecological aspects are hardly taken into account. The effects of genetic engineering on human health are generally judged in the most positive way, but their role in animal husbandry is regarded in the most negative way. The standards of judgement are, however, only rarely obvious. Ethical and moral criteria are mostly mentioned, other aspects like economic, legal, political or scientific considerations follow far behind.

The study gives above all evidence of one thing: The assertion of many scientists that reports on genetic engineering give a false image, disseminate incorrect facts and stress individual aspects wrongly, which is misleading, cannot or only under certain conditions be accepted. The impression has presumably originated from a few extremely negative reports; the average contribution is, however, presented in an adequate and balanced manner. In the meantime a positivistic tendency can be observed mainly on the economic pages in reporting about economic perspectives. Technological reports as well as particular reports on genetic engineering show parallel trends in the extra-regional press. In analogy to the growing scepticism towards technology, genetic engineering is also evaluated more critically in the media. The results demonstrate as well that in spite of the growing scepticism towards genetic engineering at the beginning of the eighties, reports about genetic engineering have never been really negative. The complaints of genetic engineering's advocates of a critical mood about genetic engineering in the German media may not be confirmed on the basis of those data (Kliment et al., 1994, p. 26).

This relatively cursory survey even shows focal points in reporting. The increase of journalistic interest is noticable when reports on approaches in human gene therapy and diagnostics as well as positive results in reproduction medicine and the Human Genome Project dominate. With the introduction of the genetic engineering law, legal and political aspects become interesting. The international situation and different national regulations are being discussed. With the increasing orientation towards application and evidently positive results in application, economic aspects are more and more in the scope of the media. The catchword deliberate release is mostly only dealt with within the discussions and actions in the sphere of up-to-date experiments. Ecological aspects of deliberate release are rarely presented comprehensively. Research on the effects of technology remains nearly ignored in reports. Differentiated scientific discussion about possible risks of deliberate release has been hardly accessible to the public so far.

Approaches to a solution

The different studies have been more and more criticized recently. Above all the Kepplinger study is accused of methodological weakness. But later studies have produced similar results as the Kepplinger study. The studies demonstrate a few essential weaknesses on the part of science as well as on the part of science reporting. Science does not make in any way sufficient effort to inform the public actively. Many organizations and institutions have no (or only a non-functioning) press department. Many scientists are not striving for popular communication with the public and meet without public admittance (Mohl, published in Flöhl and Fricke, 1987). A chance to reclaim lost confidence is given away. First of all, existing prejudices have to be removed and the dialogue between researchers and journalists has to be taken up again. Science must not bend to the needs of the population (Neidhard, 1994). Most important is, however, an improvement of the internal conditions of science journalism, so that science journalists will again become critical partners of the researchers. Therefore it is necessary to make the process of scientific work as well as the original conditions of scientific research obvious. The institutional conditions of journalistic science publishing will clearly be improved, if specific areas or departments for these reports are established and the existing areas are extended. Apart from this, regular print and distribution centres have to be created or to be expanded for science reporting (Hömberg, published in Bamme et al., 1989).

Research about technological reporting in the media as well as investigations about its influence on the public still demonstrates considerable shortcomings (Hennen, 1994). Internationally comparable studies and more exact investigations of the effects of media reporting on the public are missing. The interactions between journalism and science have only been insufficiently researched up to now. One thing, however, has become clear: better science journalism will only happen if both sides, scientists and journalists, cooperate.

References

Bamme, A., Kotzmann, E. and Reschenberg, H. (1989) *Unverständliche Wissenschaft. Probleme und Perspektiven der Wissenschaftspublizistik.* Profil Verlag, München.

Böhme-Dürr, K. and Grube, A. (1989) Wissenschaftsberichterstattung in der Presse. *Publizistik* 4: 448–466.

Bungarten, T. (1981) Wissenschaft, Sprache und Gesellschaft. *In*: T. Bungarten (ed.): *Wissenschaftssprache.* Wilhelm Fink Verlag, München, pp 14–53.

Depenbrock, G. (1976) *Journalismus, Wissenschaft und Hochschule. Eine Aussagenanalytische Studie über die Berichterstattung in Tageszeitungen.* Studienverlag Dr. N. Brockmeyer, Bochum.

Emmerich, A. (1984) *Nachrichtenfaktoren: Die Bausteine der Sensationen. Eine empirische Studie zur Theorie der Nachrichtenauswahl in den Rundfunk und Zeitungsredaktionen.* Ph.D. thesis, Saarbrücken.

Flöhl, R. and Fricke, J. (1987) *Moral und Verantwortung in der Wissenschaftsvermittlung. Die Aufgabe von Wissenschaftler und Journalist.* Neuwieder Verlagsgesellschaft, Neuwied.

Frederichs, G. (1986) *Technikmeinungen in der Bevölkerung. Zur Grundauszählung einer repräsentativen Umfrage.* Kernforschungszentrum, Abteilung für angewandte Systemanalyse, Karlsruhe.

Hayakawa, S.I. (1976) *Sprache in Denken und Handeln*, Fifth Edition. Verlag Darmstädter Blätter, Darmstadt.

Hennen, L. and Stöckle, T. (1992) *Gentechnologie und Genomanalyse aus der Sicht der Bevölkerung. Ergebnisse einer Bevölkerungsumfrage des TAB.* TAB-Diskussionspapier No. 3. (TAB) Bonn.

Hennen, L. (1994) *Ist die (deutsche) Öffentlichkeit "technikfeindlich"? Ergebnisse der Meinungs- und Medienforschung.* TAB-Arbeitsbericht No. 24. (TAB) Bonn.

Jaufmann, D., Kistler, E. and Jänsch, G. (1989) *Jugend und Technik.* Campus, Frankfurt.

Kepplinger, H.M. (1989) *Künstliche Horizonte. Folgen, Darstellung und Akzeptanz von Technik in der Bundesrepublik.* Campus, Frankfurt.

Kepplinger, H.M., Ehmig, S.C. and Ahlheim, C. (1991) *Gentechnik im Widerstreit. Zum Verhältnis von Wissenschaft und Journalismus.* Campus, Frankfurt.

Kistler, E. and Pfaff, M. (1990) Technikakzeptanz im internationalen Vergleich. *In*: E. Kistler and D. Jaufmann (eds): *Mensch, Technik, Gesellschaft.* Westdeutscher Verlag, Opladen, pp 41–70.

Kistler, E. (1991) Eurosklerose, Germanosklerose? Einstellungen zur Technik im internationalen Vergleich. *In*: D. Jaufmann and E. Kistler (eds): *Einstellungen zum technischen Fortschritt.* Campus, Frankfurt, pp 53–70.

Kliment, T., Renn, O. and Hampel, J. (1994) *Chancen und Risiken der Gentechnologie aus der Sicht der Bevölkerung.* Akademie für Technikfolgenabschätzung in Baden-Württemberg. Arbeitsbericht Nummer 29. Stuttgart.

Lohmar, U. (1972) Die Produktivkraft Wissenschaft als publizistisches Problem. *Aus Politik und Zeitgeschichte (Beilage zur Wochenzeitung das parlament)* B 21/72.

Neidhardt, F. (1994) Öffentlichkeit und Öffentlichkeitsprobleme der Wissenschaft. *In*: W. Zapf and M. Dierkes (eds): *Institutionenvergleich und Institutionendynamik.* WZB-Jahrbuch 1994. Edition Sigma, Berlin, pp 39–56.

Pörksen, U. (1986) *Deutsche Naturwissenschaftssprachen. Historische und kritische Studien.* Gunter Narr Verlag, Tübingen.

Schnabel, U. (1992) Und der Laie steht im Regen. 117. Tagung der Naturforscher und Ärzte: Die schwierige Verständigung zwischen Wissenschaft und Öffentlichkeit. *Die Zeit* 41: 45.

Thorbrietz, P. (1986) *Vernetztes Denken im Journalismus. Journalistische Vermittlungsdefizite am Beispiel Ökologie und Umweltschutz.* Niemeyer, Tübingen.

Transgenic Organisms – Biological and Social Implications
J. Tomiuk, K. Wöhrmann & A. Sentker (eds)
© 1996 Birkhäuser Verlag Basel/Switzerland

Epilogue

K. Wöhrmann, A. Sentker[1] and J. Tomiuk

Department of Population Genetics, University of Tübingen, Auf der Morgenstelle 28, D-72076 Tübingen, Germany
[1]*Eduard-Spranger-Straße 31, D-72076 Tübingen, Germany*

Methods involving gene technology have come into world-wide use in the most diverse areas of research. They have also been introduced on an increasing scale into production areas as a result of the immense economical potential of this technology. Hopes regarding the economy of using gene technology, however, still remain as hypothetical as do the possible dangers envisioned to result from its broad application. A comprehensive scenario regarding the large number of possibilities was given by von Schell and Mohr (1995). The present volume focuses in particular on two vital areas: The production of foods (Teuber, this volume) and the breeding of new plant species (Friedt and Ordon, this volume). The increasing application of gene technology in pharmaceutical industries and the possible effects of gene therapy in humans are also discussed (Leipoldt, this volume). Most of the contributions, however, concentrate on summarizing our present knowledge of the biology of organisms that are genetically modified and that may (subsequently) escape human control unintentionally or that are released deliberately. For the assessment and possible limitation of potential dangers we need to know details of the organisms in question. Knowledge of the respective genome is also important as are its ecology and population genetics.

In the discussion of profit and risks of gene technology, supporters and detractors are still irreconcilably opposed despite twenty years of heated debates (Tomiuk and Sentker, this volume). Although some questions have been resolved, many discussions simply come to a standstill, because opponents repeatedly put up arguments which have long been disproved. This phenomenon is pronounced when it comes to evaluation of the facts, and this evaluation in particular is strongly influenced by one's own capability and desire to understand the application, and one's personal interest. The present state of opinion regarding the release of transgenic organisms is most clearly illustrated by the following:

"Field release experiments with transgenic plants and microorganisms – No risk for the environment" (Casper and Landsmann, 1993), and

"However, with certain classes of transgenes for which there are no existing analogues, there will need to be greater care assessing the possible risks associated with release into environment" (Gliddon, 1994).

Some arguments repeatedly put forward in these discussions will be summarized and evaluated on the basis of our knowledge concerning the release of transgenic organisms.

Gene transfer as a tool

In plant breeding transgene technology is regarded as another possible method to produce new traits and culture lines. In the beginning only plants with the desired traits were selected. Later followed the more or less directed combination of genes (traits) of one species or of closely related species with the subsequent selection of the desired combination. Since the discovery of chemical and physical mutagenic agents science workers have tried to manipulate genes in a directed manner. Mutagenic agents, however, have unspecific effects on the whole genome and lead in most cases to negative changes. Therefore, breeding methods based on mutagenesis also had to be followed by comprehensive selection. Only the introduction of methods involving molecular genetics allows the transfer of a single gene, though positioning of this gene within the genome is still not exact. "This makes transgene technology not only a faster and more efficient technology than classical breeding, but probably also decreases the risk of unintended secondary effects" (see for discussion Meyer, this volume). This statement is certainly correct when one refers to genes that are transferred within one species or within closely related species. Genetically modified metabolisms can, however, also be achieved by the transfer of genes governing enzymatic reactions from distantly related species, e.g., from corn to petunias, from peanut into oilseed rape, or even from *Fusarium* to tobacco and from bacteria to potatoes, to mention only a few possibilities (DeLuca, 1993; Friedt and Ordon, this volume; Parker and Bartsch, this volume; Teuber, this volume). Thus, combinations of traits not occurring in the hitherto existing trait variations are established. Critics consider some of these new traits to be "risky" once the organism has unintentionally escaped into the environment. The main argument, i.e., that all possible gene combinations have already originated and been "tested" in the course of evolution, is certainly unacceptable in regard to the phylogeny of the present species spectrum. From the population biologist's and ecologist's point of view, it is not the combination of "letters" that is relevant but the combination of traits. Essential is whether an organism gains a completely new quality (Parker and Bartsch, this volume), how this trait manifests and in which combination the trait coexists with other characteristics.

Gene expression, gene interactions and gene stability

Gene technological methods – not including methods using marker genes – are used to transfer one defined gene into a host genome which already contains some thousands of genes. Methods are being developed which allow the removal of a marker gene from the genome where it is no longer necessary. It is often stated that if the effect of the transferred gene is known, no unforeseeable side-effects are to be expected, and that compared with classical breeding methods based on mutagenesis and selection the underlying genetic mechanisms are clearly better understood in transgenic plants (Meyer, this volume). In most of the cases so far examined, this kind of argument is certainly justified, but in general it cannot be accepted. Numerous examples can be found in elementary and applied genetics text-books which can clearly show that pleiotropic and epistatic effects exist because of the interactions between genes. From breeding methods based on mutagenesis it is known that the effect of a mutation ultimately depends on the genetic background from which it was induced (Röbbelen, 1995). Further supportive observations are gained from transposable elements that exist to the same extent in numerous animals and plants. Their expression and effect on the host genome depends mainly on their position in the genome (Pinsker et al., 1993; Bachmann, this volume; Miller et al., this volume). In general, it can be stated that we need to widen our very limited knowledge of genome organisation (Meyer, 1993) and that it is, e.g., an astonishing fact that the phenomenon of homology-dependent gene silencing was not recognized for a relatively long period by the scientific community (Meyer, this volume).

Transposable elements

Transposable elements have in the meantime been discovered in many plants and animals. They represent a naturally existing parallel to the artificial incorporation of genetic information (Pinsker et al., 1993; Bachmann, this volume; Miller et al., this volume), and render important indications of the effect of foreign DNA on the expression and stability of genes of the host genome. The more we learn, however, about the functional aspects of transposable elements (TEs), the clearer it becomes that it is to date impossible to explain the role of mobile elements and their homologous sequences in host evolution. It can only be speculated upon. It would seem today that the explosive expansion of the *P* elements within *Drosophila melanogaster* in America and from there to Europe and Asia within the last 40 years has not brought any visible disadvantage for the species (Bachmann, this volume). The question to what extent this model and the related knowledge can be transferred to transgenic organisms can only be answered after the acquisition of considerably more information about such naturally occurring mechanisms.

Pleiotropy

Pleiotropy refers to the production of apparently unrelated multiple effects on the phenotypic level by one particular mutant or, in our context, by one transformed gene. Numerous examples of pleiotropic effects have been described in genetics text-books. The manifold consequences are impressive, and result, for example, from a single amino acid substitution in the case of sickle cell anemia. At least 20 morphological and physiological changes can easily result from such effects.

Clear examples for pleiotropic effects have apparently not been observed in transgenic organisms. Nor – due to the limited number so far investigated – can it be expected. It is, however, important that their effects be taken into account, as the following examples demonstrate. Oilseed rape lines showed significant differences in some morphological characters. Nevertheless, the authors stated that there was no evidence that these differences were due to the presence of the introduced gene. It is assumed that they were probably caused by somaclonal variation (Dunwell et al., 1992). In *Chrysanthemus* modified plants exhibit a slower growth rate and produce fewer flowers. Despite the reduction in plant height, the weight is still higher than that of untransformed plants (Otten et al., 1992).

Horizontal gene transfer

Different biological systems allow not only the transfer of genetic information to the following generations but also the transfer between populations of different species. Thus the manipulated genetic information can move from the target organisms to, e.g., related plants and – one fears – alter their spectrum of effects and competitive behaviour in the ecosystem, i.e., lead to unpredictable and possibly negative changes.

Gene transfer in microorgansims

In bacteria, parasexual systems are known that can enable the transfer of genes between bacterial species. The first significant clue relating to the extent of such a transfer was the expansion of antibiotic resistance genes in bacteria that have already led to multiple resistances in many pathogens, along with all imaginable consequences for the medical treatment of humans. A quantitative statement on the frequency of such incidences cannot yet be made. In the meantime, however, it has been recognized that, as well as mutations, horizontal gene transfer, in combination with other recombination events, is a reason for the genetic diversity in bacteria (Lorenz and Wackernagel,

1993 and this volume). As our knowledge regarding gene transfer grows, so the arguments will also change. The existence of gene transfer between bacterial species was recognized years ago, but was still regarded as rare and therefore inadequate for making risk assessments. Today, gene transfer is common and widespread. Thus, all genetic information exists *per se* in our environment, and the release of information seems therefore not to pose any problems. The argument, however, can also be reversed: If transformed genes can spread then the expansion of a trait and its possible effects are in no way controllable.

Gene transfer in plants

In plants, systems allowing for gene flow between different species (species hybridisations, introgressions) have been described as being important for speciation (Stebbins, 1950). Hybridisations depend on the fertilization system of the plants and on the cytogenetic background (Gregorius and Steiner, 1993), and hybridisations are surely repeatedly occurring events. Here too – as in the case of bacteria – their frequency under natural conditions can hardly be determined. Only the results that have become manifest in many amphidiploid hybrids (numerous examples also exist among cultivated plants) are visible. Hybrid zones between two closely related species prove that in many cases continuous gene flow exists between the sympatric species (Barton and Clark, 1990; Jain, 1990).

The fate of genes in populations and invaders in habitats

When transgenic organisms are released into the environment, one always has to reckon with gene flow under the right biological conditions, and in particular when the release, e.g., the cultivation of transgenic plants, is carried out on an industrial scale. We can, however, not make all-encompassing statements about the fate of the hybrids or of the introduced genes. The factors responsible for the expansion of new species in a new environment and for the spread of genes in populations are too manifold. For a cautious assessment of the establishment of species or of their invasion of a new habitat, experiences and models relating to species conservation can be used, however (Tomiuk and Loeschcke, 1993). Models based on population genetics help us to understand the behaviour of genes in populations (Gabriel, 1993; Adam and Köhler, this volume).

The "cost" of additional genes

A frequent argument concerns the negative effect of genes that are not "used", i.e., that represent a "cost" for the organisms. In this model, superfluous genetic information represents a physiological strain for the organism. Many examples exist but not in every case is there a connection between usefulness and cost. In the case of transgenic sugar beets (*Beta vulgaris*) the resistance against viruses does not mean selective disadvantage in conditions with no selection pressure (Parker and Bartsch, this volume). The duplication of genetic information through polyploidisation is not necessary for the existence of a species, but opens new evolutionary prospects. The *P* elements, too, represent additional "unnecessary" information and have – as has already been mentioned – no apparent negative effect. It can also be demonstrated that genotypes and thus transgenes with reduced fitness are under certain circumstances (e.g., in small populations) able to succeed in populations (genetic drift effect) or to found new populations (founder effect).

The fate of mutated genes

For a population invading transformed genes have the same effect as spontaneous mutations. The unpredictability of a mutation's fate can be illustrated by the example of Leguminosae. Some species of this order (*Lupinus*, *Melilotus*) produce alkaloids that prevent the plants from being eaten by herbivores. Other species of the same order do not produce alkaloids (*Pisum*, *Vicia*, *Trifolium*). According to the concept of homologous variations (Vavilov, 1951) similar mutations can, however, be expected in closely related species. On the basis of this hypothesis, an alkaloid-free mutant was found in *Lupinus luteus*, the starting point for breeding different sorts of hybrids. It remains unclear why corresponding mutations manifest in one species but not in others which are closely related.

Consequences for risk assessment

Hybridisations between related species usually occur in the appropriate biological background. In most cultivated species the biology of fertilization has been well researched. With this in mind, Kareiva et al. (1994) stated that measurements of gene or pollen flow are of minor importance for any risk assessment. The absolute containment of pollen or genes is unlikely and an escape seems inevitable. The knowledge of possible hybridisation events is useful only for calculating the probability of a risk but it is not sufficient for forecasting the formation of hybrids and their possible establishment in nature.

Trait and risk

It seems therefore more important for risk assessment to take into account the altered trait. If we disregard the possible importance of the spread of antibiotic markers used for transformation procedures (Teuber, this volume), there is a broad scientific consensus that risk is a function of the traits of new organisms rather than a consequence of a specific technique for genetic modification (Tiedje et al., 1989). Furthermore, biological principles hold whether the varieties are produced by traditional breeding or by modern techniques (Dale, 1994). Most projects involving genetic engineering have not been inherently dangerous because the traits introduced into organisms were non-adaptive (but see Adam and Köhler, this volume). On the other hand, the argument that genetic engineering of plants is nothing more than an extension of conventional plant breeding and the main method used in nature, e.g., recombination and selection, is misleading. There is sometimes a difference in quality but there is always a difference in quantity. Natural selection would never cover several hundreds of acres with the same genotype. The new technology has a higher potential to produce an unexpected outcome than does traditional breeding. This may be the case if highly adaptive traits of plants, e.g., defences against insects and diseases, tolerance against stress conditions such as drought or cold resistance, or, the capacity for mineral uptake, provide them with new potential for interaction within their respective ecosystems or to broaden their available niche. In addition, modification of seed storage proteins and lipids and even changes in the mating system may have unexpected consequences for the evolutionary potential of plants (Tiedje et al., 1989; Dale et al., 1993; Darmency, 1994; Regal, 1994; Parker and Bartsch, this volume).

Ecological and evolutionary periods

The processes of gene incorporation in a population and invader establishment in a habitat are generally regarded as long-term processes on the evolutionary rather than on the ecological time-scale (Adam and Köhler, this volume; Bachmann, this volume; Miller et al., this volume; Parker and Bartsch, this volume). Short-term experiments are therefore not always suitable to detect or predict the potential spread of genetic information. *Prunus serotina* represents an impressive example for such long-term processes (Kowarik and Sukopp, 1986). In 1623 *P. serotina* was introduced to France from America. In the first 250 years effects on the ecosystem were minimal.

After 1900, however, the cultivation of *P. serotina* was prohibited and led to the plants growing wild and to the related chemical, mechanical and biological damage prevention measures. These were on the one hand able to limit the damage but in turn resulted in further negative consequences.

The need for scientific and public discussions

Despite the 20-year-old discussion of the *pros* and *cons* of gene technology, in the ongoing debate on the usefulness and risk of transgenic organisms, the positions of supporters and critics are still widely opposed. A consensus between both sides, however, is only possible in terms of continuous discussions involving experts from the related sciences. Such discussions are ideally aimed at exposing weaknesses in the arguments. Our comprehension of gene interactions (Meyer, 1993 and this volume), of the evolution and ecosystems (Adam and Köhler, this volume; Parker and Bartsch, this volume) is by far not complete. Such knowledge, however, is necessary to build up an effective monitoring system (Parker and Barsch, this volume; Smalla and van Elsas, this volume) and to extend our understanding of the interrelationships in our environment. Numerous examples relating to the "errors" of experts are known from biological pest control schemes and from species conservation. These instances testify to our limited knowledge of the interrelations between evolutionary events and ecological laws. In risk assessments regarding the release of transgenic organisms and their use, we should always keep these relationships in mind. The continuously expanding data in the respective scientifc disciplines demands an ongoing and intensive exchange as well as repeated examinations of arguments and positions.

Discussions, however, must also be demanded for the lay public. Gene technology can be used to change living standards and even lives, and a discussion of the *pros* and *cons* should therefore not be left only to the scientists and those with commercial interests. Conditions for such a discussion must be improved as well as the availability of information to outsiders (Sentker, this volume). Only then can a basis be created for decisions that are made by politicians rather than by scientists. In this context, however, Adam et al. (1993) do not sound at all hopeful:

"There is a final lesson to learn from the study of biological invasions. Even in a few cases where accurate predictions were made, and where we can argue the predictions were right by insight and not by accident, as for example in the case of mustelid introduction to New Zealand, political decision makers may choose to ignore such predictions and serve the vested interests of economically relevant groups, as happened in New Zealand in the 1880s."

References

Adam, K.D., King, C.M. and Köhler, W.H. (1993) Potential ecological effects of escaped transgenic animals: Lessons from past biological invasions. *In*: K. Wöhrmann and J. Tomiuk (eds): *Transgenic Organisms: Risk Assessment of Deliberate Release.* Birkhäuser Verlag, Basel, pp 153–147.

Barton, N. and Clark, A. (1990) Population structure and processes in evolution. *In*: K. Wöhrmann and S.K. Jain (eds): *Population Biology.* Springer-Verlag, Berlin, pp 115–173.

Casper, R. and Landsmann, J. (1993) Freisetzungsversuche mit gentechnisch veränderten Pflanzen and Mikroorganismen – Kein Risiko für die Umwelt. *Gesunde Pflanze* 45: 182–188.

Dale, P.J., Irwin, J.A. and Scheffler, J.A. (1993) The experimental and commercial release of transgenic crop plants. *Plant Breed.* 111: 1–22.

Dale, P.J. (1994) The impact of hybrids between genetically modified crop plants and their related species: General informations. *Molec. Ecol.* 3: 31–36.

Darmency, H. (1994) The impact of hybrids between genetically modified crop plants and their related species: Introgression and weediness. *Molec. Ecol.* 3: 37–40.

DeLuca, V. (1993) Molecular characterization of secondary metabolic pathways. *AGBiotech News and Information* 5: 225N-229N.

Dunwell, J.M., Lewis, G.B. and Paul, E.M. (1992) Agronomic and ecological studies of genetically modified oil seed rape. *In*: R. Casper and J. Landsmann (eds): *The Biosafty Results of Field Tests of Genetically Modified Plants and Microorganisms.* Biologische Bundesanstalt für Land- und Forstwirtschaft, Braunschweig, Germany, p 239.

Gabriel, W. (1993) Technologically modified genes in natural populations: Some sceptical remarks on risk assessment from the view of population genetics. *In*: K. Wöhrmann and J. Tomiuk (eds): *Transgenic Organisms: Risk Assessment of Deliberate Release.* Birkhäuser Verlag, Basel, pp 109–116.

Gliddon, C. (1994) The impact of hybrids between genetically modified crop plants and their related species: Biological models and theoretical perspectives. *Molec. Ecol.* 3: 41–44.

Gregorius, H.-R. and Steiner, W. (1993) Gene transfer in plants as a potential agent of introgression. *In*: K. Wöhrmann and J. Tomiuk (eds): *Transgenic Organisms: Risk Assessment of Deliberate Release.* Birkhäuser Verlag, Basel, pp 83–107.

Jain, S.K. (1990) Variation and selection in plant populations. *In*: K. Wöhrmann and S.K. Jain (eds): *Population Biology.* Springer-Verlag, Berlin, pp 199–230.

Kareiva, P., Morris, W. and Jakobi, C.M. (1994) Studying and managing the risk of cross-fertilisation between transgenic crops and wild relatives. *Molec. Ecol.* 3:15–21.

Kowarik, I. and Sukopp, H. (1986) Ökologische Folgen der Einführung neuer Pflanzenarten. *In*: R. Kollek, B. Tappeser and G. Altner (eds): *Gentechnologie – Chancen und Risiken.* J. Schweizer-Verlag, München, pp 111–135.

Lorenz, M.G. and Wackernagel, W. (1993) Bacterial gene transfer in the environment. *In*: K. Wöhrmann and J. Tomiuk (eds): *Transgenic Organisms: Risk Assessment of Deliberate Release.* Birkhäuser Verlag, Basel, pp 43–64.

Meyer, P. (1993) Expression and stability of foreign genes in animals and plants. *In*: K. Wöhrmann and J. Tomiuk (eds): *Transgenic Organisms: Risk Assessment of Deliberate Release.* Birkhäuser Verlag, Basel, pp 5–23.

Otten, A., Ackerboom, M., v.d. Meer, R., Morgan, A., Firoozabady, E., Nicholas, J., Lemieux, C., Robinson, K. and County-Gutterson, N. (1992) Field trials and further development of genetically engineered colour modified chrysanthemums. *In*: R. Casper and J. Landsmann (eds): *The Biosafty Results of Field Tests of Genetically Modified Plants and Microorganisms.* Biologische Bundesanstalt für Land- und Forstwirtschaft, Braunschweig, Germany, p 264.

Pinsker, W., Miller, W.J. and Hagemann, S. (1993) P elements of *Drosophila*: Genomic parasites as genetic tools. *In*: K. Wöhrmann and J. Tomiuk (eds): *Transgenic Organisms: Risk Assessment of Deliberate Release.* Birkhäuser Verlag, Basel, pp 25–42.

Regal, P.J. (1994) Scientific principles for ecologically based risk assesment of transgenic organisms. *Molec. Ecol.* 3: 5–13.

Röbbelen, G. (1995) Beiträge der Biotechnologie zur Verbesserung von Qualitäts- und Leistungseigenschaften landwirtschaftlicher Kulturpflanzen. *In*: T. von Schell and H. Mohr (eds): *Biotechnologie – Gentechnik. Eine Chance für neue Industrien.* Springer-Verlag, Berlin, pp 201–214.

Stebbins, G.L. (1950) *Variation and Evolution in Plants.* Columbia University Press, New York.

Tiedje, J.M., Colwell, R.K., Grossman, Y.L., Hodson, R.E., Lenski, R.E., Mack, R.N. and Regal, P.J. (1989) The planned introduction of genetically engineered organisms: Ecological consideration and recommendations. *Ecology* 70: 298–315.

Tomiuk, J. and Loeschcke, V. (1993) Conditions for the establishment and persistence of populations of transgenic organisms. *In*: K. Wöhrmann and J. Tomiuk (eds): *Transgenic Organisms: Risk Assessment of Deliberate Release.* Birkhäuser Verlag, Basel, pp 117–134.

von Schell, T. and Mohr, H. (1995) *Biotechnologie – Gentechnik. Eine Chance für neue Industrien.* Springer-Verlag, Berlin.

Vavilov, N.J. (1951) *The Origin, Variation, Immunity and Breeding of Cultivated Plants.* Waltham, Mass.

Subject index

(The page number refers to the first page of the chapter in which the keyword occurs)

B. Sobral

California Institute of Biological Research (CIBR), La Jolla, CA, USA (Ed.)

The Impact of Plant Molecular Genetics

1996. 364 pages. Hardcover.
ISBN 3-7643-3802-4

Recent years have witnessed an explosion of molecular genetic techniques and approaches that have enabled large gains in basic and applied plant genetics. These techniques span the realm of genetic transformation, which is the non-sexual introduction of new genes in plants, to DNA-marker-assisted genetic studies. DNA markers, in particular, have allowed previously intractable problems, such as genetics of polypoinds and phylogenetics and evolution of crops, to be studied in great detail. DNA markers are also largely responsible for bridging qualitative, quantitative and developmental genetics, thereby providing a rich, multidisciplinary environment for plant biology. The application of DNA markers and plant transformation technologies have serious socio-economic implications for world agriculture, especially in the developing world.

THE IMPACT OF
PLANT MOLECULAR
GENETICS

Bruno W.S. Sobral
Editor

BIRKHÄUSER

The chapters in this volume, written by international experts in diverse fields, provide not only an update of state of the art techniques in many crucial areas of plant genetics, but also look forward into the future by defining current bottlenecks and research goals. Special attention has been given to DNA markers and their applications, but a section on social and economic implications integrates laboratory science with the socio-economic realities in which they occur.

Scientists and students in both the biological and social sciences will learn a lot from this book about the recent advances and future directions of plant genetics. In addition, this book should be required reading for policymakers.

Birkhäuser Verlag • Basel • Boston • Berlin

Peter Brandt
Institut für Pflanzenphysiologie und Mikrobiologie, Freie Universität Berlin

Transgene Pflanzen

Herstellung Anwendung Risiken und Richtlinien

1995. 320 Seiten. Broschur
ISBN 3-7643-5202-7

Dieses Buch informiert über den Stand der Wissenschaft auf dem Gebiet gentechnischer Veränderungen von Pflanzen und ermöglicht jedem Leser, sich ein eigenes Urteil über dieses Teilgebiet der Gentechnologie zu bilden.

Transgene Pflanzen beschreibt umfassend die verschiedenen Methoden zur Erzeugung transgener Pflanzen und Möglichkeiten ihrer Anwendung. Ebenfalls dargestellt sind Erkenntnisse über ihre Kultivierung sowie die Bestrebungen zum Inverkehrbringen transgener Pflanzen. Darüber hinaus faßt es die wichtigsten Argumente der derzeiten Risikodiskussion, gesetzliche Regelungen zur Gentechnik sowie Überlegungen zur Verantwortung in der Wissenschaft zusammen.

Die Darstellung verschiedener Aspekte der transgenen Pflanzen macht das Buch wichtig für eine vielfältige Leserschaft: Nicht nur für NaturwissenschaftlerInnen biologischer Fachrichtung, sondern auch JuristInnen und interessierte Laien. Ihnen allen wird der Einstieg in einen Themenkomplex leichtgemacht, der immer stärker in den Brennpunkt der Diskussionen rückt.

Aus dem Inhalt:
Hybris oder Hysterie? • Begriffe und Verfahren • Anwendung • Inverkehrbringen • Tatsächliche und hypothetische Risiken • Vollzug gesetzlicher Regelungen für das beabsichtigte Freisetzen oder das Inverkehrbringen von transgenen Pflanzen • Öffentliche Meinung und Akzeptanz • Verantwortung und Wissenschaft

Birkhäuser Verlag • Basel • Boston • Berlin

G. Kjellsson / V. Simonsen,
The National Environmental Research Institute, Silkeborg, Denmark (Eds)

Methods for Risk Assessment of Transgenic Plants

I. Competition, Establishment and Ecosystem Effects

1994. 224 pages. Hardcover.
ISBN 3-7643-5065-2

The present book is a compilation of current test methods useful in risk assessment of transgenic plants. It is intended to aid the environmental researcher in finding and comparing relevant methods quickly and easily. It may also be used as a general reference work for field-ecologists, laboratory biologists and others working in plant population biology and genetics.

The major processes affecting the fate of plants are covered with emphasis on invasion, competition and establishment, e.g., seed dispersal, density-dependent competition and plant growth. Ecosystem effects and genetic structure are also covered. For each process a number of relevant test methods have been selected; in total, 84 methods for field, greenhouse or laboratory research are included, employing 51 key processwords. Each method is described and evaluated briefly and succinctly, and there are comments on assumptions, restrictions, advantages and applications. An extensive bibliography provides entry into the scientific background, and cross references make it possible to find all relevant sources quickly.

Methods to study pollination and gene transfer will be considered in a future volume.

Birkhäuser Verlag • Basel • Boston • Berlin